Low-Frequency Vibrations of Inorganic and Coordination Compounds

LOW-FREQUENCY VIBRATIONS OF INORGANIC AND COORDINATION COMPOUNDS

John R. Ferraro
Argonne National Laboratory
Argonne, Illinois

PLENUM PRESS • NEW YORK • 1971

Library of Congress Catalog Card Number 74-107528

SBN 306-30453-8

ISBN-13: 978-1-4684-1811-8 e-ISBN-13: 978-1-4684-1809-5
DOI: 10.1007/978-1-4684-1809-5

Softcover reprint of the hardcover 1st edition 1971

A Division of Plenum Publishing Corporation
227 West 17th Street, New York, N.Y. 10011

Distributed in Europe by Heyden & Son, Ltd.
Spectrum House, Alderton Crescent
London N.W.4, England

To my wife, Mary, and my children, Lawrence, Janice, and Victoria, for their patience, encouragement, and understanding during the writing of this book.

PREFACE

During the course of far-infrared investigations of inorganic and coordination compounds at Argonne National Laboratory in the years 1962–1966, it became apparent that no suitable book existed which correlated and discussed the important vibrations occurring in this region for these molecules. Early in 1967 the initial steps were taken to write such a book. Then, in 1968, an excellent text by Professor David M. Adams entitled *Metal–Ligand and Related Vibrations* was published. At this point serious consideration was given to discontinuing work on this book. However, upon examination of Adams' book, it became clear that the references covered only the period to 1966. This field of research is accelerating so tremendously, and the period 1966–1969 has seen so many new studies, that upon reconsideration it was decided to continue writing this text.

The references in this book, particularly in the last several chapters, include many papers published in 1969. However, the proliferation of the far-infrared literature has made it impossible to present all the published material that has any bearing on the subject. Many titles do not pertain primarily to the far-infrared region as such, and some of this research has been omitted for this reason. Organometallic compounds have been neglected since the author feels that adequate reviews of that subject are available. Other studies may be missing simply because, owing to space limitations, only the more important researches could be considered. Of course, "importance" may, in this case, reflect the author's interest and prejudices.

The book is intended for chemists and spectroscopists who are using the far-infrared region in studying inorganic and coordination compounds. The purpose of the book is to demonstrate the usefulness of this region for such studies. In many instances, because of the solubility problems and the amorphous nature of certain materials, far-infrared spectroscopy may be the only tool available for their study, and it is here that the method proves particularly valuable.

The book is divided into sections covering the history, instrumentation, sampling, and calibration of the far-infrared region, followed by chapters on metal–oxygen, metal–halide, metal–nitrogen, and miscellaneous M–X vibrations. The final chapter discusses other far-infrared applications.

The data presented in Chapters 5–7 are based on symmetry considerations. It is true that several different approaches of presentation were open to the author, but in the final analysis, the symmetry pathway seemed the most suitable. I hope the reader will find it agreeable from his viewpoint.

The choice of the title of the book presented some difficulty. The book discusses the region of the infrared spectrum below 650 cm^{-1}, because most of the compounds cited have rich spectra in this region. Since only the region below 200 cm^{-1} is considered the far-infrared region, a title employing the term "far-infrared" would be erroneous, and as a compromise the term "low-frequency" was selected.

Finally, it must be said that while the book attempts to show the importance of the low-frequency region of the spectrum, the correlations suggested cannot possibly be the last word. Very obviously, in a fast-moving field, there must be continuous updating. Every effort has been made in this work to stay abreast of developments to the last possible moment, up to the extreme limit where the mechanics of the publication process prohibit further revisions.

I wish to express my gratitude to Professor V. Caglioti of the Italian Research Council and Professor G. Sartori of the University of Rome for extending the invitation to spend my sabbatical leave there, and thus affording me the opportunity to start this book; to Dr. Max Matheson, Director of the Chemistry Division, Argonne National Laboratory, for allowing me to spend about one-quarter of my time to finish the book; to my wife and my children for accepting my absence during the course of this task; to Miss Mary Ellen Matthews for the monumental task of typing the manuscript; and last, but not least, I owe a debt of gratitude to Dr. Louis J. Basile of Argonne National Laboratory, who accomplished the enormous literature search for Chapters 6 and 7, and who appears as a co-author of these chapters.

Argonne, Illinois
January 1970

John R. Ferraro

CONTENTS

Chapter 1

Introduction

Chapter 2

Far-Infrared Instrumentation

Chapter 3

Sampling Techniques and Instrument Calibration in the Far-Infrared Region

Chapter 4

New Techniques Used with Far-Infrared Measurements

Chapter 5

Metal–Oxygen Vibrations

Chapter 6

Metal–Halide Vibration

Chapter 7

Metal–Nitrogen Vibration

Chapter 8

Miscellaneous Metal–Ligand Vibrations

Chapter 1

INTRODUCTION

1.1. HISTORY

The early development of far-infrared spectroscopy proceeded along the lines of reststrahlen, prism, and grating techniques. Later, developments in interferometric spectroscopy followed.

Rubens can be considered to be the father of far-infrared spectroscopy. His many classical experiments[1] served as the beginning of the technique. In 1896, Rubens[2] studied several substances to 20 μ using potassium chloride (sylvine) as the prism. In 1897, he developed and used the reststrahlen technique to conduct studies in the low-frequency region.[3] Barnes and Czerny[4] developed instrumentation in 1931 and obtained several far-infrared spectra.[5-7] To go beyond 20 μ, Strong[8] developed the residual-ray method. Barnes extended the region to 135 μ, using wire gratings with the reststrahlen method.[9-10]

Development of far-infrared instrumentation employing prisms started in 1930. The delay from the initial experiment by Rubens in 1896 was caused by the difficulty in growing a crystal large enough to serve as a prism. Strong[11] prepared the first large crystal of potassium bromide in 1951. As a consequence of this work with Randall, he constructed the first prism spectrometer which could be operated to 30 μ. Other salt crystals were developed which could serve as prism materials (e.g., AgCl, KRS-5).[12-15] In 1952, Plyler used cesium bromide[14] and in 1953 cesium iodide[15] as prism materials. With the latter he was able to reach $\sim$50 μ. Crystalline quartz could be used beyond 50 μ, but the development of the grating techniques[16] turned attention away from the prism method.

The development of grating-type instrumentation stemmed from the construction of the first echelette grating by Wood[17] in 1910. Subsequent developments were due to Randall and co-workers,[16] who introduced the first prism-grating instrument in 1918.[18] Limitations in obtaining prism materials and difficulties in the construction of gratings stymied progress. Further difficulties were caused by the lack of suitable detection

Table 1-1. Early Grating Far-Infrared Instrumentation

Source	Region covered, μ	Chief investigators	Year
University of Michigan	25–200	H.M. Randall[19-20]	1932, 1938
Johns Hopkins University	100–700	T.K. McCubbin,[21]	1950
		H.M. Randall, and	
		J. Strong[11]	1951
Ohio State University	25–400	R. A. Oetjen[22]	1952

apparatus, which did not become available until about 1935. Several universities pioneered the development of grating spectrophotometers between the thirties and fifties for use in the far-infrared region.[11,19-22] The early grating spectrophotometers are listed in Table 1-1.

The development of commercial double-beam infrared instrumentation with the capability of using suitable interchangeable prisms (e.g., KBr, CsBr, CsI) for the region to 50 μ gave tremendous impetus to far-infrared spectroscopy. Table 1-2 summarizes the early instrumentation of this type. This was followed in the sixties by the introduction of commercial filter-grating far-infrared instrumentation specifically constructed for the low-frequency region. In 1962, Perkin–Elmer introduced[28] the 301 double-beam grating spectrophotometer, and in 1964, Beckman Instruments, Inc. introduced the Beckman IR-11.[29] A thorough discussion of these instruments appears in Chapter 2.

Concurrent with the development in prism and/or grating instrumentation for far-infrared spectroscopy, spectroscopic studies using interferometric techniques were carried out.[30-35] Early experiments were concerned with the measurement of wavelength, refractive index, and surface flatness.[33-35] In 1927, Michelson[36] indicated the benefits of interferometric methods in spectroscopy. The application of these techniques to spectroscopy was realized in 1956 by Gebbie and Vanasse.[37] Because of this

Table 1-2. Early Commercial Double-Beam Instrumentation with Prism Interchange

Instrument	Region μ	Prism	Year
Perkin–Elmer No. 21	15–20	CsBr	1948
Baird 4-55[23]	to 25	KBr	1955
Perkin–Elmer No. 321[24]	to 33	CsBr	1956
Hilger-Watts[25]	to 40	KBr/CsBr	1955
Beckman IR-3[26]	to 25	KBr	1949
Beckman IR-4[27]	to 33	CsBr	1956

early work it was possible to develop commercial interferometric instruments, and these are also discussed in Chapter 2.

1.2. USEFULNESS OF THE FAR-INFRARED REGION

The term "far-infrared" is generally used to describe the region below 200 cm^{-1}. However, the low-frequency spectra of inorganic and coordination compounds below 650 cm^{-1} are so rich that this region will be the primary center of discussion in this book. In general, the distinction in nomenclature has been retained in this text, but occasionally the term "far-infrared" has been used loosely to include the 200–650 cm^{-1} region.

Randall noted the usefulness of the low-frequency region as early as 1927.[38-40] For inorganic compounds and coordination complexes the region of 650 cm^{-1} and lower has begun to be extensively examined, since it offers much promise as a tool for structural analysis for compounds for which other methods cannot be extensively used. Table 1-3 lists the types of information obtainable from this region for inorganic and coordination compounds.

1.3. PROBLEMS AND REMEDIES

Many early workers in far-infrared spectroscopy have pointed out the difficulties encountered in obtaining spectra in the low-frequency region. The problems can be considered to fall into three categories—instrumental, sampling, and interpretation. These will be discussed separately.

Table 1-3. Information Obtainable from the Far-Infrared Region for Inorganic and Coordination Compounds

1. Ionic lattice vibrations
2. Molecular lattice vibrations
3. Coordinated water modes of vibration
4. Librations in coordinated polyatomic anions (hindered rotatory motions)
5. Chain vibrations in inorganic polymers
6. Metal–ligand vibrations:
   e.g., M–O, M–N, M–S, M–X, M–C
7. Potential barriers to internal rotation
8. Potential barriers to inversion
9. Rotational spectra of gases
10. Low-lying vibrations in hydrogen-bonded compounds
11. Semiconductor studies (electronic transition)
12. Determination of stereochemistry
13. Translation and hindered rotations of gas molecules in clathrates
14. Ion-pair vibrations
15. Thermodynamic functions

Instrumental Problems

Source

The source problem has always been a serious one in far-infrared spectroscopy. The low energy available in this region from any of the sources used (Nernst glower, Globar, and mercury arc) is the result of blackbody emission characteristics. A blackbody will emit more than 5.5×10^2 times more energy at 5000 cm^{-1} than at 200 cm^{-1}. Efforts have been made to ameliorate this condition, and recently attention has turned to laser sources. A pulsed gas laser has been developed to be used with a Michelson interferometer.[41] At present, however, this source has only limited application. An excellent discussion of far-infrared sources was presented by Genzel at the NATO Advanced Study Institute on Far-Infrared Properties of Solids, Delft, Netherlands, Aug. 5–23, 1968.*

Detector

The detector problem has plagued workers in this field from the days of Rubens. Far-infrared (exclusive of the region lower than 5 cm^{-1}) detectors can be classified into two main types: (1) thermal detectors (Golay cell, bolometers); (2) photoconductive detectors (InSb detector). In the commercially available far-infrared spectrophotometers, a Golay cell with a diamond window is used. This cell was developed by Marcel Golay in 1947,[42] and is shown in Fig. 1-1. It is usually filled with xenon gas, but, for faster response, helium gas can be used. The gas is heated by the absorbed radiation and expansion occurs, which distorts the flexible plastic membrane (~100 Å thick) of the cell. The outer surface of the mirror serves as a window and reflects the beam of light to a phototube. An electric signal is thus produced by the absorption of the radiation. Unfortunately, the membrane of the cell is glued on, and, whether in use or on the shelf, it does not maintain its leakproof seal. Eppley Laboratory, Inc. in the United States manufactures the cells used in the Perkin–Elmer No. 301 and the Beckman IR-11 spectrophotometers. The cells are also manufactured by Unicam Instruments, Cambridge, England. The U.S. manufacturers guarantee the cell for six months only. Much effort has and is being expended in the development of improved detectors, and new detectors for use down to 5 cm^{-1} have been developed.[43] These detectors are the carbon-resistance,[44] the superconducting,[45] and the doped-germanium bolometers,[46] and the indium antimonide detector.[47] All of these detectors require the use of liquid helium for their response, and this item is not readily available in most laboratories. In addition, these detectors must be equipped with

* The proceedings of this institute are to be published by Plenum Press (1970).

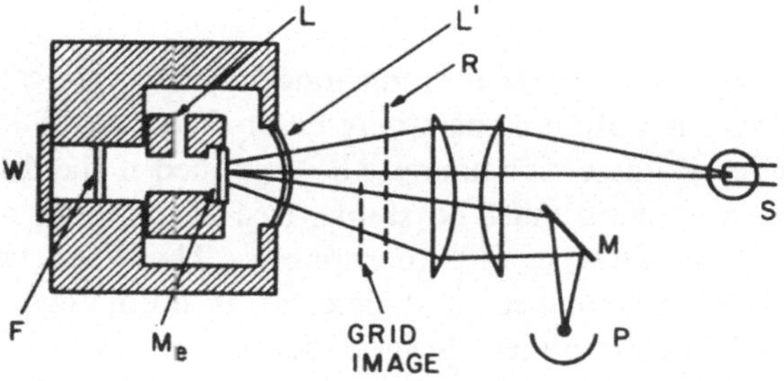

Fig. 1-1. Golay detector. W=diamond window; F=absorbing film; L=small leak; M_e=flexible membrane; L'=meniscus lens; R=grid; M=plane mirror; S=source (incandescent bulb); P=phototube. (Courtesy of Perkin–Elmer Corporation, Norwalk, Connecticut.)

a dewar for the helium and are extremely expensive. Furthermore, they suffer from an irregular frequency response and a limited frequency range. Since they are inherently more sensitive than the Golay detector and because of recent improvements in making them responsive in extended regions, they should be more extensively used in the future. Table 1-4 summarizes and compares the various detectors now in use in the far infrared.

A Rollin[48] bolometer has recently been developed and shows considerable promise for the submillimeter region up to 50–100 μ and could be used for Fourier transform spectroscopy. The spectral response is relatively uniform over the range of 1 cm to 300 μ, although some sensitivity is sacrificed in the 50–100 μ range.[49–50]

Table 1-4. Comparison of Several Detectors for Use in the Far-Infrared Region

Type	Temperature of operation, °K	Response time, sec	Spectral range, cm^{-1}
Golay cell (diamond window)	300	10^{-2}	900–15
Carbon-resistance bolometer	2.1	10^{-3}	<160
Superconducting Ge bolometer	3.7	10^{-2}	<160
Doped-Ge bolometer	2.1	10^{-3}	<160
InSb, with magnetic field	1.5	2×10^{-7}	< 40
InSb, without magnetic field	1.8	10^{-4}	< 40
InSb (Rollin type)[48–50]	1.2	3×10^{-7}	<200

Amplification

A very small electric signal is generated by most detectors, and with cooled detectors the output signals may be of the order of 10^{-6}–10^{-7} V or even lower. Clearly these signals must be amplified if the display is to be recorded. The amplifiers must be stable, free of amplifier noise, and free from electrical interference and microphony. These are very severe requirements. The performance of a detector is intimately related to the performance of the amplifier. Several successful amplifiers have been used in the low-frequency region; these are the Philips 7090, GE 7588, and RCA nuvistor tubes, all of which are of the vacuum tube type. Certain transitors are now comparable in performance to these (e.g., T.I. 2N930 and 2N918). For further discussion see Putley and Martin[51] and Conn and Avery.[52]

Miscellaneous Problems

Other instrumental problems exist, such as zero drift, stray light, fringing, overlapping orders, and water absorption. Most of these problems are substantially minimized in the commercial double-beam far-infrared spectrophotometers. The strong water vapor absorption in this region is both good and bad. Figure 1-2 shows this absorption in the region 150–350 cm^{-1} in an unpurged instrument in single-beam operation

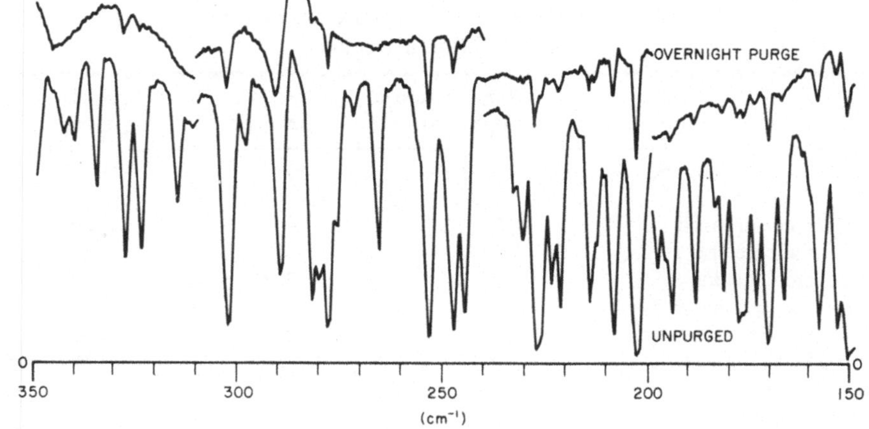

Fig. 1-2. Water absorptions in the 150–350 cm^{-1} range. (Courtesy of Beckman Instruments, Inc., Fullerton, California.)

and the improvement by going to double beam. If some compartment of a far-infrared instrument must be opened for change or repair, the purge is broken, and it is then necessary to repurge to remove the accumulated water vapor in the instrument. This may result in a delay of several hours. However, the strong water vapor absorption allows one to use these absorptions as calibration points for far-infrared instruments. This will be discussed further in Chapter 3.

Sampling Problems

To obtain good far-infrared spectra for inorganic solids is not an easy chore. For the most part, inorganic solids are hard to grind, and it is very difficult to obtain a good dispersion in a matrix material. It has been found that some hydrated materials and polymeric substances present problems. Coupling of vibrations is strong in the low-frequency region, and this can cause broad bands which can hide weak peaks. Certain vibrations are probably weak because of the lack of a significant change in dipole moment occurring between the atoms involved in the vibration. An example of such a weak vibration in the low-frequency region is the metal–nitrogen stretching (ν_{MN}) vibration. Symmetry may also play an important role in determining the intensity of a vibration. For example, it is difficult to assign a ν_{CoN} stretching vibration in $Co(NH_3)_6Cl_3$, because of its low intensity. However, in $[Co(NH_3)_5Cl]Cl_2$ the ν_{CoN} stretching vibration is more readily observed.[53]

These sampling problems are hardly solved at present, and continue to hinder the far-infrared spectroscopist. However, certain measures do offer some help. Low-temperature studies to 100°K have been shown to

Table 1-5. Factors Determining Position of Metal–Ligand Stretching Vibrations

Oxidation number of metal	⟶	The higher the oxidation number, the higher the frequency
Mass of metal and ligand	⟶	The larger the mass, the lower the frequency
Coordination number of metal	⟶	The higher the CN, the lower the frequency
Stereochemistry of complexes	⟶	The frequency decreases from a T_d to O_h structure
Basicity of ligand	⟶	The higher the basicity, the higher the frequency[a]
Counter-ion effect	⟶	The larger the counter-ion, the lower the frequency
Bridging or nonbridging	⟶	Nonbridging ligands cause vibrations to occur at higher frequency
High ligand field stabilization energy	⟶	The higher the energy, the higher the frequency

[a] Only when sigma bonding is involved.

produce better quality spectra for polymers and particularly for hydrated materials.[54] Other techniques will be discussed in Chapter 4.

Interpretative Problems

Interpretation of spectra for solids in the far-infrared range can be very difficult. The spectra of inorganic and coordination compounds are particularly difficult to interpret. This is due in part to the multiplicity of bands involved in the low-frequency range, such as metal–ligand, metal–water, and ligand vibrations, differentiation of which can present a problem. Coupling of vibrations is very important in this region and is a further source of trouble. The differentiation between a lattice vibration (movement of groups of atoms) and an internal vibration (movement of atoms in a group) in a solid is also difficult. The many factors which determine the position of a metal–ligand vibration in inorganic and coordination complexes have been cited by Clark.[55] These factors, which are impossible to assess and screen, are listed in Table 1-5. Further, the interpretation becomes more difficult as the compound becomes more complex.

Certain aids to interpretation which can be used in the far-infrared region are listed in Table 1-6. These aids are not without difficulties. For example, the deuteration of some hydrogen-bonded compounds does not shift the low-lying $\nu_{O.H.O}$ vibration. However, deuteration studies have helped assign the librational modes of water coordinated to the transition metals[56] and the lanthanides.[57] The conversion of a chloride complex to a heavier halide complex, such as the bromide or iodide, doesn't always result in a shift toward lower frequency, as expected for the metal–halide stretching (ν_{MX}) vibration. Examples are now available in which the chloride complexes are of octahedral symmetry with chloride bridging,[55,58] while the bromide complex involves tetrahedral symmetry with no bromide bridging. Thus, the ν_{MBr} stretching vibration may be found at a higher frequency than the ν_{MCl} stretching vibration. The use of low-temperature techniques to distinguish between lattice modes and internal vibrations is ineffectual in some cases, as the shifts of the lattice vibrations are minimal. Some benefits from a qualitative aspect can be obtained from the observa-

Table 1-6. Measures Used to Aid Far-Infrared Interpretation

1.	Deuteration studies in hydrogen-containing compounds
2.	Converting a chloride complex to a bromide and/or iodide complex
3.	Low-temperature studies
4.	Observation of breadth and intensity of ligand and metal–ligand absorptions
5.	Slow-neutron scattering studies of hydrated materials
6.	High-pressure studies
7.	Complementation with Raman spectroscopy

tion of breadth and intensities of low-frequency absorption bands. Ligand peaks are usually sharp whereas metal–ligand bonds are broad. Generally, ν_{MO} stretching vibrations are more intense than ν_{MN} stretching vibrations, and ν_{MX} stretching vibrations are usually equal to or of greater intensity than ν_{MN} stretching vibrations.

The use of slow-neutron scattering studies and high-pressure measurements as aids to the interpretation problem in the low-frequency region of the infrared will be discussed in Chapter 4. Raman scattering experiments made in conjunction with low-frequency infrared studies offer some promise in aiding far-infrared interpretation. For example, the distinction between a ν_{MN} stretching vibration and a ν_{MX} stretching vibration (in particular the ν_{MCl} vibration) can be made in some cases by obtaining the Raman spectrum of the complex. Whereas in the low-frequency spectrum of a complex the infrared intensity of the ν_{MX} stretching vibration is greater than that of the ν_{MN} stretching vibration, the reverse may be true in the Raman spectrum. Thus, two low frequencies lying close to one another and corresponding to the ν_{MN} and ν_{MX} stretching vibrations may be possibly distinguished.

For some recent reviews on far-infrared spectroscopy, see Bentley and co-workers,[59-60] Adams,[61] Lord,[62] Stewart,[63] Wood,[64] Wilkinson *et al.*,[65] and Brasch *et al.*[66] For a recent review on infrared detectors, see Levinstein.[67]

BIBLIOGRAPHY

1. H. Rubens and co-workers, papers from 1889 to 1922, cited in *A Far Infrared Bibliography* by E. D. Palik, U. S. Naval Research Laboratory (1962).
2. H. Rubens, *Verhandl. Deut. Phys. Ges.* **15,** 108 (1896).
3. H. Rubens and E. F. Nichols, *Ann. Phys. Chem.* **60,** 418 (1897).
4. R. B. Barnes and M. Czerny, *Z. Physik* **72,** 447 (1931).
5. R. B. Barnes, W. S. Benedict, and C.M. Lewis, *Phys. Rev.* **47,** 129 (1935).
6. R. B. Barnes, *Phys. Rev.* **47,** 658 (1935).
7. R. B. Barnes, W. S. Benedict, and C.M. Lewis, *Phys. Rev.* **47,** 918 (1935).
8. J. Strong, *Phys. Rev.* **37,** 1565 (1931).
9. R. B. Barnes, *Rev. Sci. Instr.* **5,** 237 (1934).
10. R. B. Barnes, *Phys. Rev.* **39,** 562 (1932).
11. J. Strong, *Phys. Today* **4,** 4 (1951).
12. E. K. Plyler, *J. Res. Natl. Bur. Std.* **41,** 125 (1948).
13. L. W. Tilton, E. K. Plyler, and R. E. Stephens, *J. Res. Natl. Bur. Std.* **43,** 81 (1949).
14. E. K. Plyler and N. Acquista, *J. Res. Natl. Bur. Std.* **49,** 51 (1952).
15. D. E. Mann, N. Acquista, and E. K. Plyler, *J. Chem. Phys.* **21,** 1949 (1953).
16. H. M. Randall, *J. Opt. Soc. Am.* **44,** 97 (1954).
17. R. W. Wood, *Phil. Mag.* **20,** 770 (1910).
18. H. M. Randall, *J. Appl. Phys.* **10,** 768 (1939).
19. H. M. Randall and F. A. Firestone, *Rev. Sci. Instr.* **9,** 404 (1938).

20. H. M. Randall, *Rev. Sci. Instr.* **3,** 196 (1932).
21. T. K. McCubbin and W. M. Sinton, *J. Opt. Soc. Am.* **40,** 537 (1950).
22. R. A. Oetjen, W. H. Haynie, W. M. Ward, R. L. Hansler, H. E. Schauwecker, and E. E. Bell, *J. Opt. Soc. Am.* **42,** 559 (1952).
23. Preliminary Tech. Bull. 43A, Baird-Atomics, Cambridge, Mass.
24. E. H. Siegler and T. F. Flynn, Abstract of Columbus Symposium on Molecular Spectroscopy, June, 1956, *Perkin-Elmer Instrument News* **7** (3): 7 (1956).
25. *News Letter*, Jarrell-Ash Co., Newtonville, Mass., May (1955).
26. Beckman Bulletin No. 273.
27. R. H. Muller, *Anal. Chem.* **28,** 73A (1956).
28. C. C. Helms, H. W. Jones, A. J. Russon, and E. H. Siegler, *Spectrochim. Acta* **19,** 819 (1963).
29. G. T. Keahl, H. J. Sloane, and M. Lee, 15th Pittsburgh Conference in Analytical Chemistry, March (1964).
30. A. A. Michelson, *Phil. Mag.* **5,** 131, 256 (1891).
31. H. Rubens and R. W. Wood, *Phil. Mag.* **21,** 249 (1911).
32. H. Rubens and von Baeyer, *Phil. Mag.* **21,** 689 (1911).
33. H. Kuhn, *Rept. Prog. Phys.* **14,** 64 (1951).
34. P. Jacquinot, *Rept. Prog. Phys.* **23,** 267 (1960).
35. R. S. Longhurst, *Geometrical and Physical Optics*, John Wiley and Sons, Inc., New York (1967), p. 159.
36. A. A. Michelson, *Studies in Optics*, Univ. of Chicago Press, Chicago (1927) p. 45.
37. H. A. Gebbie and G. A. Vanasse, *Nature* **178,** 432 (1956).
38. H. M. Randall, *Science* **65,** 168 (1927).
39. H. M. Randall, *Rev. Mod. Phys.* **10,** 72 (1938).
40. H. M. Randall, *J. Appl. Phys.* **10,** 768 (1939).
41. H. A. Gebbie, *Mol. Spectry. Rep. Conf.* **8,** 577 (1965).
42. M. S. E. Golay, *Rev. Sci. Instr.* **18,** 357 (1947).
43. P. L. Richards, *Rev. Sci. Instr.* **18,** 535 (1947).
44. P. L. Richards and M. Tinkham, *Phys. Rev.* **119,** 575 (1960).
45. D. H. Martin and D. Bloor, *Cryogenics* **1,** 159 (1961).
46. F. J. Low, *J. Opt. Soc. Am.* **51,** 1300 (1961).
47. D. G. Avery, D. W. Goodwin, and A. E. Rennie, *J. Sci. Instr.* **34,** 394 (1957).
48. B. V. Rollin, *Proc. Phys. Soc.* **77,** 1102 (1961).
49. M. A. Kinch and B. V. Rollin, *Brit. J. Appl. Phys.* **15,** 672 (1963).
50. M. A. Kinch, *Appl. Phys. Letters* **12,** 78 (1968).
51. E. H. Putley and D. H. Martin, in *Spectroscopic Techniques* (D. H. Martin, Ed.) John Wiley and Sons, Inc., New York (1967).
52. G.K.T. Conn and D. G. Avery, *Infrared Methods, Principles, and Applications*, Academic Press, New York (1960).
53. K. Nakamoto, *Infrared Spectra of Inorganic and Coordination Compounds*, J. Wiley and Sons, New York (1963).
54. J. E. Katon, J. T. Miller, Jr., and F. F. Bentley, *Arch. Biochem. Biophys.* **121,** 798 (1967).
55. R. J. Clark and C. S. Williams, *Inorg. Chem.* **4,** 350 (1965).
56. I. Nakagawa and T. Shimanouchi, *Spectrochim. Acta* **20,** 429 (1964).
57. C. Postmus and J. R. Ferraro, *J. Chem. Phys.* **48,** 3605 (1968).
58. J. R. Ferraro, W. Wozniak, and G. Roch, *Ri. Sci.* **38,** 433 (1968).
59. F. F. Bentley, E. F. Wolfarth, N. E. Srp, and W. R. Powell, *Analytical Applications*

of Far Infrared Spectra. I. Historical Review, Apparatus and Techniques, WADC TR 57-359, Wright Air Development Center, September (1957).
60. F. F. Bentley and E. F. Wolfarth, *Spectrochim. Acta* **18,** 165 (1959).
61. D. M. Adams, *Metal–Ligand and Related Vibrations*, Edward Arnold Publishers, Ltd. (1967).
62. R. C. Lord, *Investigations of the Far Infrared*, Contract AK33 (616)-5578, WADC TR59-498, Wright Air Development Center, February (1960).
63. J. E. Stewart, in *Interpretative Spectroscopy* (S. K. Freeman, editor) Reinhold Publishing Corp., New York (1965), p. 131.
64. J. L. Wood, *Quart. Rev. (London)* **17,** 362 (1963).
65. G. R. Wilkinson, S. A. Inglis, and C. Smart, in *Spectroscopy* (H. J. Wells, Ed.,) Institute of Petroleum, London (1962) p. 157.
66. J. W. Brasch, Y. Mikawa, and R. J. Jakobsen, *Applied Spectroscopy Reviews*, Vol. I, M. Dekker, New York (1968).
67. H. Levinstein, *Anal. Chem.* **41,** 81A (1969).

Chapter 2

FAR-INFRARED INSTRUMENTATION

2.1. INTRODUCTION

The commercially available far-infrared instrumentation will be discussed and summarized. The prism instruments will only be briefly mentioned, since the emphasis has now passed to the grating instruments. As a result, the Beckman IR-11 and the Perkin–Elmer No. 301 will be comprehensively discussed. Commercial interferometric instruments will also be discussed, the conventional spectrometers will then be compared with interferometers, and, finally, trends in far-infrared instrumentation will be discussed.

The present commercial instruments capable of obtaining spectra at frequencies below 650 cm^{-1} are listed in Table 2-1. The table includes both prism and/or grating and interferometric instruments.

2.2. PRESENT COMMERCIAL GRATING/PRISM INSTRUMENTS

Beckman Prism and/or Grating Instruments

IR-11 Infrared Spectrophotometer

The optical diagram for the IR-11 spectrophotometer is illustrated schematically in Fig. 2-1. Figure 2-2 is a photograph of the instrument. The IR-11 is an optical-null instrument when in double-beam operation. The practical application of the principle of optical-null instrumentation was developed by Wright.[1] The IR-11 covers the region of 12.5–300 μ (800–33 cm^{-1}). It consists of a single monochromator having nine transmission filters and four gratings, which allows for maximum wavelength coverage. Suitable combinations of one of the four diffraction gratings with one or more of the nine transmission filters provides high energy with low stray light. The filter/grating combinations are illustrated in Table 2-2. The filters are mounted on a turret located beyond the exit slit. The sample and reference beams enter the monochromator where they are combined by a chopper. The bilateral slits are continuously adjustable from 0 to 12

Table 2-1. The Contemporary* Far-Infrared Commercial Instruments

Prism and/or Grating-Type Spectrophotometers	
Beckman instruments	FIR limit, cm^{-1}
IR–11	33
IR–12	200
IR–9	400
IR–4	285 (with prism interchange)
IR–20	250
Perkin–Elmer instruments	
301	14
621	200
521	250
421	250 (with grating interchange)
457	250
337	400
225	200
Hitachi FIS-3	30
Bausch and Lomb—270 IR	400
Unicam (SP1200)	400
Unicam (SP100G)	375
Grubb Parsons DM4	222/200
Grubb Parsons GM3	65
Grubb Parsons Spectromaster MR. 3	400
Grubb Parsons Spectromajor	400
Interferometers	
Interferometer (RIIC-Subsidiary of Beckman Instrument Inc.)	
FS–820	10
FS–720	10
LR–100	3
Grubb Parsons/N.P.L. Cube MK 11	10
Grubb Parsons Iris Interferometer	10
Dunn Associates Inc., Block FTS-16	10 (standard) 5 (optional)

* As of August 1968

mm. Slit-programming control is obtained through a conventional padding-potentiometer system. The radiation source used throughout the range is the high-pressure 100-W AH-4 mercury lamp with the outer glass envelope removed, since glass is not transparent to radiation of wavelengths beyond 3 μ. The detector is a diamond-window Golay cell manufactured by Eppley Laboratory, Inc. A linear 10-in. strip-chart flat-bed recorder is used. The instrument can be kept free of moisture either by use of a pneumatic air dryer or by purging with dry nitrogen. Most of the controls are external and many of the functions are automated, so that opening the

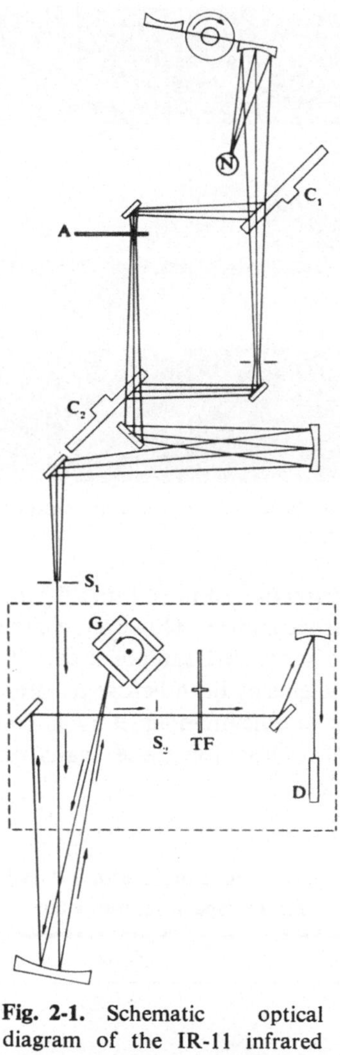

Fig. 2-1. Schematic optical diagram of the IR-11 infrared spectrophotometer. (Courtesy of Beckman Instruments, Inc., Fullerton, California.)

instrument, with a subsequent loss of purge, is unnecessary. The entire range can be scanned as rapidly as nine minutes or as slowly as several weeks. The instrument may be used as a single-beam instrument by blocking the reference beam. Several automatic features of the instrument are tracking accuracy control, repetitive scan, and scale expansion.

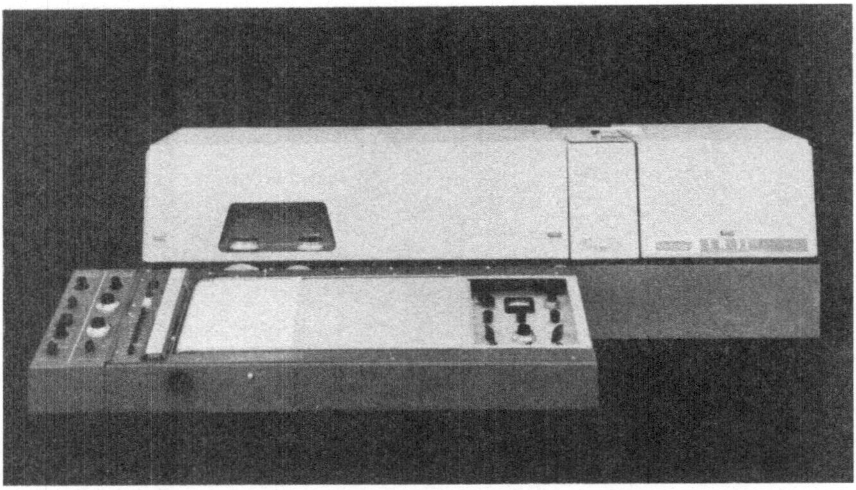

Fig. 2-2. Photograph of the IR-11 infrared spectrophotometer. (Courtesy of Beckman Instruments, Inc., Fullerton, California.)

The sample compartment is 5 in. × 9 in. × 12 in. and is rather small for dewars and other special accessories. However, a supersize instrument is now available which has an increased sample area.[2] It should be pointed out that the IR-11 chops the energy both before it enters and after it leaves the sample, and operating in this manner it is not suited for nonambient-temperature spectra. However, for these scans one can stop the after-chopper of the instrument.

Table 2-2. Filter/Grating Combinations Used in the Beckman IR-11 Spectrophotometer

Range	Filter	Grating[a]
33–59 cm^{-1}	1	1
59–80	2	1
70–95	2	2
92–120	3	2
120–150	4	2
140–200	5	3
200–250	6	3
250–300	7	3
300–350	8	3
350–500	8	4
500–800	9	4

[a] (1) 3 lines/mm blazed at 165 μ. (2) 8 lines/mm at 112 μ. (3) 20 lines/mm at 45 μ. (4) 50 lines/mm at 19.98 μ. Area is 64 × 64 mm.

IR-12, IR-9, and IR-20 Infrared Spectrophotometers

The IR-12 spectrophotometer is similar in design to the IR-11. Its optical diagram is illustrated in Fig. 2-3. It has a single monochromator with 4 gratings and 8 transmission filters and is fully automatic.

The IR-9 optical diagram is illustrated in Fig. 2-4. It has a double monochromator with a prism and two gratings.

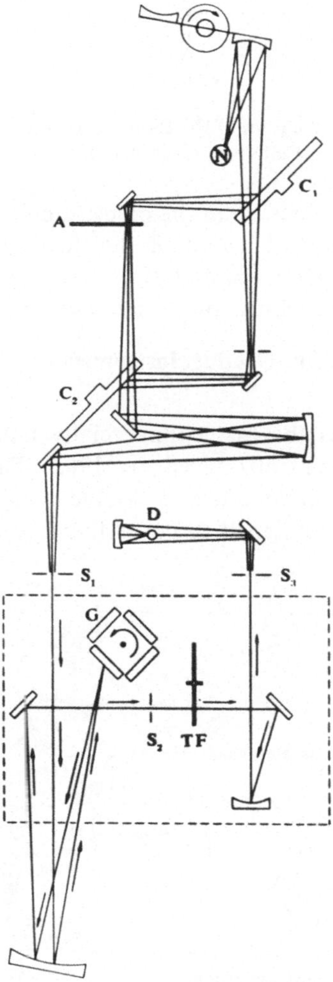

Fig. 2-3. Schematic optical diagram of the IR-12 infrared spectrophotometer. (Courtesy of Beckman Instruments, Inc.. Fullerton, California.)

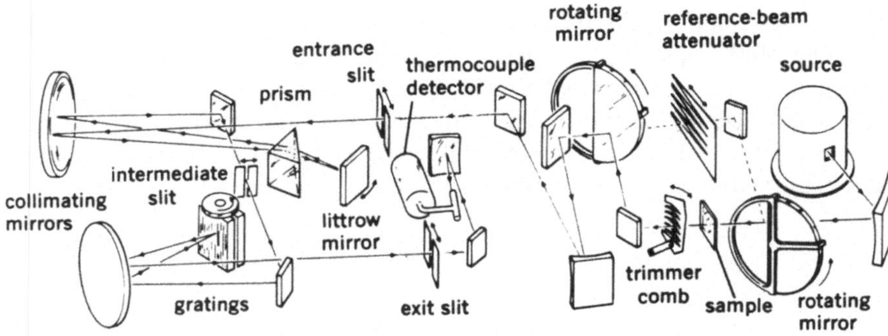

Fig. 2-4. Schematic optical diagram of the IR-9 infrared spectrophotometer. (Courtesy of Beckman Instruments, Inc., Fullerton, California.)

The IR-20 is one of the newest in the collection of Beckman instruments. Figure 2-5 shows the optical layout of the instrument and Fig. 2-6 is a photograph of it. The instrument covers the range from 4000 to 250 cm^{-1} and utilizes two gratings with a single monochromator.

Perkin–Elmer Prism and/or Grating Instruments

Model 301

The Model 301 was the first commercially built far-infrared spetrophotometer, having been introduced in 1962. The optical diagram is shown in Fig. 2-7. The Model 301 is a double-beam ratio-recording spectrophotometer primarily designed to cover the region between 660 and 60

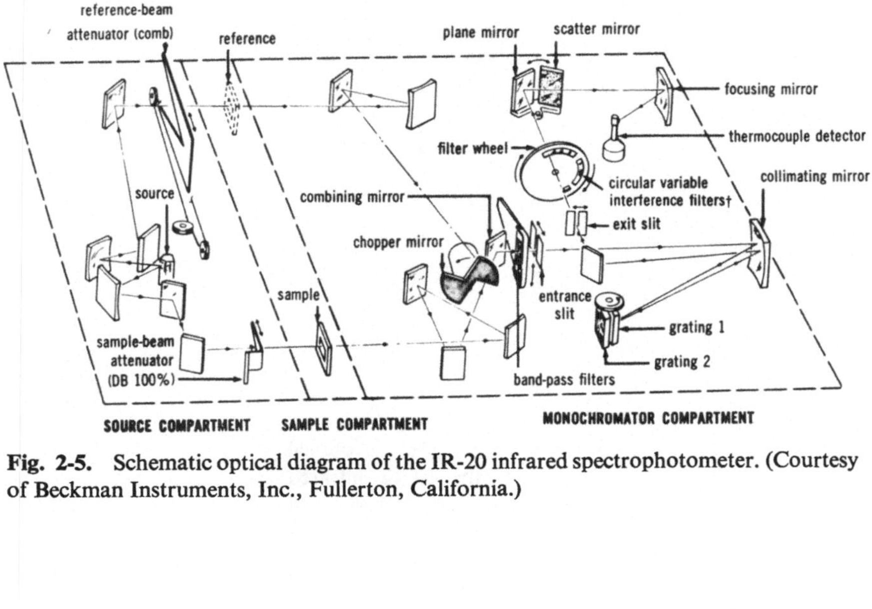

Fig. 2-5. Schematic optical diagram of the IR-20 infrared spectrophotometer. (Courtesy of Beckman Instruments, Inc., Fullerton, California.)

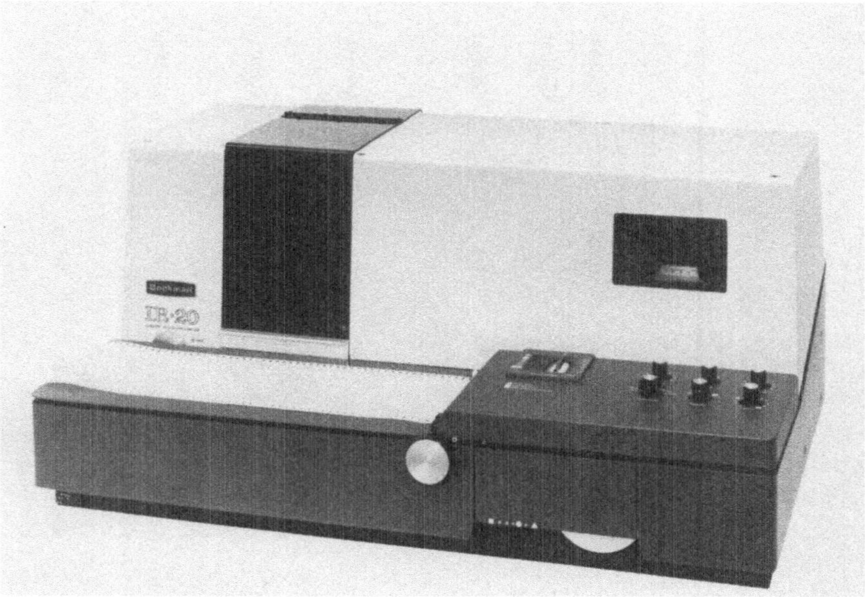

Fig. 2-6. Photograph of the IR-20 infrared spectrophotometer. (Courtesy of Beckman Instruments, Inc., Fullerton, California.)

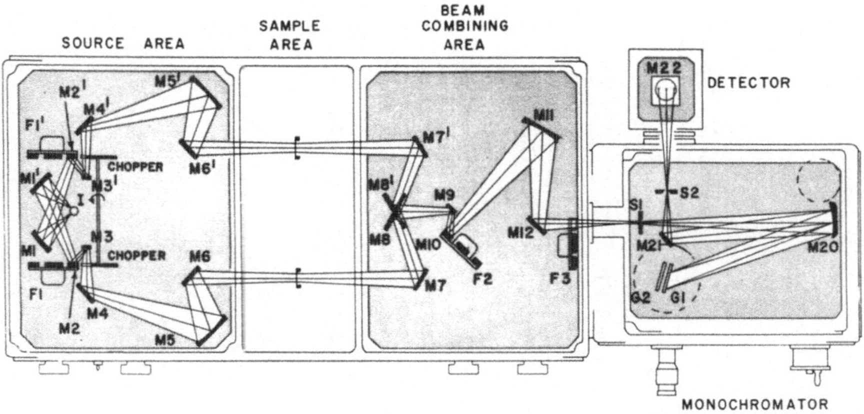

Fig. 2-7. Schematic optical diagram of the Perkin–Elmer Model No. 301 spectrophotometer. (Courtesy of Perkin–Elmer Corporation, Norwalk, Connecticut.)

Table 2-3. Operating Conditions for No. 301 Spectrophotometer

Wavelength or frequency range	M2 and M2′ on filter wheel F1′ and F1	Chopper	M4 and M4′	M10 on filter wheel F2	M12	F3 filter wheel	Mono blaze, 1st order	Grating ruling, 1/mm	Source
900–667 cm^{-1}	Pos. 1 mirror	Opaque	Mirror	LiF pos. 1	Mirror	Open pos. 2	7.5 μ	101	Globar
667–400 cm^{-1} 15–25 μ	Pos. 1 mirror	BaF_2	Fine scatter plate	LiF pos. 1	Fine scatter plate	Open pos. 2	22.5 μ	40	Globar
400–333 cm^{-1} 25–30 μ	Pos. 4 NaF	BaF_2	Fine scatter plate	LiF pos. 1	Fine scatter plate	Open pos. 2	22.5 μ	40	Globar
333–222 cm^{-1} 30–45 μ	Pos. 4 NaF	BaF_2	Fine scatter plate	BaF_2 pos. 2	Fine scatter plate	Open pos. 2	45 μ	20	Globar
222–167 cm^{-1} 45–60 μ	Pos. 5 NaCl	BaF_2	Fine scatter plate	BaF_2 pos. 2	Fine scatter plate	Open pos. 2	45 μ	20	Globar
167–130 cm^{-1} 60–77 μ	Fine scatter plate pos. 2	CsI	Coarse scatter plate	KCl pos. 3	Coarse scatter plate	Open pos. 2	90 μ	10	Hg

130–100 cm^{-1} 77–100 μ	Fine scatter plate pos. 2	CsI	Coarse scatter plate	KBr pos. 4	Coarse scatter plate	Open pos. 2	90 μ	10	Hg
100–80 cm^{-1} 100–125 μ	Fine scatter plate pos. 2	CsI	Coarse scatter plate	CsBr pos. 5	Coarse scatter plate	Open pos. 2	90 μ	10	Hg
80–69 cm^{-1} 125–145 μ	Coarse scatter plate pos. 3	CsI	Coarse scatter plate	CsBr pos. 5	Coarse scatter plate	Open pos. 2	180 μ	5	Hg
69–41 cm^{-1} 145–240 μ	Coarse scatter plate pos. 3	CsI	Coarse scatter plate	KRS-5 pos. 6	Coarse scatter plate	Filter No. 1 pos. 4	180 μ	5	Hg
41–32 cm^{-1} 240–310 μ	Coarse scatter plate pos. 3	CsI	Coarse scatter plate	Mirror pos. 1 (remove LiF)	Coarse scatter plate	Filter No. 2 pos. 5	360 μ	2.5	Hg
.32–20 cm^{-1} 310–500 μ	Coarse scatter plate pos. 3	CsI	Coarse scatter plate	No. 145 mesh screen pos. 1 (remove LiF)	Coarse scatter plate	Filter No. 2 pos. 5	360 μ	2.5	Hg
20–17.5 cm^{-1} 500–570 μ	No. 145 mesh screen pos. 6	CsI	Coarse scatter plate	No. 145 mesh screen pos. 1 (remove LiF)	Coarse scatter plate	Filter No. 2 pos. 5	720 μ	1.25	Hg
17.5–14 cm^{-1} 570–714 μ	Coarse scatter plate pos. 3	CsI	Coarse scatter plate	No. 70 mesh screen pos. 1 (remove LiF)	Coarse scatter plate	Filter No. 2 pos. 5	720 μ	1.25	Hg

cm^{-1}. However, both the high- and low-frequency regions can be extended. For example, in the low-frequency range, the instrument is capable of reaching 714 μ (14 cm^{-1}).

The instrument can best be described by discussing each of its five components individually. The five components are the source, the sample area, the beam-condensing area, the monochromator, and the detector chamber.

Source Area. The source area consists of the source, which can be either a water-cooled 200-W Globar or a high-pressure mercury AH-4 lamp with the outer glass envelope removed. The inner quartz bulb is usually dimpled to prevent fringing. Two toroidal mirrors, which magnify the image 2×, direct the beam to a filter wheel. The filter wheel has six positions for reflection filters. Two crystal choppers present in this housing chop the beam at 13 Hz. Kinematically mounted scatter plates are also present. The proper source, filter, and chopper are used for a desired frequency range, as indicated in Table 2-3. The reference beam follows a similar path, as observed in Fig. 2-7. The filters can be changed by external knobs. The chamber is sealed from the air and can be purged by nitrogen, but it must be opened to air when replacing the choppers, scatter plates, and source.

Sample Area. The sample area is accessible from the rear and front and can be used with attached rubber gloves to prevent moisture from entering the compartment. The area is sufficiently large to accommodate dewar flasks and is about 6 in. wide.

Beam-Condensing Area. The beam-condensing area contains toroidal mirrors which reflect the sample and reference beams and directs them to a half-aperature plane mirror. By means of this mirror, the two beams are combined in the same vertical plane with only the top and bottom halves being utilized. The beam is the directed to a reststrahlen filter wheel having six positions for different filters. These can be changed with external manual controls. Scatter plates, which can be changed depending on the far-infrared region used, are also mounted in this housing. The light beam leaves this compartment by going through a filter on a filter wheel with six positions—again externally controlled. The chamber can be sealed and nitrogen purged but must be opened to replace the scatter plates.

Monochromator. The monochromator is a Littrow-type *f*/3.8, designed for far-infrared operation and has a 21° off-axis parabolic mirror which collimates the beam. Two gratings, placed back-to-back on a turntable, are 68.6 mm square and can be positioned externally. The grating drive turns the grating as a function of the cosecant of the incidence angle and achieves a scan that is linear with respect to wave number. A special slit mechanism permits a slit opening from 0 to 10 mm. The monochromator is airtight and can be purged with nitrogen. If the grating is to be changed, the purge must be broken and the compartment opened.

Detector Compartment. The detector compartment contains the detector, a 90° total off-axis ellipsoid, and a specially designed preamplifier. The detector is a Golay pneumatic cell, from Eppley Laboratory, with a 3/16-in.-diameter target and a diamond window. The ellipsoid is used to focus the light to the detector.

A drum turn disk which controls the rotation of the grating is tied in with the grating mechanism. The drum turn disk is marked in drum turns of 0 to 24, in 1/10 units. The drum turn readings must be calibrated in wavenumber (cm^{-1}) units. A different calibration chart is necessary for each grating (see Chapter 3 for methods of calibration). The instrument also contains an electronics rack which has a Leeds and Northrup Model G recorder of 10 in. size.

The Model 301 can also be used in single-beam operation and this is recommended for the region below 32 cm^{-1}. Table 2-3 lists the recommended filters, choppers, scatter plates, gratings, and sources to be used with the 301 from 900 to 14 cm^{-1}.

Model 621

The Perkin–Elmer Model 621 is a double-beam optical-null filter/grating instrument capable of reaching 200 cm^{-1}. It uses a fast $f/4.5$ single monochromator. Two gratings are used mounted back-to-back, a Nernst glower serves as the source, and a thermocouple serves as the detector. It is capable of reaching 200 cm^{-1}. Figure 2-8 depicts the optical layout of the 621.

Model 225

The Perkin–Elmer Model 225 is a double-beam optical-null spectrophotometer manufactured in Germany. It has a range of 5000 to 200

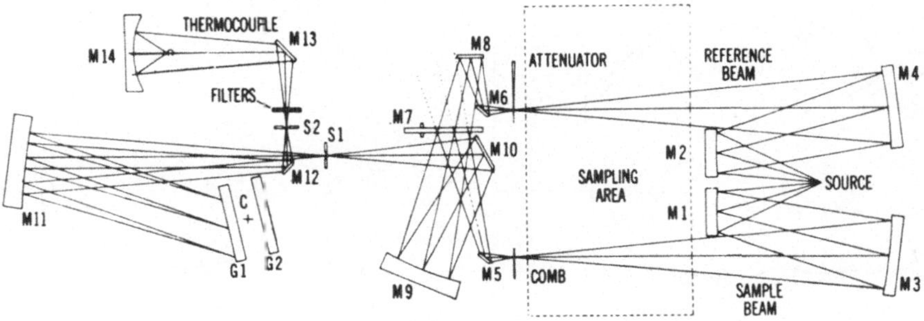

Fig. 2-8. Schematic optical diagram of the Perkin–Elmer Model No. 621 spectrophotometer. (Courtesy of Perkin–Elmer Corporation, Norwalk, Connecticut.)

cm^{-1}. An Ebert grating monochromator is used. A thermostatted KBr prism predispenses the energy. Two gratings are used in the instrument. An air-cooled Globar is the source, and a Golay cell with a CsI window serves as the detector.

Hitachi FIS-3

The Hitachi FIS-3 covers the range from 400 to 30 cm^{-1} and is manufactured by Hitachi Ltd. in Japan. It is marketed in the U.S.A. by the Perkin–Elmer Corporation and replaces the Model 301. It is a double-beam single-pass optical-null instrument. The light source is a Globar or high-pressure mercury lamp, and a Golay cell serves as the detector. It uses three gratings in the first order, ruled at 20, 8, and 4 lines/mm respectively, along with eight transmission filters. The monochromator is a Littrow type. The instrument can be used in single-beam operation if desired and is designed so that the entire optical system may be either evacuated or purged. It scans automatically. (For further information see Ref. 3.) Figure 2-9 illustrates the schematic optical diagram for the Hitachi FIS-3.

Grubb Parsons Prism and/or Grating Instruments

Grubb Parsons DM 4

The DM 4 instrument is manufactured by Grubb Parsons in Newcastle upon Tyne, England. It is available in two ranges: 667–222 cm^{-1} and 500–200 cm^{-1}. The instrument utilizes a high-energy double monochromator (*f*/5 aperture) and is a double-beam optical-null spectrophotometer. It has

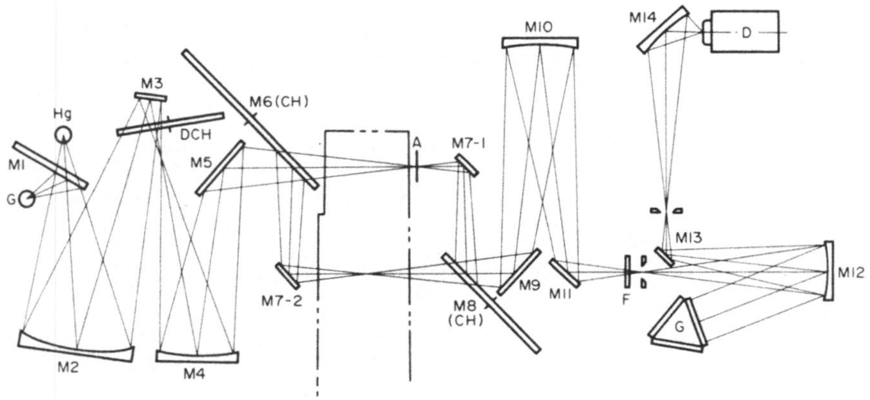

Fig. 2-9. Schematic optical diagram of the Hitachi FIS-3 spectrophotometer. (Courtesy of Perkin–Elmer Corporation, Norwalk, Connecticut.)

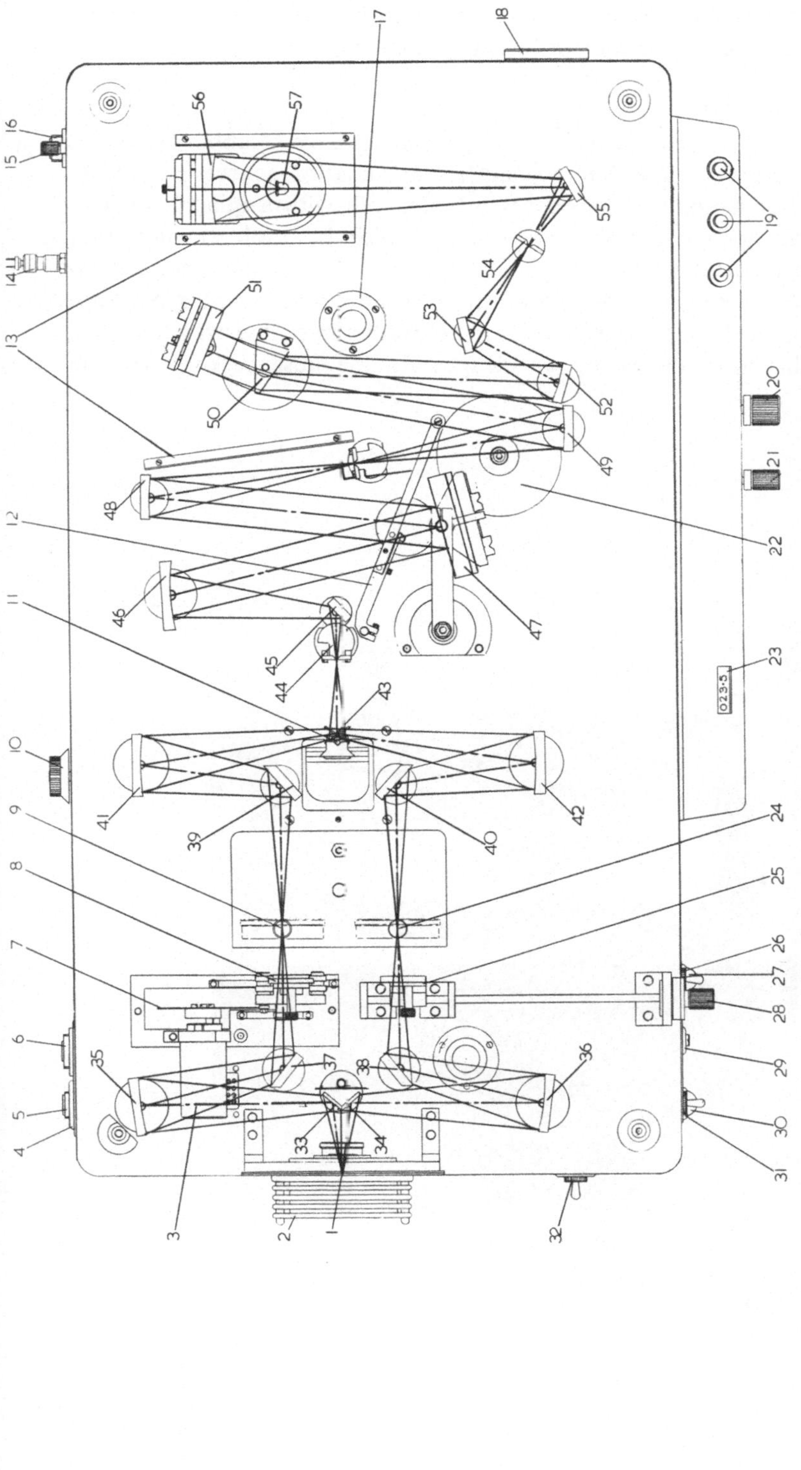

Fig. 2-10. Optical diagram of the Grubb Parsons DM 4 instrument.

1) Position of Nernst; 2) Nernst housing (air cooled); 3) servo motor; 4) mains input socket; 5) barretter socket; 6) electronics socket; 7) comb cam; 8) null-balance comb; 9) reference cell position; 10) "phasing adjustment"; 11) reciprocating mirror mech; 12) slit mech. lever arm; 13) stray light shields; 14) dry air inlet; 15) earthing connection; 16) detector signal output socket; 17) dry air filters; 18) thermostat control; 19) wave number selectors; 20) gear change control (scan speed); 21) wave number select (hand control); 22) slit cam; 23) wave number indicator; 24) sample cell position; 25) trimmer comb (100%T); 26) instrument on–off switch; 27) heater indicator lamp; 28) trimmer comb control; 29) fuse; 30) mains indicator lamp; 31) servo on–off switch; 32) electronics on–off switch; 33, 34) plane mirrors; 35, 36) spherical mirrors; 37–40) plane mirrors; 41, 42) spherical mirrors; 43) plane mirrors (reciprocating); 44) slits (variable); 45) plane mirror; 46) spherical mirror (collimating); 47) diffraction grating; 48, 49) spherical mirrors; 50) prism (CsI); 51) Littrow mirror; 52) spherical mirror; 53) plane mirror; 54) slit (fixed); 55) plane mirror; 56) on-axis ellipsoid; 57) detector. (Courtesy of Hilger & Watts Ltd., London, England.)

a Nernst source, a thermocouple detector, and uses two gratings in the first order, ruled at 30 and 20 lines/mm, together with a CsI afterprism. Figure 2-10 shows the optical layout of the instrument.

Grubb Parsons GM 3

The GM 3 is also manufactured in England by Grubb Parsons. It covers the range from 200 to 65 cm^{-1}. The radiation source is a 125-W mercury vapor lamp. The detector is a Golay cell with a quartz window. It has provisions for continuous purging, which is absolutely necessary since the instrument is operated on single beam. To obtain spectra comparable in appearance to double-beam recordings, the coupled entrance and exit slits of the monochromator are controlled in such a way that they open together with increasing wavelength. In the absence of a sample

Fig. 2-11. Photograph of the Grubb Parsons GM 3 instrument. (Courtesy of Hilger & Watts Ltd., London, England.)

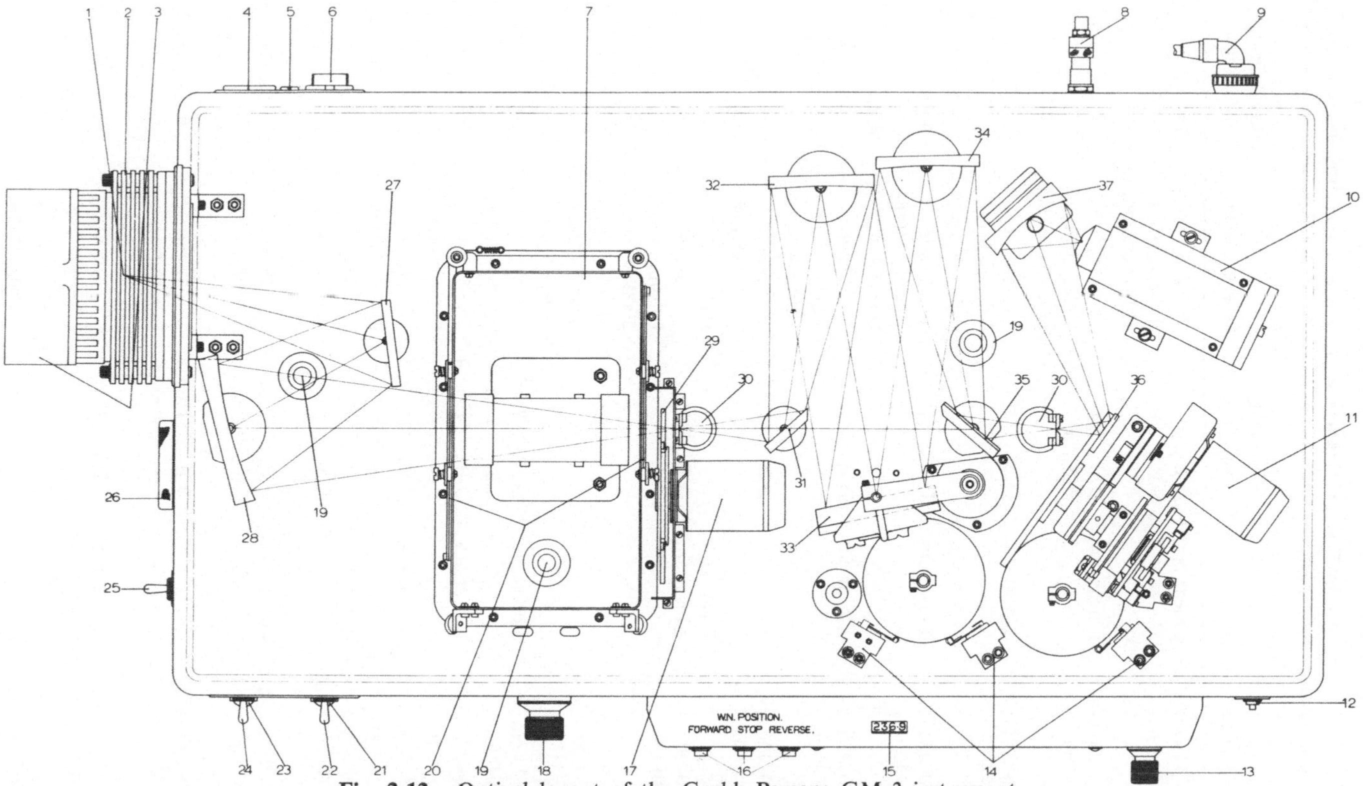

Fig. 2-12. Optical layout of the Grubb Parsons GM 3 instrument.

1) Position of IR source (mercury lamp); 2) source housing (air cooled); 3) source housing cooling fan; 4) electronics socket; 5) fuse; 6) mains socket; 7) sample box; 8) dry air connection (inlet); 9) detector signal output; 10) detector (Golay) with quartz windows; 11) reststrahlen filter drive mech; 12) wave number pause press button switch; 13) gearchange knob (scan speed); 14) filter-change switches; 15) wave number indicator; 16) wave number selector controls; 17) chopper drive mech; 18) sample box shutter-control knob; 19) dry air inlet filters; 20) sample box shutters; 21) mains indicator lamp; 22) instrument on–off switch; 23) heater indicator lamp; 24) mercury lamp on–off switch; 25) electronics on–off switch; 26) thermostat control; 27) scatter grating; 28) spherical mirror; 29) KBr chopper blades (3); 30) slits; 31) scatter grating; 32) spherical mirror; 33) diffraction grating; 34) spherical mirror; 35) reflecting filter; 36) reststrahlen filters (NaCl, KBr, KCl, KRS-5); 37) off-axis ellipsoid. (Courtesy of Hilger & Watts Ltd., London, England.)

Table 2-4. Comparison of Various Grating Instruments

Instrument	Range	Presentation	Operating modes	Monochromator	Filter ranges	Gratings
P-E 301	665–14 cm^{-1} (15–715 μ)	Linear cm^{-1}	D.B., %T S.B., %T	*Single* Filter/grating (Littrow mount)	13	6
P-E 621	4000–200 cm^{-1} (2.5–50 μ)	Linear cm^{-1}	D.B., %T D.B., A	*Single* Filter/grating (Littrow mount)	8	2
P-E 521	4000–250 cm^{-1} (2.5–40 μ)	Linear cm^{-1}	D.B., %T	*Single* Filter/grating (Littrow mount)	7	2
P-E 421	4000–250 cm^{-1} (2.5–40 μ)	Linear cm^{-1}	D.B., %T	*Single* Filter/grating (Littrow mount)	4/4 in interchange	2
P-E 457	4000–250 cm^{-1}[a] (2.5–40 μ)	Linear cm^{-1}	D.B., %T	*Single* Filter/grating (Littrow mount)	7	2
P-E 337	4000–400 cm^{-1} (2.5–25 μ)	Linear λ or linear cm^{-1}	D.B., %T	*Single* Filter/grating (Littrow mount)	4	2
P-E 225	5000–200 cm^{-1} (2–50 μ)	Linear cm^{-1}	D.B., %T	*Single* Prism/grating 5000–400 cm^{-1}, filter/grating 450–200 cm^{-1} (Ebert type)	2	2
IR-11	800–33 cm^{-1} (125–300 μ)	Linear cm^{-1}	D.B., %T or A S.B., %T or A	*Single* Filter/grating	9	4
IR-12	4000–200 cm^{-1} (2.5–50 μ)	Linear cm^{-1}	D.B., %T or A S.B., %T or A	*Single* Filter/grating	8	4
IR-9	4000–400 cm^{-1} (2.5–25 μ)	Linear cm^{-1}	D.B., %T or A S.B., %T or A	*Double* Prism/grating	–	2
IR-5A	900–285 cm^{-1} (11–35 μ)	Linear λ	D.B., %T or A S.B., %T or A	*Single* Prism	–	–
IR-4	10,000–285 cm^{-1} (1–35 μ)	Linear λ	D.B., %T or A S.B., %T or A	*Double* Prism	–	–
IR-20	4000–250 cm^{-1} (2.5–40 μ)	Linear cm^{-1}	D.B., %T or A S.B., %T or A	*Single* Filter/grating	6	2
Hitachi FIS-3	400–30 cm^{-1} (25–333 μ)	Linear cm^{-1}	D.B., %T	*Single* Filter/grating	8	3
Unicam SP 100G	8950–375 cm^{-1} (1.1–26.6 μ)			Prism/grating		

Capable of Low-Frequency Measurements

Source	Detectors	Stray light[d]	Wave number accuracy, cm^{-1}	Resolution, cm^{-1}	Recorder	Price[e]
Globar, AH-4 Hg lamp	Golay	<1% from 665–32 cm^{-1}	0.5	0.7 in entire range	L & N Model "G"	$33,700
Nernst glower	TC[c]	<2% from 400–200 cm^{-1}	±0.5	0.3 at 1000 cm^{-1}	Drum	$22,250
Nernst glower	TC	<2% at 250 cm^{-1}	±0.5	0.3 at 1000 cm^{-1}	Drum	$21,250
Nernst glower	TC	<0.1 to 550 cm^{-1} <2% 400–250 cm^{-1} (interchange)	1	0.3 at 1000 cm^{-1}	Drum	$19,250
Nernst glower	TC	3% at 250 cm^{-1}	±2 (2000–250 cm^{-1})	2 at 1000[b] cm^{-1}	Flow-chart	$ 8,800
Nernst glower	TC	<4% at 400 cm^{-1}	±2 (1333 to 400 cm^{-1})	Better than 2 at 1000 cm^{-1}	Drum	$ 7,500
Globar	Thermopile (Golay on request)	<2.5 from 260–200 cm^{-1}	±0.5	0.33 at 2400 cm^{-1}, 0.16 at 1034 cm^{-1}, 0.41 at 286 cm^{-1}	Drum	$28,500
AH-4 Hg	Golay	0.4% at 200 cm^{-1}	1	0.5	Flat-bed strip	$36,000
Nernst glower	TC	<1% 500–200 cm^{-1}	0.3 at 400 cm^{-1}	0.25	Flat-bed strip	$22,750
Nernst glower	TC	<0.1% at 700 cm^{-1}	0.2 at 400 cm^{-1}	0.25	Flat-bed strip	$20,750
Nichrome wire	TC	< 2% at 400 cm^{-1}	0.05 μ from 30–35 μ	6 at 400 cm^{-1} (25 μ)	Flat-bed strip	$ 6,400
Nernst glower	TC	0.1% at 700 cm^{-1}	.008 μ entire range	0.01 μ at 1000 cm^{-1} (10 μ)	Flat-bed strip	$23,115
Nichrome wire	TC	1% 600–250 cm^{-1}	2 2000–250cm^{-1}	1 2000–200 cm^{-1}	Flat-bed strip	$ 8,500
Globar or Hg lamp	Golay		±0.5	0.5	Flat-bed strip	$32,500
	Golay					

Table 2-4

Instrument	Range	Presentation	Operating modes	Monochromator	Filters ranges	Gratings
Unicam SP 1200	4000–400 cm^{-1} (2.5–25 μ)			Filter/grating		
Grubb Parsons DM 4	667–222 cm^{-1} (15–45 μ)	Linear cm^{-1}	D.B., %T	*Double* Prism/grating		2
Grubb Parsons GM 3	200–65 cm^{-1} (50–153 μ)	Linear cm^{-1}	S.B., %T	Filter/grating	4	1
Grubb Parsons Spectromaster		Linear λ	D.B., %T	*Double* Prism/grating		2

[a] With interchange.
[b] Resolution pertains to survey scan slit runs in general; better resolution is obtained with narrower slits.
[c] TC = thermocouple.
[d] Stray light figures are indicated for only low-frequency ranges.
[e] Prices as of November 20, 1969.[4]

the signal remains at a substantially constant level throughout the entire range of the instrument. The instrument employs a grating used in the first order, ruled at 8 lines/mm, and four reststrahlen filters. Figure 2-11 is a photograph of the instrument and Fig. 2-12 shows its optical layout.

Table 2-4 compares the present prism grating instruments capable of reaching the low-frequency region.

2.3. PRESENT COMMERCIAL INTERFEROMETRIC INSTRUMENTS

Introduction

Spectroscopy using interferometric methods is called Fourier transform spectroscopy. The principles of this type of spectroscopy date back to Michelson's studies.[5] His research on the interference of light led to a device called a "differential refractometer." Later the instrument became known as a interferometer. The theory of the interferometer is well known today. For reviews on the theory see Strong and Vanasse,[6] Jacquinot,[7] Genzel,[8] and Connes.[9]

Conventional far-infrared instruments employ prisms and/or gratings for selecting the monochromatic radiations desired. Much of the energy of the source is lost because of the dispersion and diffraction occurring. This loss, in addition to the fact that the source energy is low at long wave-

(Continued)

Source	Detectors	Stray light[b]	Wave number accuracy, cm^{-1}	Resolution, cm^{-1}	Recorder	Price[e]
						$ 5,800
Nernst glower	TC	<2%	1	2.5	L & N strip	
Hg lamp	Golay	<2% at 70 cm^{-1}	1	>3	L & N strip	
Nernst glower	TC	<0.1% at 700 cm^{-1}		3	L & N strip	

lengths, presents an energy problem further complicated by insensitivities of the detectors toward the low energies obtained in this region. Wider slit programs are used to optimize the signal-to-noise ratio, but in order to do this, resolution is sacrificed. Decreasing slit widths further compounds the energy problem. It is for this reason that the far-infrared spectroscopist has turned his attention toward the interferometric technique in the past ten years. The interferometric spectrophotometer is considerably more efficient in its light-gathering power than grating-type instruments, since it does not enploy gratings, prisms, or slits, and, in addition, affords one high resolution. For reviews on the subject of interferometers see Martin[10] and Hurley.[11]

Many developments have occurred since the first Michelson interferometer. As a result, commercial instruments are now available, and it is these instruments that will be discussed.

Types of Interferometers

Two types of interferometers have been found to be most useful for the far-infrared region. These are the lamellar grating interferometer[12] and the Michelson interferometer.[5] (A third type, the Fabry–Perot interferometer, will not be discussed. See Renk and Genzel[13] for a discussion of this type of interferometer.) The optical diagram of a lamellar grating interferometer is illustrated in Fig. 2-13, and in Fig. 2-14 a typical Michelson

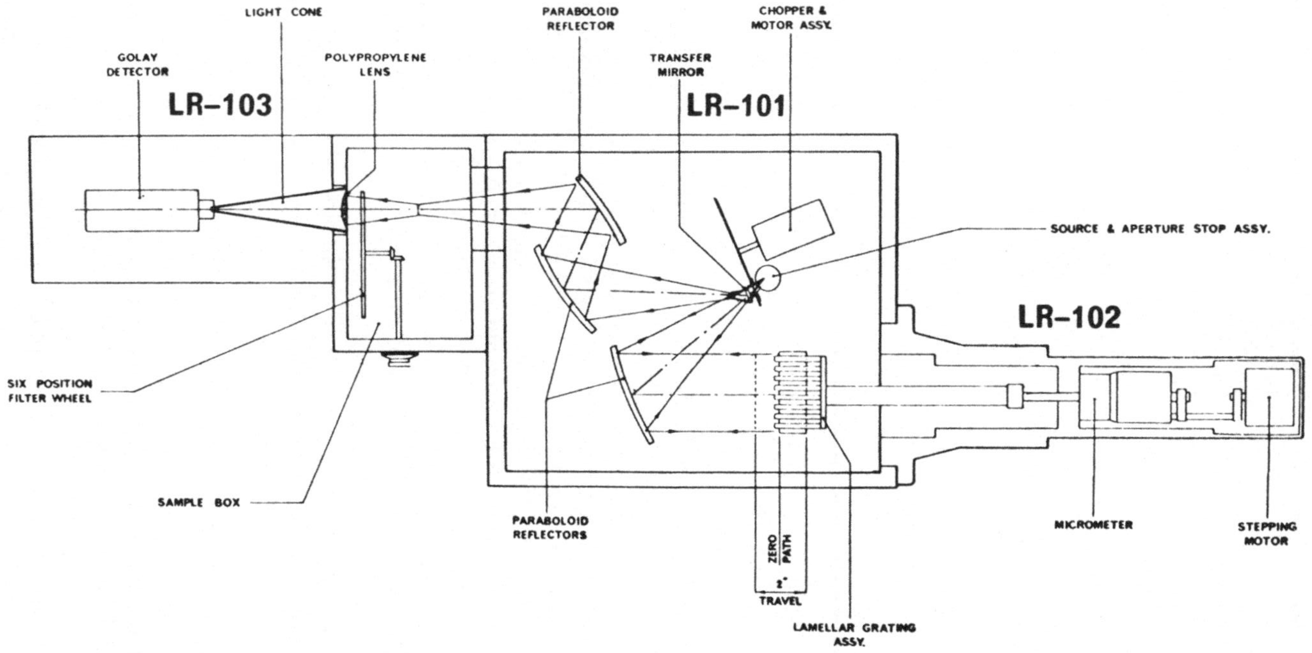

Fig. 2-13. Optical diagram of a lamellar grating interferometer, the LR-100. (Courtesy of Beckman Instruments, Inc., Fullerton, California.)

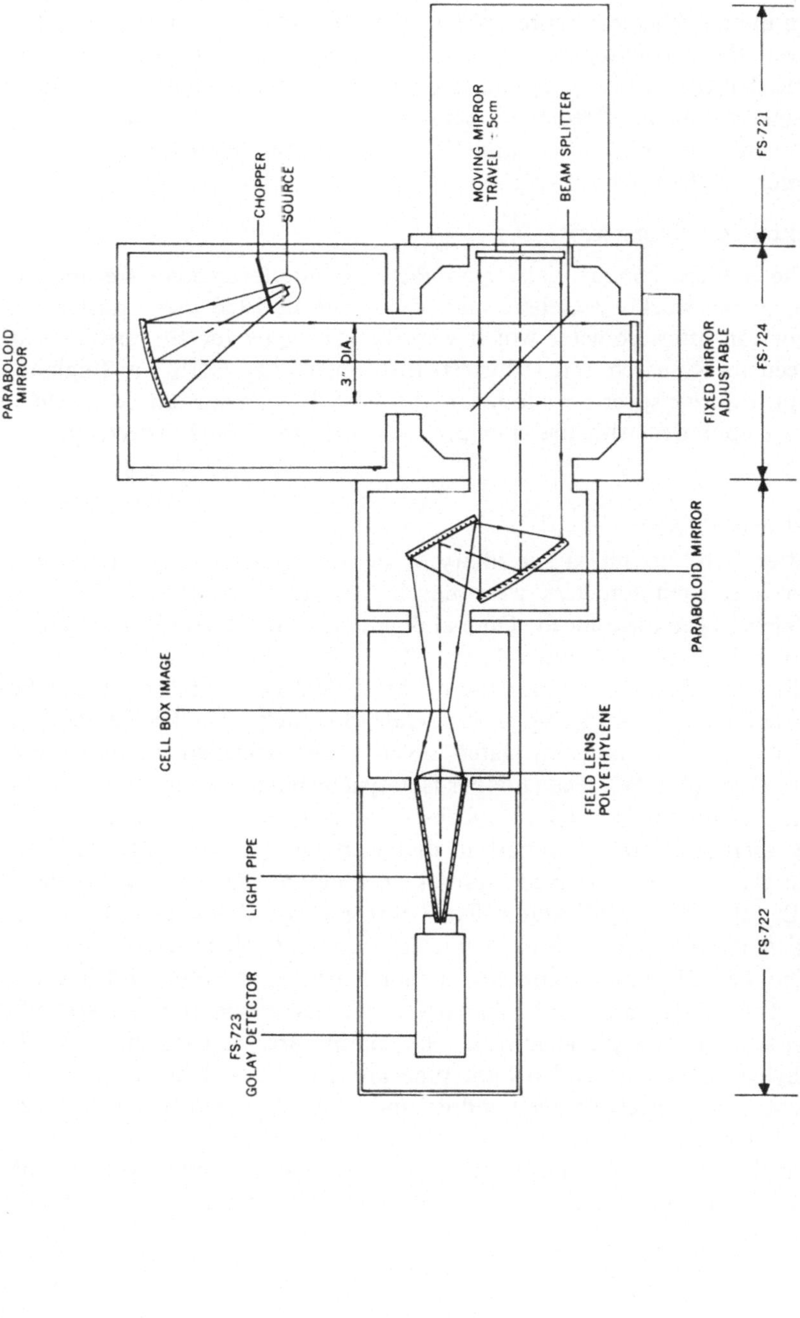

Fig. 2-14. Optical diagram of a Michelson interferometer, the FS-720. (Courtesy of Beckman Instruments Inc., Fullerton, California.)

interferometer is illustrated. The limiting feature of the Michelson interferometer is the beam splitter. The commonly used materials for beam splitters are a Mylar polyester film and metal mesh. Because of the limitation of the beam-splitting efficiencies of these materials, the lamellar grating interferometer is more efficient than the Michelson interferometer. However, the Michelson can be constructed for less cost, and it is quite possible that the two types of interferometers can complement one another. The lamellar-type interferometer, being more sensitive in the low-frequency region, can be used between 3 and 100 cm^{-1}, and the Michelson-type interferometer can be used above 100 cm^{-1}.

Michelson Interferometers

The modern commercially manufactured interferometers are an outgrowth of the work by Gebbie and Vanasse.[14] In 1956 they constructed a Michelson interferometer which was the prototype for the instruments produced by Research and Industrial Instruments Co. (RIIC) in England. The first Fourier spectrophotometer, the FS-520, was introduced in 1963. This was upgraded with the later production of the FS-620 spectrophotometer.

FS-720 and FS-820

After RIIC became a subsidiary of Beckman Instruments, additional Fourier spectrophotometers were developed. Recently, the FS-720 and FS-820 have been introduced. Optical diagrams of the FS-720 and FS-820 are shown in Figs. 2-14 and 2-15.

The FS-720 is intended for use in the 10–500 cm^{-1} region. It can be best described by considering each module separately. The FS-724 module (see Fig. 2-14) contains a water-cooled quartz-jacketed high-pressure mercury lamp (125 W). The chopper is a synchronous motor driven at 15 Hz. The collimator is a $f/1.5$ surface-aluminized off-axis paraboloid. The source aperature can be varied in diameters of 3, 5, and 10 mm. The module also contains the beam splitter on a mount made of polyethylene terephthalate and an adjustable fixed mirror (3 in. diameter), also on a mount. A moving mirror (3 in. diameter) is present which can travel ± 5 cm. The FS-721 module contains a moire grating monitor and a drive motor. The FS-722 module contains a condenser which is a surface off-axis paraboloid and plane mirror. The sample area is a $5\frac{3}{4}$-in. cube. A polyethylene field lens and a light pipe are also housed in this module. The light pipe is electroformed copper and transfers the energy to the detector, which is a Golay cell with a diamond window.

The FS-820 consists essentially of the source and interferometer of the FS-720. In place of the moire drive a stepping drive is used. In addi-

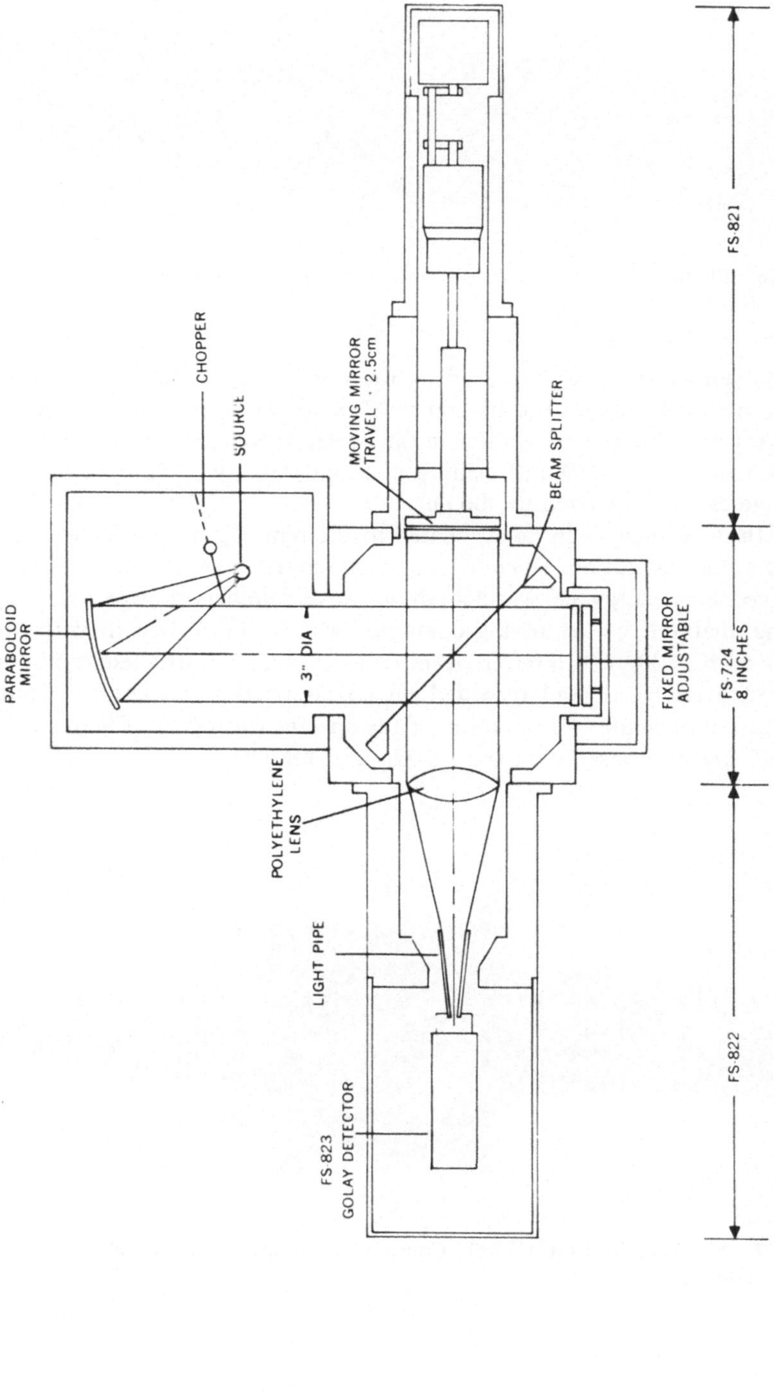

Fig. 2-15. Optical diagram of a Michelson interferometer, the FS-820. (Courtesy of Beckman Instruments, Inc., Fullerton, California.)

Fig. 2-16. Photograph of the FS-720. (Courtesy of Beckman Instruments Inc., Fullerton, California.)

tion, polyethylene lenses are used as the condensing optics. The instrument is intended only for the 10–200 cm^{-1} region because of the polyethylene lenses and the quartz window in the detector. By substituting a FS-722 module from the FS-720 and using a diamond-windowed Golay detector, the range can be increased to 400 cm^{-1}.

Both instruments can be evacuated to 0.1 mm Hg to eliminate atmospheric water absorption. Since the results are in terms of an interferogram, which contains energy *vs.* wavelength or wave number information, they must be analyzed by an analog computer which can be tied in with the FS-720 or FS-820. The interferogram can then be reconstructed in analog form, wave analyzed, and recorded on an attached x–y recorder. Thus, an almost immediate presentation of the spectra can occur. Figures 2-16 and 2-17 are photographs of the FS-720 and FS-820.

Fig. 2-17. Photograph of the FS-820. (Courtesy of Beckman Instruments, Inc., Fullerton, California.)

Grubb Parsons/N.P.L. Cube Interferometer MK 11

This instrument was developed by the National Physical Laboratory in England and is manufactured by Grubb Parsons. It is composed of five basic components and is the smallest commercially available interferometer. The source is a high-pressure mercury lamp, and the detector is a Golay cell with a quartz or diamond window. It operates from 10 to 200 cm^{-1} in standard operation but can be extended to 500 cm^{-1} with suitable accessories.

Grubb Parsons Iris Interferometric Spectrometer

This instrument employs Melinex beam splitters and is capable of operation from 10 to 500 cm^{-1}. A high-pressure mercury lamp serves as the source, and the detector is a Golay cell. The detector can be obtained with a diamond or quartz window. The instrument can be flushed with nitrogen or can be evacuated.

Dunn Associates, Inc., Block FTS-16

The Block FTS-16 is a Michelson-type interferometer of cube construction, utilizing a high-pressure mercury arc discharge lamp and a Golay cell (standard—quartz window; optional—diamond window). It can scan from 10 cm^{-1} to 200 cm^{-1} in standard operation, although it can be operated optionally from 5 cm^{-1} to 650 cm^{-1}. The resolution is claimed to be 0.5 cm. The instrument may be used for liquids, gases, or solids and may be purged or evacuated.

Dunn Associates, Inc., Block FTS-14

At the 20th Mid-America Spectroscopy Symposium held in Chicago, May 12–15, 1969, Dunn Associates introduced a new interferometer called the FTS-14. This instrument is capable of scanning 2.8–25 μ and can scan 1–1000 μ by simply changing a beam splitter. This is the first interferometer that can combine scans of the mid-infrared and far-infrared regions. Survey scans of one second may be taken giving resolutions of $\sim$4 cm^{-1}. Resolutions of $\sim$0.5 cm^{-1} in 8 sec. are also possible. A fast automatic Fourier transform computer achieves computation of a survey spectrum in 30 sec, and a high-resolution spectrum in 100 sec. Figure 2-18 shows a photograph of the basic components of the Block instrument, and Fig. 2-19 illustrates the optical schematic of the Block interferometer.

The FTS-14 incorporates a Michelson interferometer, a stabilized infrared source, a new infrared detector, appropriate signal electronics, a general-purpose digital computer, programs to control the instrument and perform data reduction and analysis, a Teletype, a high-speed digital plotter, and a variety of sampling accommodations.[15] The instrument employs a

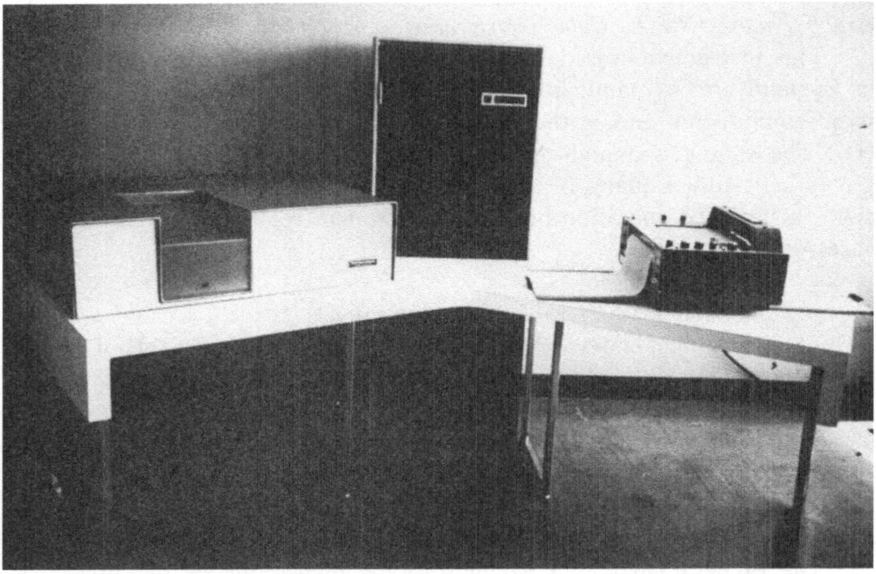

Fig. 2-18. Photograph of basic components of the Block interferometer. (Courtesy of Dunn Analytical Instruments, Silver Spring, Maryland.)

double-beam system which produces spectra at speeds up to 300 times that of grating instruments. The infrared source is a glower operated at about 1100°C, stabilized by an electroservo to limit any infrared intensity fluctuations to within $\pm 0.2\%$ at 2 μ. The detector is a triglycine sulfate (TGS) pyroelectric bolometer, which operates at room temperature.

Coderg Co. MIR 2

Another new interferometer has recently been introduced by the Coderg Co. is France. It is the MIR 2 of the Michelson type which is capable of obtaining spectra from 800 to 10 cm^{-1} at high resolution and in a relatively short time. It is a double-beam interferometer employing a Unicam Golay detector.

Lamellar Grating Interferometers

Beckman Instruments Co. LR-100

At the Pittsburgh–Cleveland Conference in Cleveland, held in March 1968, a new interferometer was introduced by the Beckman Instruments Co. This instrument is the LR-100 lamellar grating interferometer and is the first commercially available instrument of this type. The interference modulator in the lamellar instrument consists of 4.75-mm Al plates which

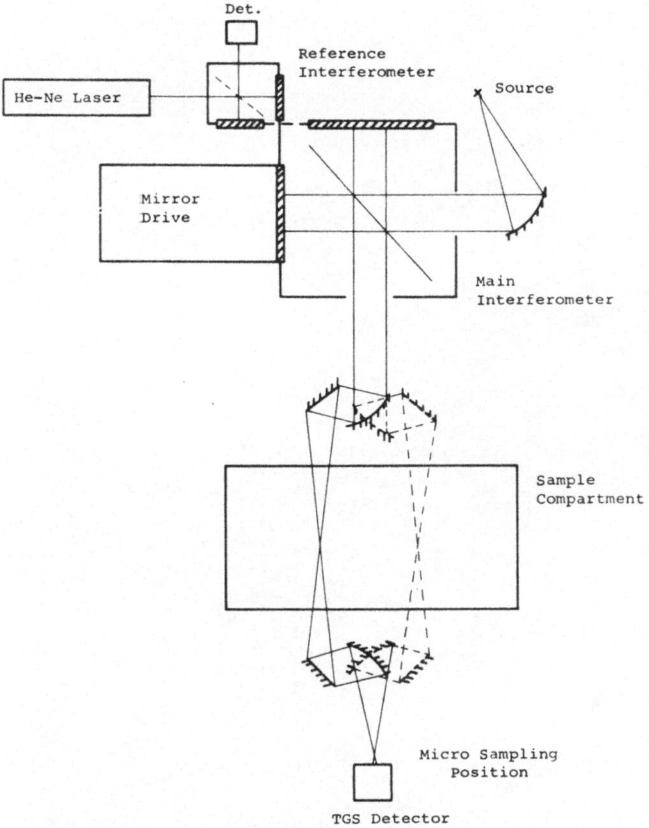

Fig. 2-19. Schematics of optical layout of the Block interferometer. (Courtesy of Dunn Analytical Instruments, Silver Spring, Maryland.)

act as a wavefront divider. Strong and Vanasse[6] designed the first lamellar interferometer. The instrument covers the range 3–70 cm^{-1}. The theoretical maximum resolutions are 0.1 cm^{-1} for a single-sided interferogram and 0.2 cm^{-1} for a double-sided interferogram. The resolution is observed to be better than a Michelson instrument at the low-frequency end. The optical diagram of the LR-100 is shown in Fig. 2-13.

Perkin–Elmer Corp. Model 180

Despite the competition received from the interferometric instruments, instrument makers are busy attempting to improve the conventional grating-type spectrophotometers.

At the Eastern Analytical Symposium held in New York, November

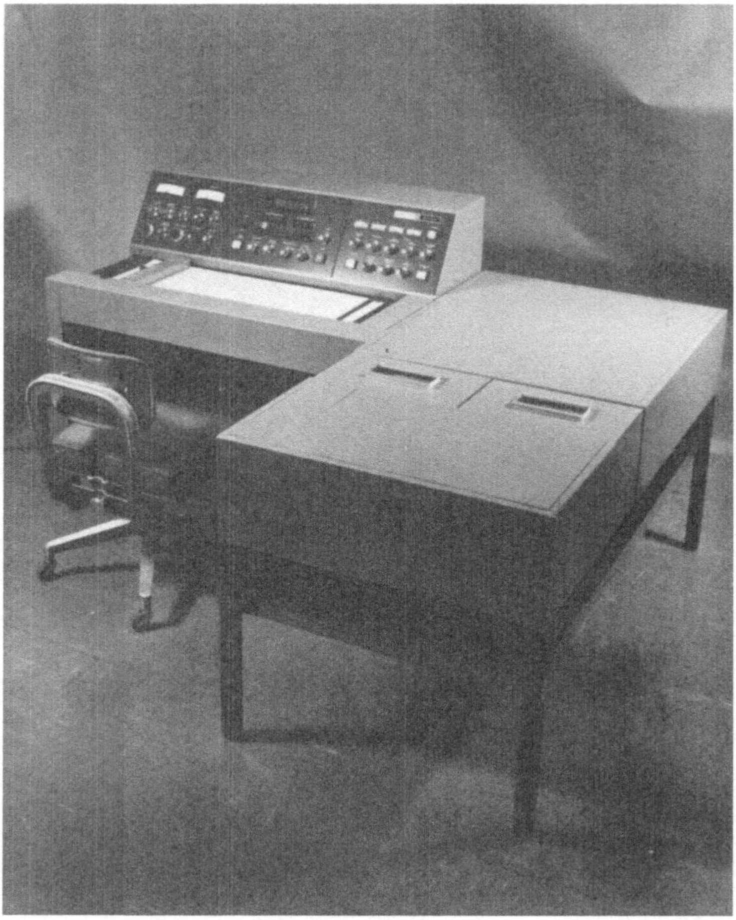

Fig. 2-20. The Perkin–Elmer Model 180 infrared spectrophotometer. (Courtesy of Perkin–Elmer Corp., Norwalk, Conn.)

19–21, 1969, the Perkin–Elmer Corporation introduced a new instrument, the Model 180. The instrument is shown in Fig. 2-20 and its optical schematic appears in Fig. 2-21.

The Model 180 uses a full-aperature time-shared electrical-null ratio-recording system. A 500-mm focal length Ebert monochromator is employed, which uses five 84×84 mm diffraction gratings, each used in the first order only. The instrument may be operated in three modes—constant I_0, programmed slits, and manual slits. The source is an air-cooled Globar. A thermopile serves as the detector. The basic range of the instrument under survey conditions is 4000 to 180 cm^{-1}.

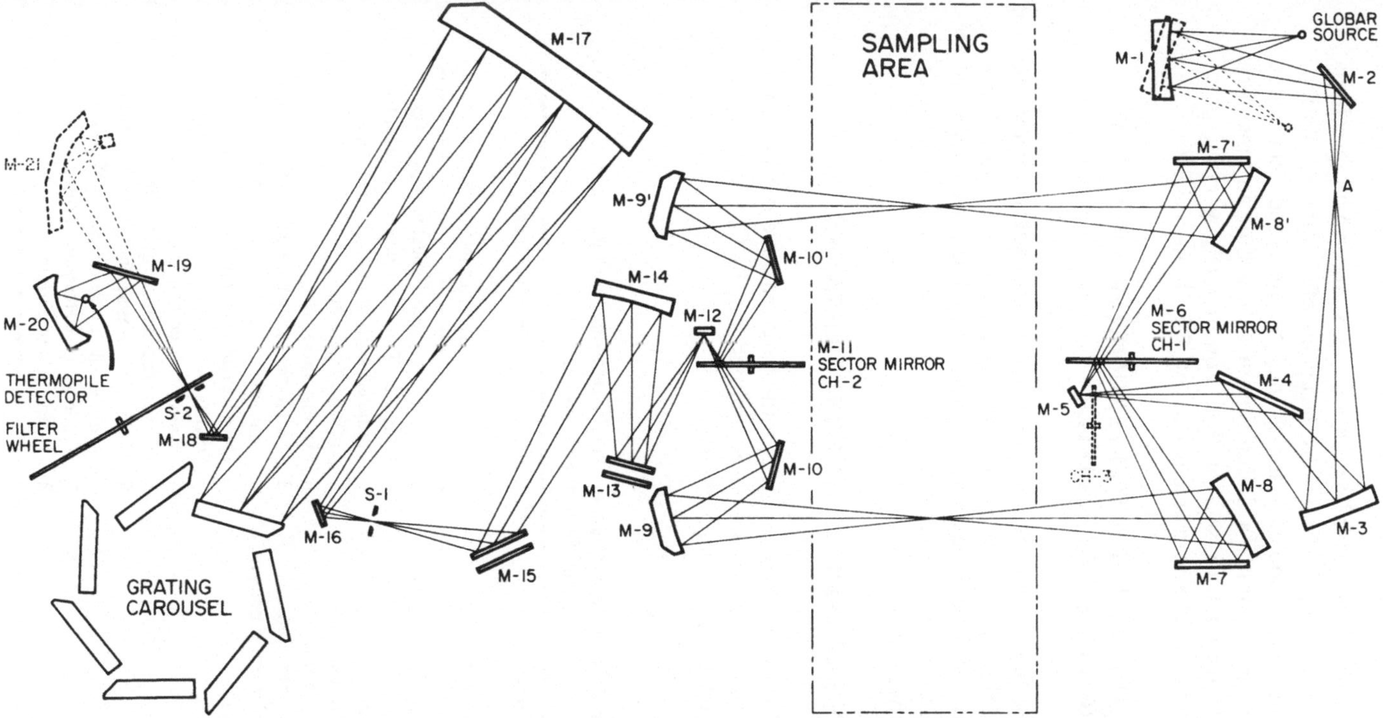

Fig. 2-21. Optical schematic of the Perkin–Elmer Model 180 Infrared Spectrophotometer.

M-1) Toroid; dotted lines near M-1 show location of alternate source—in this case, a mercury arc for planned far-infrared extension; M-2) flat; A) source focus—alternate sources could be located here if it is desired to study their spectral properties; M-3) toroid; M-4) flat; CH-3) alternate 180° single beam chopper—used optionally in place of M-6 and M-11 for high-energy single-beam operation; M-5) field toroid; M-6) 15-Hz chopper; M-7) flat; M-7′) flat; M-8) toroid; M-8′) toroid; M-9) toroid; M-9′) toroid; M-10) flat; M-10′) flat; M-11) 30-Hz chopper; M-12) field toroid; M-13) flat mirror on one side, coarse diffraction grating on back side (mount/rotates introducing grating at 250 cm^{-1}; grating is used to scatter short-wavelength energy); M-14) toroid; M-15) flat mirror on one side, coarse diffraction grating on back side (mount rotates introducing grating at 250 cm^{-1}; grating is used to scatter short-wavelength energy); S-1) monochromator entrance slits; M-16) flat; M-17) spherical Ebert monochromator mirror; grating carousel with positions for 7 gratings—5 used to cover standard range; 2 positions for gratings to accomplish planned far-infrared extension; M-18) flat; S-2) monochromator exit slits; 14-position filter wheel; 9 filters used over standard frequency range; 5 positions reserved for filters to be used with planned far-infrared extension; M-19) flat; M-20) on-axis ellipsoidal mirror; 2-element thermopile detector with flat cesium iodide window; M-21) 90° off-axis ellipsoidal mirror. (This is shown dotted to indicate that it is a part of the planned far-infrared extension and not included with standard-range instrument. When used, mirror M-19 swings out of the path, and beam from S-2 passes through filter and directly to M-21, which focuses energy onto Golay detector. Golay detector is part of planned far-infrared extension and not included with standard instrument.) (Courtesy of the Perkin–Elmer Corp., Norwalk, Conn.)

The full-aperature ratio-recording chopper system, which prechops the IR beam, is a feature which eliminates the effects of sample reradiation. Thus, spectra of heated or cooled samples may be measured accurately.

The instrument incorporates a digital output with a built-in "memory bank" capable of storing four sets of four continuously varying parameters for instant recall. The instrument will sell for about $33,000.

2.4. COMPARISON OF THE INTERFEROMETER WITH THE CONVENTIONAL SPECTROMETER

Since both the interferometer and the conventional spectrometer are commercially available, a comparison of the relative merits of both types of instrumentation becomes necessary. Comparison of the two types of instruments has been made by Richards,[16] Hurley,[11] and Miller.[17] Some of the points of comparison are listed in Table 2-5.

2.5. TRENDS IN FAR-INFRARED INSTRUMENTATION

The need for more powerful sources and more sensitive detectors for use in the low-frequency region has been pointed out in Chapter 1. The developments in this area were also cited.

Table 2-5. Comparison of the Interferometer with the Spectrophotometer[17]

Advantages of interferometer	Disadvantages of interferometer
1. Higher energy, particularly at lower frequencies.	1. Time delay until spectrum is reproduced. This is being constantly reduced with computers tied directly into instrument.
2. Higher resolution potential if long running times are used.	2. Computation costs.
3. Lower frequency not limited	3. Single-beam instrument. To get an equivalent of a double-beam scan requires more time and is more expensive.
4. Convenience of operation.	4. Beam splitter changes are necessary.
5. Easily adapted for other detectors.	5. Fewer accessories available.
	6. Difficult to determine if one has poor sample or instrumental difficulties.
	7. Difficult to determine noise.
	8. Intensities uncertain.

An additional trend in far-infrared instrumentation has been the use of multiple-scan interferometry. Applications for this technique can be found in absorption, reflection, or emission infrared spectroscopy.[18-21] The sum of the numerous individual interferograms is stored digitally in the core memory of the computer. The built-up signal is converted automatically into a normal spectrum. The use of multiple-scan interferometry increases the signal-to-noise ratio. Speed is not sacrificed, for a signal scan may take only about 0.1 sec. With the commercial availability of interferometers, it can be expected that more applications in chemistry and physics will develop using this type of spectroscopy.

BIBLIOGRAPHY

1. N. Wright and L. W. Herscher *J. Opt. Soc. Am.* **37,** 211 (1947).
2. L. Basile, C. Postmus, and J. R. Ferraro, *Spectry. Letters* **1,** 189 (1968).
3. *Instr. News* **19,** 1 (1968).
4. *Ind. Res., 1970 Instruments Specifics Annual*, p. 48, November 20 (1969).
5. A. A. Michelson, *Studies in Optics*, Univ. of Chicago Press, Chicago, Illinois (1927).
6. J. D. Strong and G. A. Vanasse, *J. Opt. Soc. Am.* **50,** 113 (1960), **49,** 844 (1959); *J. Phys. Radium* **19,** 192 (1958).
7. P. Jacquinot, *Rep. Prog. Phys.* **23,** 267 (1960).
8. L. Genzel, *J. Mol. Spectry.* **4,** 24 (1960).
9. J. Connes, *Rev. Optique* **40,** 45 (1961).
10. P. L. Richards, in *Spectroscopic Techniques* (D. H. Martin, Ed.,) John Wiley and Sons, Inc., New York (1967).
11. W. J. Hurley, *J. Chem. Ed.* **43,** 236 (1966).
12. J. D. Strong and G. A. Vanasse, *J. Phys. Radium* **19,** 192 (1958).
13. K. F. Renk and L. Genzel, *Appl. Opt.* **1,** 643 (1962).
14. H. A. Gebbie and G. A. Vanasse, *Nature* **178,** 432 (1956).
15. S. T. Dunn and M. J. Block, *Amer. Lab.*, **52,** November (1969).
16. P. L. Richards, *J. Opt. Soc. Am.* **54,** 1474 (1964).
17. F. A. Miller, Institute Symposium on Far-Infrared Transpose Spectroscopy (MISFITS), June (1965).
18. *Chem. Eng. News* **43,** 40 (1965).
19. M.J.D. Low, *Science and Technology*, February (1967).
20. M.J.D. Low, R. Epstein, and A. C. Bond, *Chem. Commun.*, 226 (1967).
21. M.J.D. Low, *Appl. Opt.* **6,** 1503 (1967).

Chapter 3

SAMPLING TECHNIQUES AND INSTRUMENT CALIBRATION IN THE FAR-INFRARED REGION

3.1. INTRODUCTION

Materials used as windows in the far-infrared region will be discussed, including plastics, salt crystals, and semiconductors. Next, solvents useful in the far infrared will be mentioned, and then sampling techniques. Finally, calibration of instruments will be discussed.

3.2. WINDOW MATERIALS FOR CELLS

Windows must, of course, be transparent in the region to be probed. However, a good window material must also be rigid, tough, and relatively cheap, and, in addition, must be moisture, solvent, and chemical resistant. It should be able to be used at nonambient temperatures. At present, hardly any material will fulfill all these requirements and more research is required in this area. Table 3-1 lists several materials which have been used as materials in the far infrared. They will be discussed below.

Plastic Windows

Many synthetic polymers have been studied for far-infrared transparency.[3] Of these, high-density polyethylene offers many advantages and appears to be the best suited for the far infrared. It is rigid and inexpensive and is transparent below 650 cm^{-1}, although it does show an absorption at about 75 cm^{-1}. It resists solvents fairly well and has been used for solutions involving fluorine compounds.[4–6] The presence of the fluorine converts the polyethylene surface to a very thin film of the fluorinated polyethylene, which shows only weak absorptions in the far-infrared region. Polyethylene does show undesirable properties when heated. For work at nonambient temperatures, other polymeric materials can be used. Teflon, polypropylene,

Table 3-1. **Window Materials for Far-Infrared Spectroscopy**

Material	Practical frequency limit, cm^{-1}	Remarks
KBr	250	
TlBr	400	
CsBr	200	
CsI	145	
KRS-5	225	High refractive index—high reflection losses.
Polyethylene	10	Very versatile at ambient temperatures.
Polypropylene	*a*	Can be used to 150°C.
4-Methylpentene-1	*a*	Can be used to 200°C.
Teflon	*a*	Can be used to 200°C.
Polystyrene	40	
Kel-F[b]		
Fluorothene[c]	<100	For fluorine compounds.
Germanium		High refractive index.
Silicon[1,2]	*a*	High refractive index, shows conduction bands of impurities at −196°C.
Diamond	5	Can be used at ambient and nonambient temperatures.
Quartz (crystal)	<150	
Mica	< 50	

[a] Has far-infrared absorptions.
[b] Chlorotrifluoroethylene-vinylidene fluoride copolymer.
[c] Formed by putting fluorine compounds in polyethylene-fluorthene formed on surface.

and poly 4-methylpentene-1 have been used for high-temperature spectra. However, all these materials have some absorptions in the far-infrared region.

Salt-Type Windows

From 650 to 200 cm^{-1} KBr, TlBr, CsBr, CsI, and KRS-5 (thallium bromide–iodide) windows can be used. With the exception of TlBr, all of these materials are moisture sensitive and must be kept under conditions of low humidity. There are also other disadvantages: salt windows are easily broken, KBr scratches easily, TlBr is toxic, and KRS-5 is soft.

Silicon, Germanium, and Diamond Windows

The semiconducting materials such as silicon and germanium show promise as far-infrared windows. However, they have high refractive indices, and, therefore, show high reflective losses. Diamond appears to be the most versatile far-infrared window available. It is hard, chemically resistant, transparent, and can be used at temperatures of 300°C or higher. Unfortunately, it is expensive.

Quartz and Mica

Crystalline quartz up to 1 mm thick can be used in the region below 150 cm^{-1}. Here again, depending on the type of quartz, some low-frequency absorptions are found. Mica has also been used in this region.

At present, it may be necessary to use several types of windows to obtain spectra in the far infrared. For example, polyethylene can be used for room-temperature spectra and quartz for high-temperature spectra. Dia-

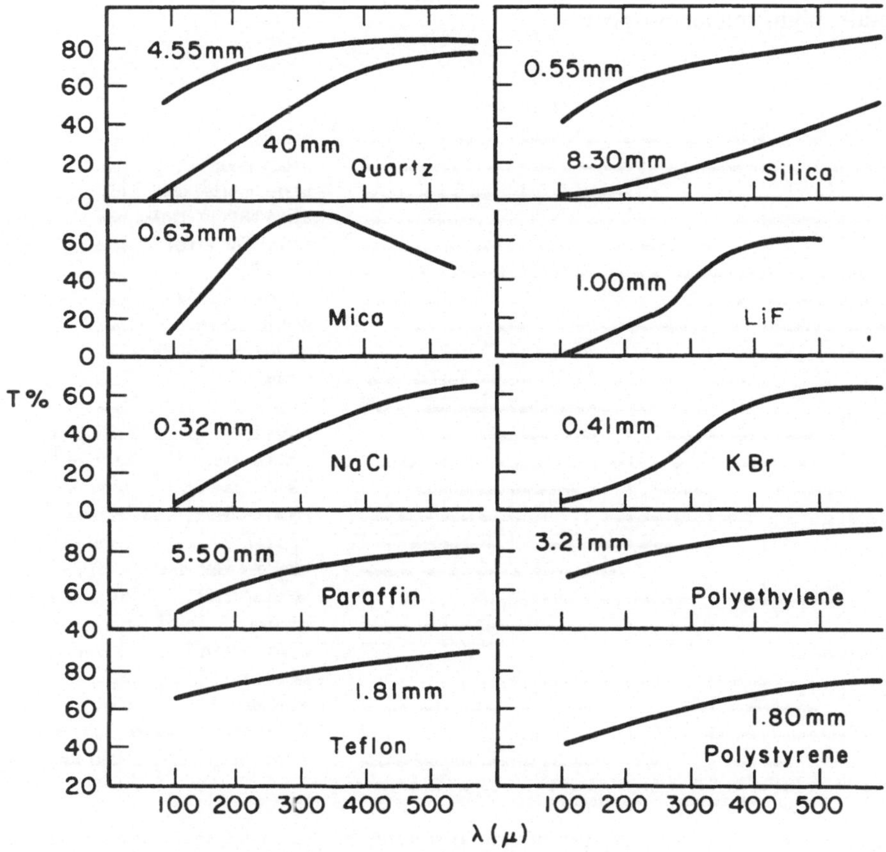

Fig. 3-1. Transmission of various window materials in the far-infrared region. (Courtesy of Pergamon Press, New York. From A. Hadni.[7])

mond windows, on the other hand, can be used for ambient and nonambient temperatures.

Figure 3-1 shows the transmission properties of various optical materials.

3.3. FAR-INFRARED SOLVENTS

Suitable solvents are available for use in the far-infrared region. Several studies to determine the transparency of organic liquids in the far infrared have been made.[8] Figure 3-2 illustrates the transparency characteristics of several solvents. Nujol is transmissive and very useful as a mulling agent. Benzene has an absorption at 300 cm^{-1} and is quite useful as a far-infrared solvent, as are cyclohexane (weak band at 240 cm^{-1}) and carbon disulfide (medium band at 260 cm^{-1}). In some cases one can use several solvents to obtain high-frequency spectra.

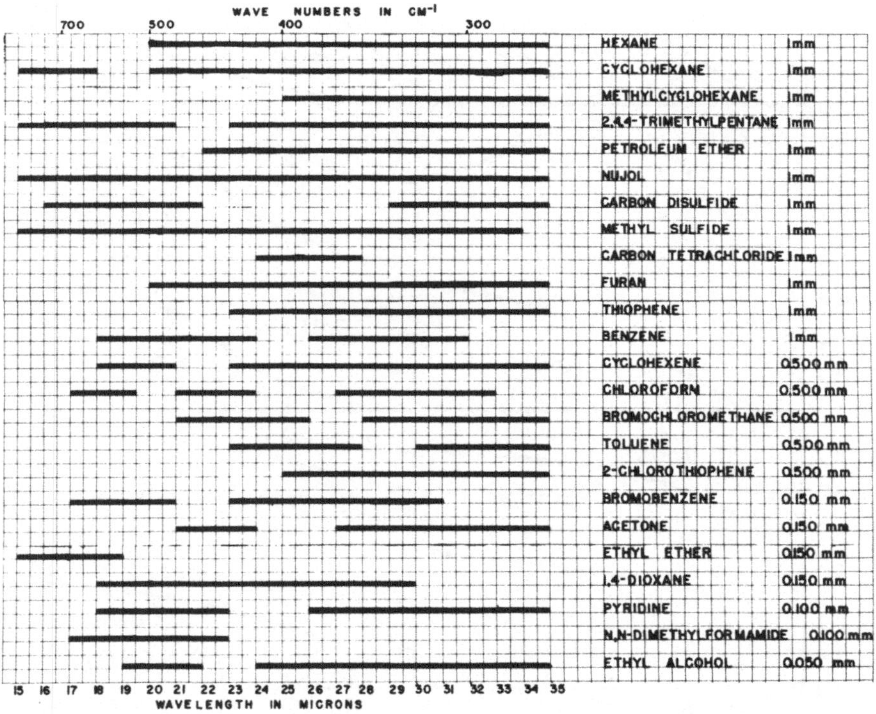

Fig. 3-2. Transmission properties of several solvents in the far-infrared region. The black lines represent useful regions. (Courtesy of F. F. Bentley, Wright-Patterson Air Force Base, Ohio)

Table 3-2. Sampling Techniques in the Far Infrared

Solids	*Remarks*
Nujol mull	Use polyethylene windows
KBr pellet	
CsBr pellet	
CsI pellet	
Polyethylene pellet	
Between diamond windows	No matrix necessary
Molten spectra	Quartz, CsI, or teflon windows
Solid on polyethylene sheet	Rub solid into polyethylene
Liquids	
Liquid film	Use polyethylene windows
Solution	
Dissolve in appropriate solvent	Use polyethylene windows

3.4. SAMPLING TECHNIQUES

In the far-infrared region one is usually concerned with spectra of solids, solutions, and liquids. Table 3-2 lists the possible techniques available for sampling of those materials. The choice of method depends on the particular compound whose spectra is to be determined.

3.5. CALIBRATION OF FAR-INFRARED INSTRUMENTS

As in the conventional infrared region, it is necessary to calibrate far-infrared instruments. In common use are the calibration of instruments with gases, orders of mercury, and mercuric oxide.

Rotation Spectra of Gases

Far-infrared instruments can be calibrated by the use of known frequencies for the rotational spectra of simple gases. A number of gases can be used, such as HCl, HBr, DBr, HI, CO, HF, HCN, N_2O, NH_3, and H_2O. Calibration with water vapor or the orders of mercury is preferred, as the other gases are either toxic or corrosive and must be placed in an infrared gas cell. On the other hand, the instruments may be easily exposed to the open atmosphere containing water vapor, and the calibration made with the numerous water vapor absorption bands appearing in the far-infrared spectrum. Figure 1-2 shows the pure rotational spectrum of water vapor between 150 and 350 cm^{-1}. Wavelength calibrations from 30 to 1000 μ using HCN, CO, N_2O, and H_2O vapor have been cited by Rao *et al.*[9]

Table 3-3. Observed Transitions for Rotational Spectrum of HF

Transition $J \rightarrow J'$	Vacuum wave numbers (observed), cm^{-1}
$0 \rightarrow 1$	41.30 ± 0.71
$1 \rightarrow 2$	82.35 ± 0.25
$2 \rightarrow 3$	122.83 ± 0.28
$3 \rightarrow 4$	163.92 ± 0.13
$4 \rightarrow 5$	204.50 ± 0.12
$5 \rightarrow 6$	244.97 ± 0.16
$6 \rightarrow 7$	284.98 ± 0.13
$7 \rightarrow 8$	324.52 ± 0.22
$8 \rightarrow 9$	363.89 ± 0.18
$9 \rightarrow 10$	402.77 ± 0.18
$10 \rightarrow 11$	441.05 ± 0.29

Rothschild[10] has measured the pure spectrum of HF vapor from 22 to 250 μ. Table 3-3 shows the transitions for the rotacion spectrum of HF. Figure 3-3 shows typical calibration data obtained for a Perkin–Elmer No. 301 spectrophotometer using H_2O vapor and the orders of mercury for calibration in the various grating regions.

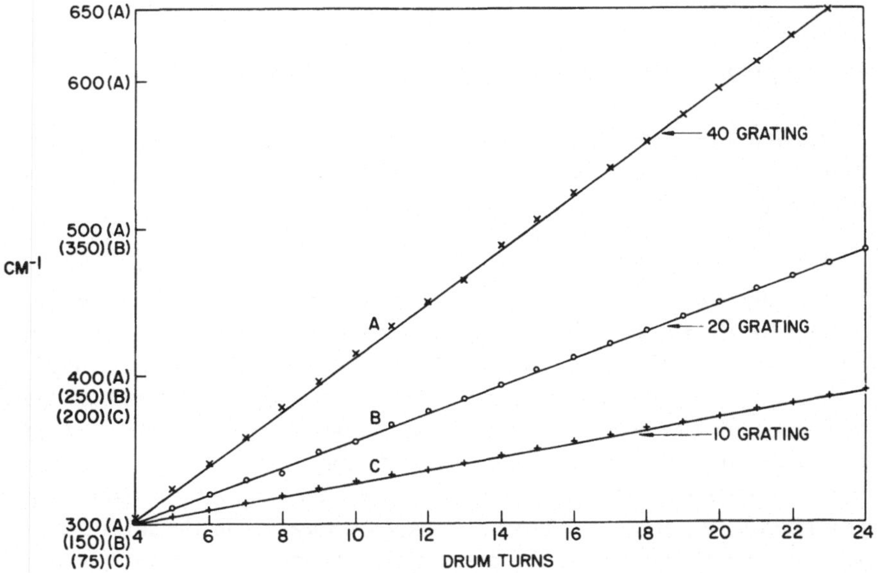

Fig. 3-3. Typical calibration data obtained for the Perkin–Elmer No. 301 spectrophotometer in the far-infrared region.

Table 3-4. Typical Data Obtained with the Orders of Hg for the Perkin–Elmer No. 301 Spectrophotometer

μ	Order	cm^{-1}	μ	Order	cm^{-1}
22.3890	41	446.65	41.5016	76	240.95
22.9351	42	436.01	42.0477	77	237.83
23.4812	43	425.87	42.5938	78	234.78
24.0272	44	416.19	43.1398	79	231.80
24.5733	45	406.95	43.6859	80	228.91
25.1194	46	398.10	44.2320	81	226.08
25.6655	47	389.63	44.7781	82	223.32
26.2115	48	381.51	45.3241	83	220.63
26.7576	49	373.73	45.8702	84	218.01
27.3037	50	366.25	46.4163	85	215.44
27.8498	51	359.07	46.9623	86	212.94
28.3958	52	352.16	47.5084	87	210.49
28.9419	53	345.52	48.0545	88	208.10
29.4880	54	339.12	48.6006	89	205.76
30.0341	55	332.96	49.1466	90	203.47
30.5801	56	327.01	49.6927	91	201.24
31.1262	57	321.27	50.2388	92	199.05
31.6723	58	315.73	50.7840	93	196.91
32.2184	59	310.38	51.3309	94	194.81
32.7644	60	305.21	51.8770	95	192.76
33.3105	61	300.21	52.4231	96	190.76
33.8566	62	295.36	52.9692	97	188.79
34.4026	63	290.68	53.5152	98	186.86
34.9487	64	286.13	54.0613	99	184.98
35.4948	65	281.73	54.6074	100	183.13
36.0409	66	277.46	55.1535	101	181.31
36.5869	67	273.32	55.6995	102	179.53
37.1330	68	269.30	56.2456	103	177.79
37.6791	69	265.40	56.7917	104	176.08
38.2252	70	261.61	57.3377	105	174.41
38.7712	71	257.92			
39.3173	72	254.34			
39.8634	73	250.86			
40.4095	74	247.47			
40.9555	75	244.17			

Mercury Lines

The mercury lines from a mercury lamp can be used for calibration in the far-infrared region. Table 3-4 shows typical data obtained for the Perkin–Elmer No. 301 spectrophotometer.

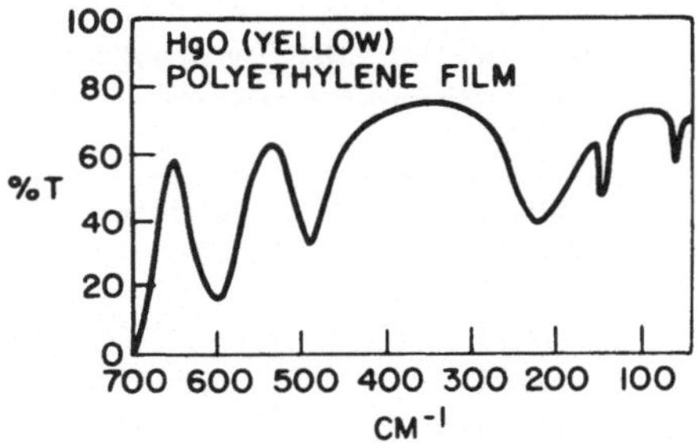

Fig. 3-4. Spectrum of HgO from 50 to 750 cm^{-1}. (Courtesy of *Journal of the Optical Society of America*, American Institute of Physics. From McDevitt and Davidson.[11])

Mercuric Oxide

Mercuric oxide is substance that is as versatile for calibration in the far infrared as polystyrene is in the conventional region.[11] The material has several bands which serve as calibration points (596, 492, 225, 145, 60 cm^{-1}). Figure 3-4 illustrates the spectrum from 750 to 50 cm^{-1} for HgO.

BIBLIOGRAPHY

1. W. Fateley, R. E. Witkowski, and G. L. Carlson, *Appl. Spectry.* **20,** 190 (1966).
2. J. R. Durig, S. F. Rush, and F. G. Baglin, *Appl. Spectry.* **22,** 212 (1968).
3. H. A. Willis, R. G. Miller, D. M. Adams, and H. A. Gebbie, *Spectrochim. Acta* **19,** 1457 (1963).
4. H. H. Claassen and H. Selig, *J. Chem. Phys.* **43,** 103 (1965).
5. H. H. Claassen and H. Selig, *J. Chem. Phys.* **44,** 4039 (1966).
6. B. Frlec and H. H. Claassen, *J. Chem. Phys.* **46,** 4603 (1967).
7. A. Hadni, *Essentials of Modern Physics Applied to the Study of the Infrared*, Pergamon Press, New York (1967).
8. H. A. Willis, R. G. J. Miller, D. M. Adams, and H. A. Gebbie, *Spectrochim. Acta* **19,** 1457 (1963).
9. K. N. Rao, R. V. de Vore, and S. K. Plyler, *J. Res. Natl. Bur. Std.* **67A,** 351 (1963).
10. W. G. Rothschild, *J. Opt. Soc. Am.* **54,** 20 (1964).
11. N. T. McDevitt and A. D. Davidson, *J. Opt. Soc. Am.* **55,** 1695 (1965).

Chapter 4

NEW TECHNIQUES USED WITH FAR-INFRARED MEASUREMENTS

4.1. INTRODUCTION

Recently, several new techniques have been used with far-infrared measurements. That the use of other physical tools complements measurements made in the mid-infrared region has been well documented in the literature. Examples of the use of diffraction techniques, nuclear magnetic resonance, chromatographic methods, Raman scattering, visible and ultraviolet spectroscopy, mass spectroscopy, electron spin resonance, and others in conjunction with infrared spectroscopy are numerous. No attempt will be made to discuss these. Only the complementation of the newest techniques, such as the slow-neutron scattering (SNS) and high-pressure methods with far-infrared spectroscopy, will be discussed in this chapter.

4.2. THE SLOW-NEUTRON SCATTERING (SNS) TECHNIQUE

The use of Raman scattering as a complementary tool to infrared (or *vice versa*) is long known. These two techniques work together so well because of the different selection rules obtained for each method. Whereas a dipole moment change is necessary for an infrared absorption to occur, a polarizability change is needed in Raman scattering.

By comparison, slow-neutron scattering requires no selection rules at all and is thus complementary to both infrared and Raman spectroscopy. It is particularly attractive for use with inorganic compounds such as hydrates and hydrogen-bonded molecules. The type of information that can be obtained with the SNS method is shown in Table 4-1. The method offers much promise in the area of molecular and lattice dynamics and is rapidly taking a respected place with the Raman and infrared tools in molecular spectroscopy. Table 4-2 lists some advantages and disadvantages of the method.

Table 4-1. Information Obtained from SNS Method[a]

Neutron wavelength (Å)

2.5	2.0	1.5		1.0
		Vibrational molecular levels		
Rotational molecular levels				
		Optical lattice vibrations, light elements		
Acoustic lattice vibrations				

20	40	60	80	100

E, meV

meV	cm^{-1}
100	800
63	500
13	100
1.3	10

[a] After Ferraro.[1]

Theory

A discussion of the theory of slow-neutron scattering is beyond the scope of this book. The interested reader is referred to Lomer and Low[2] and Blume.[3]

Table 4-2. Advantages and Disadvantages of SNS Method[a]

Advantages	Disadvantages
1. Nondestructive technique.	1. Large-sized sample needed (e.g., ~5 g for hydrate).
2. No selection rules. All hydrogen motions observed.	2. Limited to low-frequency range (~800 cm^{-1}—energy gain experiment used—higher states unpopulated at room temperature. Can be counteracted by doing energy loss experiments).
3. Aids far-infrared interpretations.	3. Poor resolution.
4. Potential observation of wide ranges of dispersion of optic modes.	4. Expensive—need reactor to produce neutrons.
5. For liquids, direct observation of diffusive motion.	5. Complexity of interpretations.

[a] After Ferraro.[1]

Instrumentation

Several experimental techniques are used in SNS, and these are listed in the book by Egelstaff.[4] Briefly, a source of neutrons is provided by an experimental reactor. (The need for a reactor for this type of research has been the major reason for the delay in the development of the SNS method.) The neutron beam is monochromatized either by causing Bragg scattering through a crystal such as copper, lead, aluminum, quartz, or germanium, by passing it through a mechanical velocity selector, or by use of a beryllium filter technique. The latter technique is perhaps the most common. The beryllium acts as a low-pass filter, being transparent for neutrons of energy $<40\ cm^{-1}$ and opaque for higher energies. The low-energy neutrons are then scattered with great efficiency by the sample studied. (A schematic for such an instrument is shown in Fig. 4-1.)

The gain or loss in energy of the neutrons is measured after scattering from the sample. In order to make the energy analysis of the neutrons, they are passed through a chopper which gives a pulsation of the beam. They are then detected by $^{10}BF_3$ counters, and the velocity distribution of the scattered neutrons is measured by conventional time-of-flight techniques, using a multichannel analyzer.

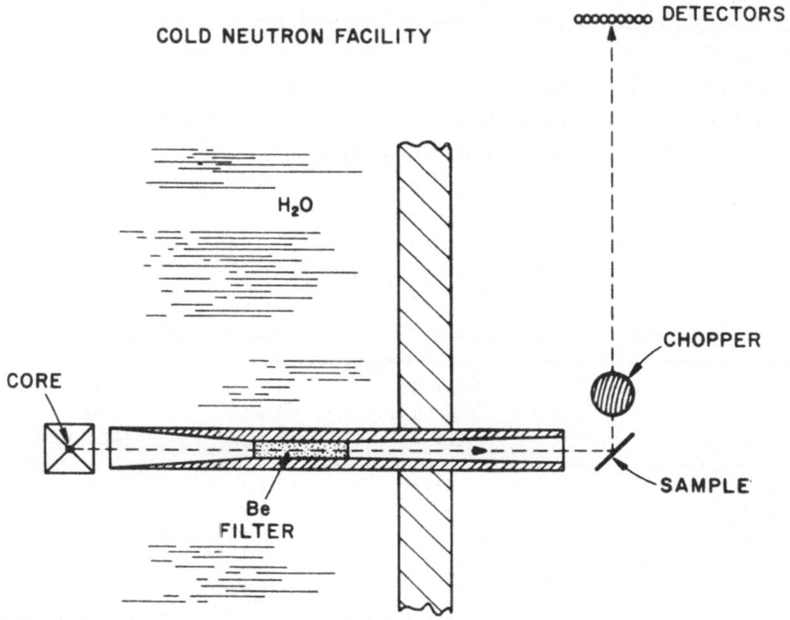

Fig. 4-1. Schematic diagram for the SNS instrumentation. (Courtesy of *Applied Spectroscopy*, American Institute of Physics. From Treviño.[5])

Applications

The SNS method has been used to measure the hindered internal rotational frequencies of solids containing the ammonium ion.[6-8] Table 4-3 shows some results for the ammonium salts. The results indicate that there is an increase in the freedom of rotation of the ammonium ion from $NH_4F \rightarrow NH_4Cl \rightarrow NH_4Br \rightarrow NH_4I$.

The scattering of low-energy neutrons by solid hydrates has been used to study librational and translational motions of the water molecules.[9,10] Boutin *et al.*[10] have studied several hydrates by the SNS method. Comparisons of these results with infrared measurements are shown in Table 4-4. Similar studies have been made by Rush, Ferraro, and Walker,[15] and these results are tabulated in Tables 4-5 and 4-6.

Point-for-point agreements with infrared results are not to be expected since the SNS method will study modes not allowed in the infrared, such as acoustic modes. Furthermore, one obtains the frequency for a wave vector equal to or very close to zero in infrared or Raman spectroscopy, while in SNS spectra energy transfers over a wide range of wave vectors are observed. Because of this and the poor resolution obtained, peak energies in SNS are broad. It is possible that the band envelope may include several modes of vibrations. Further differences are possible, since some nonhydrogen motions active in infrared may be seen only as weak bands in the SNS technique. Despite this, fair agreement is observed for the librational frequencies of water in hydrated compounds. The complementation of the SNS method with far-infrared spectroscopy may aid in far-infrared interpretations in some cases.

Table 4-3. SNS Results for Several Ammonium Salts

Compound	Torsional frequency, cm^{-1}	Slope,[a] barns/Å	Barrier to rotation, kcal/mol
NH_4F	523	2.8	10
NH_4Cl	359	4.8	5.2
NH_4Br	311	5.7	4
NH_4I (phase II)	279	6.5	2.7
NH_4I (phase I)		11.3	0.4 (freely rotating)

[a] Slope of a plot of scattering cross section per proton *vs.* wavelength of neutrons.

Table 4-4. Vibrational Frequencies and Tentative Assignments*

Salt		H_2O librations	M–O stretch	H-bond stretch	Latice modes
$CoCl_2 \cdot 6H_2O$	I	776-*R*[a] 656, 560, 464	341, 288[b]	216, 184, 136[c]	104, 80
	II	781-*R*[11] 775-*R*[12]			
$NiCl_2 \cdot 6H_2O$	I	728-*R* (650-*W*)[d], 584, 480, 396[c]	(350),[d] 296[b]	200, 136[c]	80, 66, 56
	II	763-*R*,[11] 755-*R*,[13] 640-*W*[13]			
$CaCl_2 \cdot 6H_2O$	I	656, 480	380	240, 192	136, 100, 80
	II	719-*R*[11]			
$SrCl_2 \cdot 6H_2O$	I	656, 500	(350),[d] 312[b]	224, 160	112, 80
	II	690-*R*[11]			
$CrCl_3 \cdot 6H_2O$	I	730-*R*, 590-*W*, 384-*T*	464	248, 160[e]	120, 72, 44, 32
	II	844-*R*,[11] 800-*R*,[13] 541-*W*[13]	490[13]		
$AlCl_3 \cdot 6H_2O$	I	790-*R*, 570-*W*, 300-*T*[c]	465[c]	216,[c] 168[c]	128, 110, 65, 44
$MgSO_4 \cdot 7H_2O$	I	665-*R*, 595-*T*, 485-*W*	380[c]	247, 165[e]	102, 70
	II	749-*R*,[11] 460-*W*[13]	310[13]		
$MgCl_2 \cdot 6H_2O$	I	615, 536, 464	384[c]	200, 162[e]	112
	II	714-*R*[11]			
$FeCl_2 \cdot 4H_2O$	I	725, 552, 456	379, 320[c]	184, 141	76, 64
	II	750[14]			
$Al_2(SO_4)_3 \cdot 18H_2O$	I	810-*R*, 585-*W*, 320-*T*[c]	440[c]	220, 160[e]	< 160
$Al(NO_3)_3 \cdot 9H_2O$	I	790-*R*, 595-*W*, 310-*T*[c]	475[c]	250, 176	100
$KAl(SO_4)_2 \cdot 12H_2O$	I	810-*R*, 585-*W*, 310-*T*, 710	480[c]	140, 196, 165[e]	107

* From Ferraro.[1] I. SNS results (see reference 14); II. Infrared results.
[a] *R*=rocking, *T*=twisting, *W*=wagging.
[b] M–Cl stretching mode.
[c] See discussion in text of reference 14.
[d] Very weak or unresolved transition.
[e] Possibly OMO deformation.

Table 4-5. Comparison of Neutron Peak Energies with Far-Infrared
Results (900–400 cm^{-1})*

Compound	Neutron	Infrared
$CuSO_4 \cdot H_2O$	860[a]	863,[a] 800,[a] 500[a]
	450[b]	410[b]
$CuSO_4 \cdot 5H_2O$	600[a]	870[a]
	400[b]	442[b]
$Co(NO_3)_2 \cdot 2H_2O$	650[a]	650[a]
	475[b]	410[b]
$Co(NO_3)_2 \cdot 6HO_2$	630[a] ⎱	Too diffuse
	450[b] ⎰	
$Cu(NO_3)_2 \cdot 3H_2O$	735,[a] 620[a]	895,[a] 580[a]
	470[b]	460[b]
$UO_2(NO_3)_2 \cdot 6H_2O$	535[a]	570,[a] 555[a]
	420[b]	425[b]
$HCrO_2$	1070,[e] 475[g]	1200,[e] 620,[g] 520[g]
$DCrO_2$	826[b], 460[g]	840,[b] 635,[g] 505,[g] 470,[g] 430[g]

* From Ferraro.[1]
Legend for Tables 4–5 and 4–6
[a] H_2O rocking and wagging modes.
[b] M–OH_2 stretch and H_2O torsional mode in neutron results only.
[c] Hydrogen bond stretching regions.
[d] H_2O–Cu–OH_2 deformation modes+contribution from lattice-vibrational modes+oscillations of nitrate or sulfate+acoustic vibrations (in neutron results only).
[e] O–H–O bending vibrations.
[f] O–D–O bending vibrations.
[g] Lattice optic modes, $\nu_{OH \cdots O}$.
[h] Acoustic vibrations (neutron results only).
[i] ν_{MO} in $M\!\!\begin{smallmatrix}O\\ \diagdown\\ \diagup\\ O\end{smallmatrix}\!\!SO_2$; $M\!\!\begin{smallmatrix}O\\ \diagdown\\ \diagup\\ O\end{smallmatrix}\!\!NO$

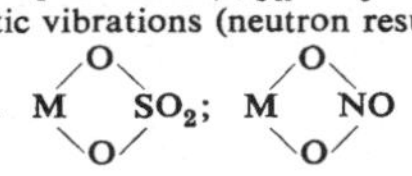

Table 4-6. Comparison of Neutron Peak Energies with Far-Infrared
Results (400–50 cm^{-1})*

Compound	Neutron	Infrared
$CuSO_4 \cdot HO_2$	243[c]	299,[i] 197[g]
	208[c]	147,[d] 101,[d] 96[d]
$CuSO_4 \cdot 5H_2O$	253[c] ⎱	Too diffuse
	133,[d] 70[d] ⎰	
$Co(NO_3)_2 \cdot 2H_2O$	280,[c] 200[c]	308,[i] 200[d]
	165[d]	125[d]
$Co(NO_3)_2 \cdot 6H_2O$	228[c] ⎱	Too diffuse
	120[d] ⎰	
$Cu(NO_3)_2 \cdot 3HO_2$	240[c]	325,[i] 257,[d] 253[d]
	202,[c] 98[d]	180,[d] 160,[d] 120,[d] 104[d]
$UO_2(NO_3)_2 \cdot 6H_2O$	250[c]	348,[b] 260,[i] 249[i]
	162,[d] 98[d]	140,[d] 122,[d] 94[d]
$HCrO_2$	226,[g] 110,[b] 65[h]	nil
$DCrO_2$	180,[g] 62[h]	nil

* From Ferraro.[1]

4.3. HIGH-PRESSURE TECHNIQUES IN THE FAR-INFRARED REGION

Speciroscopic measurements at nonambient pressures were first made by Drickamer.[16] His studies covered the regions of ultraviolet, visible, and near infrared.[17] Extension to 35.0 μ (285 cm^{-1}) was made by Weir, Van Valkenburg, and Lippincott.[18] The capability of reaching 200 μ (50 cm^{-1}) was the result of research by Ferraro, Postmus, and Mitra.[19,20] The use of interferometers extended this to 250 μ (40 cm^{-1})[21] and 1000 μ (10 cm^{-1}).[22]

Instrumentation

The method utilizes the opposed diamond anvil cell developed by Weir, Van Valkenburg, and Lippincott[18,23–24] and marketed by High Pressure Diamond Optics, Inc., McLean, Virginia. The cell can be used with a grating spectrophotometer[19,20,23] or an interferometer.[21,22] A quartz anvil cell with an interferometer has also been used. For use with a grating spectrophotometer, a beam condenser is absolutely essential. Furthermore, it is necessary to use a microscope to determine whether one has a good solid load between the diamonds and whether a phase transition is occurring under pressure.

Table 4-7 summarizes the high-pressure apparatus in current use. Figure

Table 4-7. High-Pressure Apparatus Currently Used for Low-Frequency Studies*

Workers	Spectrophotometer or interferometer	Wavelength range, μ	Optical cell
Weir, Van Valkenburg, and Lippincott[18,23,24]	Commercial double-beam spectrophotometer with beam condenser	2–35	Diamond anvil
Jacobsen and Brasch[26,27]	Perkin–Elmer No. 521[a]	2–35	Diamond anvil
Ferraro, Mitra, and Postmus[19–20]	Perkin–Elmer No. 301[a]	16–200	Diamond anvil
	Beckman IR-11[a]	16–200	Diamond anvil
	Beckman IR-12[a]	2–40	Diamond anvil
McDevitt, Witkowski, and Fateley[21]	FS-520 interferometer	to 250	Diamond anvil
Bradley, Gebbie, *et al.*[22]	Michelson interferometer	50–1000	Anvil, quartz window

* From Ferraro.[25]

[a] With 6 × beam condenser.

Note: Foroperation to 200 μ with a grating spectrophotometer a beam condenser and diamond cell are needed, at a cost of about $6000.

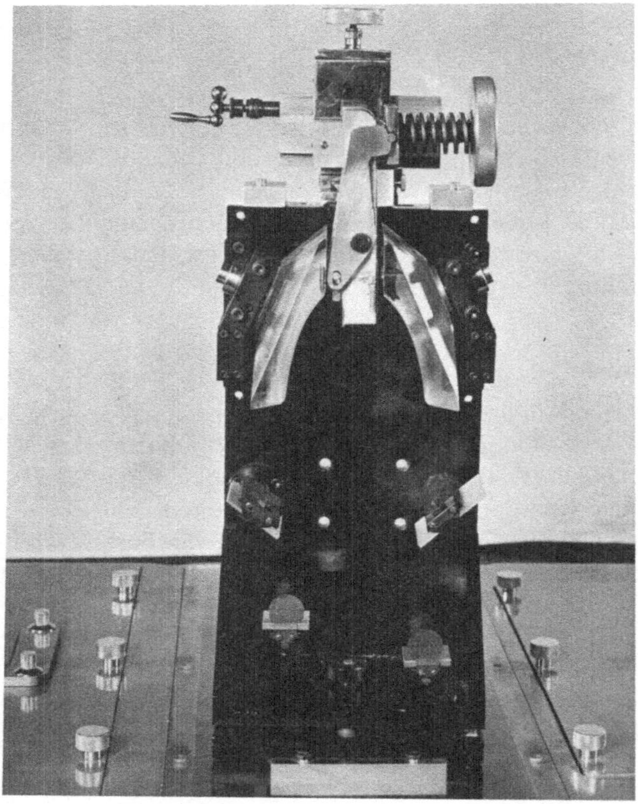

Fig. 4-2. Diamond cell with 6× beam condenser for Perkin–Elmer No. 301 spectrophotometer.[19]

4-2 shows the diamond cell in place in the 6× beam condenser of the Perkin–Elmer No. 301 spectrophotometer.

The diamond cell can be calibrated by any one of three methods:

(1) The compression of the spring can be measured by means of a Dillon force gauge. The contact area of the diamond is determined by means of microphotographs. This gives force per unit area or the pressure of the cell, relative to the spring compression.

(2) The use of solids which undergo phase transformations at known pressures can be used (e.g., KBr, 18 kbar; KCl, 20 kbar).

(3) Calibrations can be made by using nickel dimethylglyoxime which shows a change in spectral properties with pressure.[28,29]

Table 4-8. Lattice Vibrations for Alkali Halides,* cm^{-1}

Material	ν_{TO}
LiF	306
LiCl	191
LiBr	159
NaF	244
NaCl	164
NaBr	134
NaI	117
KF	190
KCl	146
KBr	113
KI	101
RbF	156
RbCl	116
RbBr	88
RbI	77
CsCl	99
CsBr	73
CsI	62

* From Mitra.[30]

Applications

*Lattice Vibrations**

Table 4-8 lists some ionic lattice frequencies. It can be observed that the tranverse optical lattice modes (ν_{TO}) are located below 300 cm^{-1}. The used of the high-pressure technique in the far-infrared region, has made possible the study of these modes under pressure for the first time. Several ionic salts (mostly of the cubic type) have now been studied. Table 4-9 lists the pressure dependences of the ν_{TO} vibration for several solids.[31,32] It can be observed that blue shifts are obtained for these ionic lattice vibrations with increasing pressure. The shifts may be considerable and certainly are of a greater magnitude than those observed at low temperatures. However, not all ionic lattice vibrations will show dramatic shifts since the compressibility of the solid is involved. The relationship between the change in frequency with pressure for simple ionic salts of the type AB, is given by the relationship

$$\gamma\chi\nu = \left(\frac{\partial \nu}{\partial p}\right)_T \quad 4(1)$$

* Motion of a group of atoms with respect to another group of atoms or an ion with respect to another ion.

Table 4-9. **Pressure Dependences of Several Transverse Optical Lattice Vibrations,* cm^{-1}**

P, kbar[a]	NaF	LiF	CsBr	ZnS
5	250	319	78	280
10	262	324	83	281
15	270	329	88	285
20	278	333	93	290
25	284	337	98	292
30	290	341	—	295
40	298	349	—	301

* From Postmus *et al.*[20] and Mitra *et al.*[31]
[a] 1 kbar = 1000 atm

where γ is the Grüneisen parameter, χ is the isothermal compressibility of the solid, and ν is the frequency of the particular lattice mode. For solids which are relatively difficult to compress, it is possible that only small shifts will occur. For example, the lattice modes of zirconia and hafnia failed to show significant shifts at 40 kbar.[21]

Pressure dependences of two phases of a solid can be studied by these techniques. Such studies have been made with KBr[32] and KCl.[21,32] It is also possible to study mixed ionic crystals using the diamond cell technique.[33] Certain molecular lattice modes were investigated by McDevitt *et al.*[21] using interferometer techniques and by others using the Raman exrpeiment.[34] For these materials the experiments are more difficult to perform, since a thicker sample is necessary and gaskets must be employed. Molecular lattice vibrations have been observed that show similar shifts toward higher frequency with increasing pressure. It would be expected that pressure dependences of molecular lattice modes would be greater than ionic lattice modes due to their greater compressibility, and this expectation has been realized.[34]

As a result of the sensitivity to pressure of lattice modes, the technique may serve to distinguish between lattice vibrations and internal vibrations in the far-infrared region, the internal vibrations being relatively insensitive of shifts in frequency with pressure.

*Internal Vibrations**

Whereas certain compressible ionic solids show shifts of lattice modes toward higher frequencies with pressures, internal modes in other compounds show very little pressure shift. Nevertheless, certain intensity changes are possible under pressure. For example, coordination complexes having two resolved vibrations involving metal-halide or metal–

* Movement of an atom with respect to another atom within a group of atoms.

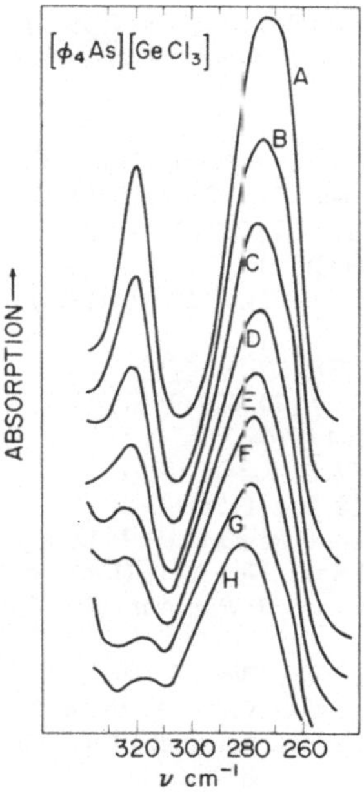

Fig. 4-3. Low-frequency spectra of [φ_4As] [GeCl$_3$] with and without pressure.

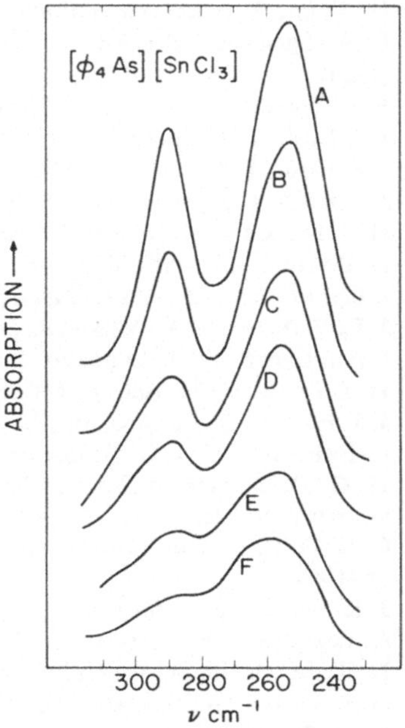

Fig. 4-4. Low-frequency spectra of [φ_4As] [SnCl$_3$] with and without pressure.

nitrogen stretches (asymmetric and symmetric modes) show pressure dependences involving changes in intensities. Figures 4-3 and 4-4 illustrate the effect of pressure on the asymmetric and symmetric stretching vibration of [φ_4As][GeCl$_3$] and [φ_4As][SnCl$_3$].[35] It is observed that both peaks diminish in intensity and broaden, but the symmetrical vibration (a_1 species) is the most sensitive to pressure. The determination in these compounds as to which is the symmetric mode and which is the antisymmetric mode was made by Raman polarizability data in solution.[36,37] It is important to be able to distinguish between these modes in making correct far-infrared interpretations. The technique enables one to accomplish this using powdered or polycrystalline materials. In attempting to make this distinction by dichroism infrared measurements, single crystals are necessary; by Raman polarization methods, single crystals or solubility in a solvent is required.

BIBLIOGRAPHY

1. J. R. Ferraro, *Anal. Chem.* **40,** 24A (1968).
2. W. M. Lomer and G. G. Low, in *Thermal Neutron Scattering* (P.A. Egelstaff, Ed.), Academic Press, New York (1965).

3. M. Blume, Brookhaven National Laboratory Report 940 (C-45), p. 1 (1965).
4. P. A. Egelstaff (Ed.), *Thermal Neutron Scattering*, Academic Press, New York (1965).
5. S. F. Treviño, *Appl. Spectry.* **22,** 659 (1968).
6. J. J. Rush, T. J. Taylor, and W. W. Havens, *Phys. Rev. Letters* **5,** 507 (1960).
7. J. J. Rush, T. J. Taylor, and W. W. Havens, *J. Chem. Phys.* **35,** 2265 (1961).
8. J. J. Rush, T. J. Taylor, and W. W. Havens, *J. Chem. Phys.* **37,** 234 (1962).
9. H. J. Prask and H. Boutin, *J. Chem. Phys.* **45,** 699 (1966).
10. H. Boutin, G. J. Stafford, and H. R. Danner, *J. Chem. Phys.* **40,** 2670 (1964).
11. I. Gamo, *Bull. Chem. Soc. Japan* **34,** 760 (1961).
12. J. R. Ferraro and A. Walker, *J. Chem. Phys.* **42,** 1278 (1965).
13. I. Nakagawa and T. Shimanouchi, *Spectrochim. Acta* **20,** 429 (1964).
14. H. J. Prask and H. Boutin, *J. Chem. Phys.* **45,** 3284 (1966).
15. J. J. Rush, J. R. Ferraro, and A. Walker, *Inorg. Chem.* **6,** 346 (1967).
16. E. Fishman and H. G. Drickamer, *Anal. Chem.* **28,** 804 (1956).
17. H. G. Drickamer, in *Progress in Very High Pressure Research* (E. P. Bundy, W. R. Hibbard, and H. M. Strong, Eds.), J. Wiley and Sons, New York (1961), p. 16.
18. C. E. Weir, A. Van Valkenburg, and E. R. Lippincott, *J. Res. Natl. Bur. Std.* **63A,** 55 (1959).
19. J. R. Ferraro, S. S. Mitra, and C. Postmus, *Inorg. Nucl. Chem. Letters* **2,** 269 (1966).
20. C. Postmus, J. R. Ferraro, and S. S. Mitra, *Inorg. Nucl. Chem. Letters* **4,** 55 (1968).
21. N. T. McDevitt, R. E. Witkowski, and W. G. Fateley, Abstract 13th Colloquium Spectroscopium Internationale, June 18–24, 1967, Ottawa, Canada.
22. C. C. Bradley, M. A. Gebbie, A. C. Gilby, V. V. Kelchin, and J. H. King, *Nature* **211,** 839 (1966).
23. E. R. Lippincott, F. W. Welsh, and C. E. Weir, *Anal. Chem.* **33,** 137 (1961).
24. E. R. Lippincott, C. E. Weir, A. Van Valkenburg, and E. N. Bunting, *Spectrochim. Acta* **16,** 58 (1960).
25. J. R. Ferraro, in *Far-Infrared Properties of Solids* (S. S. Mitra and S. Nudelman, Eds.), Plenum Press, New York (1970).
26. J. W. Brasch and R. Jakobsen, *Spectrochim. Acta* **21,** 1183 (1965).
27. J. W. Brasch, *J. Chem. Phys.* **43,** 3473 (1965).
28. J. C. Zahner and H. G. Drickamer, *J. Chem. Phys.* **33,** 1625 (1960).
29. H. W. Davies, *J. Res. Natl. Bur. Std.* **72A,** 149 (1968).
30. S. S. Mitra, in *Optical Properties of Solids* (S. S. Mitra and S. Nudelman, Eds.), Plenum Press, New York (1969).
31. S. S. Mitra, C. Postmus, and J. R. Ferraro, *Phys. Rev. Letters* **18,** 455 (1967).
32. C. Postmus, J. R. Ferraro, and S. S. Mitra, *Phys. Rev.* **74,** 983 (1968).
33. J. R. Ferraro, S. S. Mitra, C. Postmus, C. Hoskins, and E. C. Siwiec, *Appl. Spectry.* **24,** 187 (1970)
34. O. Brafman, S. S. Mitra, R. K. Crawford, W. B. Daniels, C. Postmus, and J. R. Ferraro, *Solid State Commun.* **7,** 449 (1969).
35. C. Postmus, K. Nakamoto, and J. R. Ferraro, *Inorg. Chem.* **6,** 2194 (1967).
36. L. A. Woodward and M. J. Taylor, *J. Chem. Soc.*, 407 (1962).
37. D. F. Shriver and M. P. Johnson, *Inorg. Chem.* **6,** 1265 (1967).

Chapter 5

METAL-OXYGEN VIBRATIONS

5.1. METAL-OXYGEN VIBRATIONS IN HYDRATES

There are several types of water in inorganic salts. Those water molecules that are coordinated to the metal and retain their identity in solution —e.g., aquo-ions of the type $[M(H_2O)_a]^b$—are termed coordinated water, and those water molecules that are hydrogen bonded to other water molecules or to the anion or to both are termed lattice water. The distinction, however, is not clear-cut, for some coordinated water may also engage in hydrogen bonds within the crystal. Confirmation of structures of this type has come from neutron diffraction studies of hydrates. Studies by Taylor *et al.*[1] with $Th(NO_3)_4 \cdot 5H_2O$ showed coordinated water,[1] water–water hydrogen bonds, and water hydrogen bonded to the nitrate. Figure 5-1 illustrates the

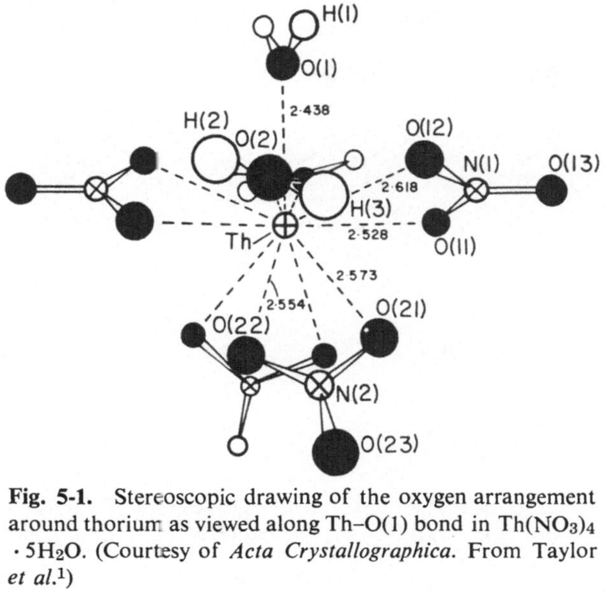

Fig. 5-1. Stereoscopic drawing of the oxygen arrangement around thorium as viewed along Th–O(1) bond in $Th(NO_3)_4 \cdot 5H_2O$. (Courtesy of *Acta Crystallographica.* From Taylor *et al.*[1])

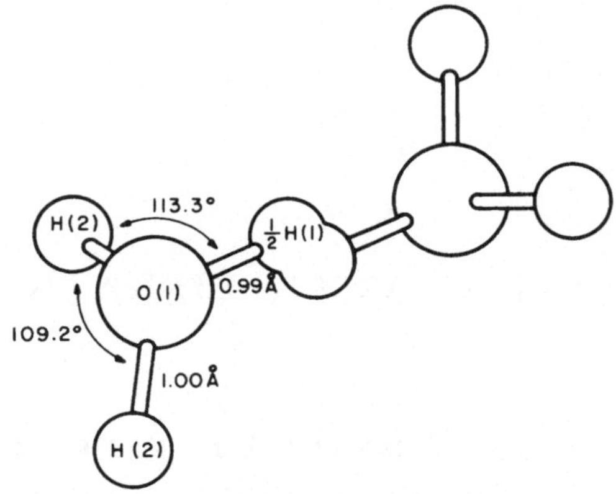

Fig. 5-2. Hydrogen bonding in *trans*-diaquohydrogen ion. (Courtesy of Pergamon Press, New York. From Williams.[8])

structure of $Th(NO_3)_4 \cdot 5H_2O$. Uranyl nitrate hexahydrate exhibits similar types of environments for the water molecules.[2] Stable solids containing the oxonium ion (H_3O^+) have also been reported.[3–7] Recently, single-crystal neutron diffraction evidence for the diaquohydrogen ion in *trans*-$[Co(en)_2Cl_2]^+Cl^-(H_5O_2)^+Cl^-$ has been found.[8] The hydrogen bonding in the diaquohydrogen ion (*trans*) is illustrated in Fig. 5-2. Complex structures

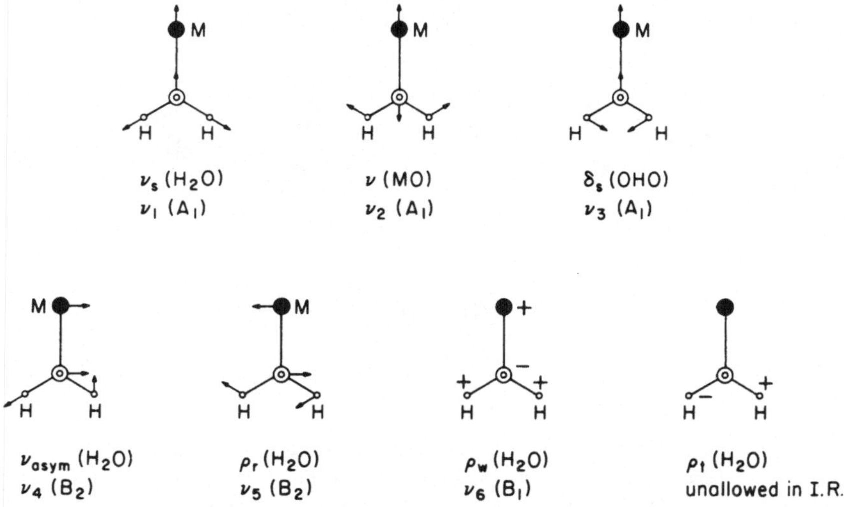

Fig. 5-3. Infrared vibration of MOH_2 moiety.

of the aforementioned types (as illustrated by $Th(NO_3)_4 \cdot 5H_2O$ and $[Co(en)_2Cl_2]^+Cl^-(H_5O_2)^+Cl^-$) should be considered to be typical of hydrates rather than exceptional. Chidambaram *et al.*[8a] have proposed a classification of hydrates based on the type of coordination of the lone pair orbitals of water. For an extension of this classification see Hamilton and Ibers.[8b]

Both the lattice water and coordinated water will illustrate the librational modes of water molecules, and these frequencies will be found below 800 cm^{-1}. Figure 5-3 depicts the vibrations possible for a MOH_2 moiety. Sartori and co-workers[9] used such a model in making a normal coordinate treatment for coordinated water in inorganic solids. Besides the metal–oxygen stretch vibration, two librational modes of the coordinated water will be found in the low-frequency infrared. These are the water rocking mode and the water wagging mode. The water twisting vibration is active only in the Raman.

Table 5-1. Librational Vibrations for Lattice-Type Water[10]

		Frequency, cm^{-1}*	
Hydrate	Type of H_2O	H_2O	D_2O
$NaBr \cdot 2H_2O$	Br...HOH...Br	590sh	
		405vs	
	O...HOH...Br	625vs	
		470vs	
$NaI \cdot H_2O$	I...HOH...I	510sh	350vs
		385vs	
	O...HOH...I	585vs	430vs
		450vs	
$NaI \cdot 2H_2O$	I...HOH...I	470vs	
		400vs	
	O...HOH...I	615vs	
		470vs	
$SrCl_2 \cdot H_2O$	Cl...HOH...Cl	585vs	435vs
		395vs	
$SrCl_2 \cdot 2H_2O$	Cl...HOH...Cl	710vs	520vs
		480vs	365vs
$BaCl_2 \cdot H_2O$	Cl...HOH...Cl	560vs	405vs
		410vs	
$BaCl_2 \cdot 2H_2O$	Cl...HOH...Cl	690vs	515vs
	O...HOH...Cl	520sh	370sh
$BaBr_2 \cdot H_2O$	Br...HOH...Br	550vs	415vs
		335vs	
$BaBr_2 \cdot 2H_2O$	Br...HOH...Br	550vs	415vs
		445vs	

* Not investigated below 300 cm^{-1}.
Legend: sh = shoulder, vs = very high intensity.

Table 5-2. Librational Modes of Water Coordinated to Transition Metals*

	Frequency, cm^{-1}			
	Rock (H_2O)	Wag (H_2O)	Rock (D_2O)	Wag (D_2O)
Anion—SiF_6^{2-}				
Cd $(H_2O)_6^{2+}$	867			
Mn $(H_2O)_6^{2+}$		560		
Fe $(H_2O)_6^{2+}$		575, 540		
Ni $(H_2O)_6^{2+}$		645		450
Anion—Cl^-				
Al $(H_2O)_6^{3+}$	835	595		
Cr $(H_2O)_6^{3+}$	820, 810, 800	541, 550		
$[Cr(H_2O)_5Cl]^{2+}$		547		
$[Cr(H_2O)_4Cl]^+$	609	495		
$[Cr(H_2O)_4(NH_3)_2]^{3+}$	820			
Anion—SO_4^{2-}				
Ni $(H_2O)_6^{2+}$	765	625		450
Mg $(H_2O)_6^{2+}$		460	474	391
Zn $(H_2O)_6^{2+}$	621	541, 555	467	392
Cu $(H_2O)_6^{2+}$	887, 885	535		
Mn $(H_2O)_6^{2+}$		560		
Fe $(H_2O)_6^{2+}$		575		

* Transition metals primarily.[11]

Typical of lattice-type water librations were those studied by Van der Elskin and Robinson.[10] Results with alkali and alkaline earth metal halide hydrates showed two absorptions in the 335–700 cm^{-1} region, which could be associated with the two types of water (i.e., X. . .HOH. . .X and O. . .HOH. . .X). Both of these bands were found to shift upon deuteration. Table 5-1 tabulates the low-frequency lattice-type water librations.

A number of studies have now been made in attempts to locate coordi-

Table 5-3. The Metal-Oxygen Stretching Vibration in Transition Metal-Aquo Complexes*

Complex	$\nu_{M\leftarrow OH_2}$, cm^{-1}	Complex	$\nu_{M\leftarrow OD_2}$, cm^{-1}
$Cr(H_2O)_6^{3+}$	490		
$Ni(H_2O)_6^{2+}$	405	$Ni(D_2O)_6^{2+}$	389
$Mn(H_2O)_6^{2+}$	395		
$Fe(H_2O)_6^{2+}$	389		
$Cu(H_2O)_4^{2+}$	440		
$Zn(H_2O)_6^{2+}$	364	$Zn(D_2O)_6^{2+}$	358
$Mg(H_2O)_6^{2+}$	310		

* All complexes contain sulfates as the anion.[11]

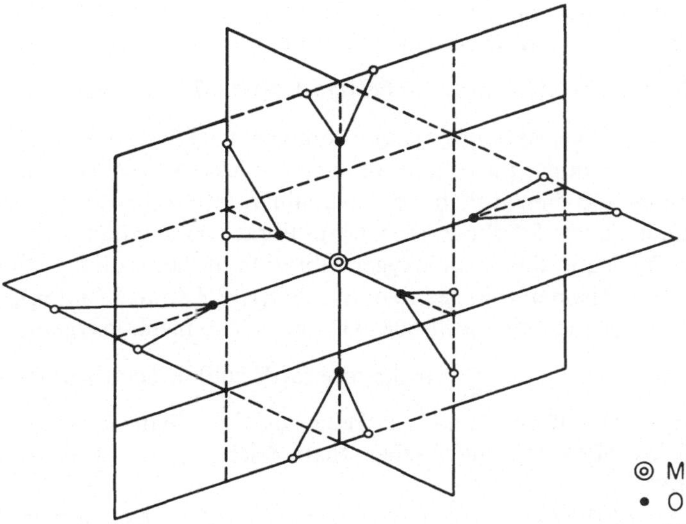

Fig. 5-4. Model used for the $M(H_2O)_6$ ion. (Courtesy of Pergamon Press, New York. From Nakagawa and Shimanouchi.[11])

nated water librations. Probably the most extensive investigation of the transition metal–aquo complexes has been made by Nakagawa and Shimanouchi.[11] Deuteration studies were found to be very helpful. Tables 5-2 and 5-3 summarize the data accumulated for the coordinated water vibrations. It can be observed that the metal–oxygen vibration is little changed by deuterium substitution, as expected. Figure 5-4 shows the model used by Nakagawa and Shimanouchi for the $M(H_2O)_6$-type ion. Table 5-4 illustrates the normal vibrations for the $M(H_2O)_6$-type ion of O_h symmetry, which are active in the infrared. Of the 51 total vibrations $(3 \cdot 19 - 6)$, only $8 f_u$ species will be infrared active. The 3 a_g, 3 e_g, and 5 f_g species are Raman active. The

Table 5-4. Infrared Allowed Vibrations for $M(H_2O)_6$-Type Ion (O_h)

Species	Vibrational mode
f_u	OH stretch (sym)
"	OH stretch (asym)
"	OH_2 scissor
"	OH_2 rock
"	OH_2 wag
"	MO stretch
"	OMO deformation
f_u	OMO deformation

a_u and the e_u species are inactive. From these studies it has been demonstrated that the metal–oxygen vibration varies as follows:

$$\text{Cr(III)} > \text{Ni(II)} \sim \text{Mn(II)} \sim \text{Fe(II)} > \text{Cu(II)} \sim \text{Zn(II)} > \text{Mg(II)}$$

and that metal–oxygen bonds in aquo complexes are less covalent than those found in the cyanide, nitro, and amine complexes of the same metals.

Postmus and Ferraro[12] have made similar studies with hydrated rare earth sulfates. Table 5-5 shows the various librational modes for coordinated water (or D_2O) and the metal–oxygen vibration in these sulfates. Figure 5-5 shows a comparison of the spectra of $Sm_2(SO_4)_3 \cdot 8H_2O$ and $Sm_2(SO_4)_3 \cdot 8D_2O$ in the low-frequency region of 666–333 cm^{-1}. The metal–oxygen stretching vibration to water $\left(\nu_{MO}\!<^{H}_{H}\right)$ for the rare earth sulfate octahydrates appears in the range 431–437 cm^{-1} and is at frequency lower than $Cr(H_2O)_6^{3+}$, about equal to $Cu(H_2O)_4^{2+}$, and higher than $Ni(H_2O)_4^{2+}$, $Fe(H_2O)_6^{2+}$, and $Zn(H_2O)_6^{2+}$.

In view of the fact that some water molecules are hydrogen bonded to the anion, it would appear that the librational water modes should be dependent on the anion present. Preliminary studies in the infrared by Ferraro and

Table 5-5. Coordinated H_2O Vibrations vs. Coordinated D_2O Vibrations for Rare Earths,[12] cm^{-1}

Sulfate	$M{-}OH_2$ rock	$M{-}OD_2$ rock	$M{-}OH_2$ wag	$M{-}OD_2$ wag	$(M{-}O)H_2$ stretch	$(M{-}O)D_2$ stretch
		567	495			
$Sc_2(SO_4)_3 \cdot 6H_2O$	767	540	471	(?)	433(?)	
$Y_2(SO_4)_3 \cdot 8H_2O$	745	544	495	367	442	426
					434	
$La_2(SO_4)_3 \cdot 9H_2O$		540	469		433	
$Ce_2(SO_4)_3 \cdot 5H_2O$		534		359		406
$Pr_2(SO_4)_3 \cdot 8H_2O$	780	534	488	359	437	413
	740					
$Nd_2(SO_4)_3 \cdot 8H_2O$	740	536	502	360	434	415
$Sm_2(SO_4)_3 \cdot 8H_2O$	740	536	490	370	435	416
$Eu_2(SO_4)_3 \cdot 8H_2O$	740		498		433	
$Gd_2(SO_4)_3 \cdot 8H_2O$	735	540	488	356	435	422
$Tb_2(SO_4)_3 \cdot 8H_2O$	740	540	484	361	434	418
$Dy_2(SO_4)_3 \cdot 8H_2O$	742	(?)	492	334(?)	434	422
$Ho_2(SO_4)_3 \cdot 8H_2O$	735	542	498	363	435	422
$Er_2(SO_4)_3 \cdot 8H_2O$	745	542	484	361	434	422
$Tm_2(SO_4)_3 \cdot 8H_2O$	750	545	480	361	431	423
$Yb_2(SO_4)_3 \cdot 11H_2O$	750	547	484	362	434	421
$Lu_2(SO_4)_3 \cdot 8H_2O$	758	555	488	359	433	433

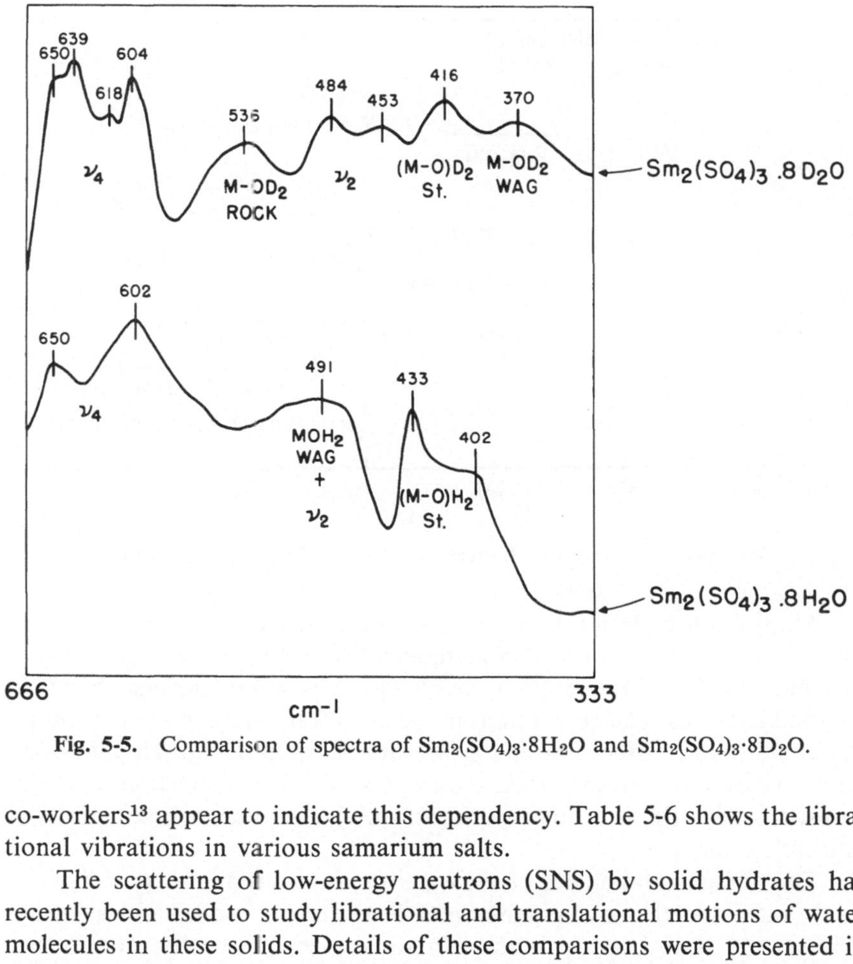

Fig. 5-5. Comparison of spectra of $Sm_2(SO_4)_3 \cdot 8H_2O$ and $Sm_2(SO_4)_3 \cdot 8D_2O$.

co-workers[13] appear to indicate this dependency. Table 5-6 shows the librational vibrations in various samarium salts.

The scattering of low-energy neutrons (SNS) by solid hydrates has recently been used to study librational and translational motions of water molecules in these solids. Details of these comparisons were presented in Chapter 4. The agreement is only fair. Only when the energy resolution of the slow-neutron scattering method approaches the resolution of optical methods can better agreement be expected.

Table 5-6. Comparison of Librational Modes of Water in $SmCl_3$, $Sm(NO_3)_3$, $Sm_2(SO_4)_3$ Hydrates,[13] cm^{-1}

Compound	(H_2O) ν_{rock}	(D_2O) ν_{rock}	(H_2O) ν_{wag}	(D_2O) ν_{wag}	$\nu_{(MO)H_2}$	$\nu_{(MO)D_2}$
$SmCl_3$	627	462	470	375		
$Sm(NO_3)_3$	750	580	645	506	437	440
$Sm_2(SO_4)_3$	740	536	490	370	435	416

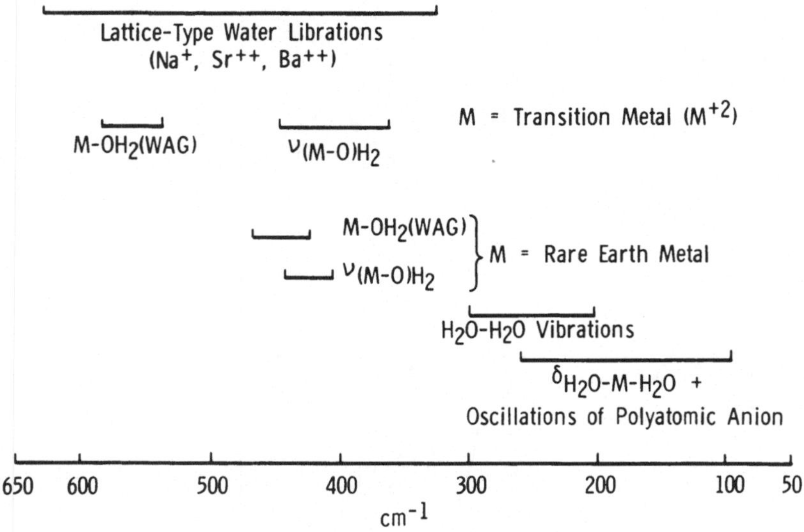

Fig. 5-6. Characteristic frequencies observed for metal hydrates.

Many of the hydrates show strong absorptions in the far-infrared spectra and strong scattering in the SNS method between 200 and 300 cm^{-1}. These have been attributed to vibrations involving water–water hydrogen bonds in the solid hydrates. The absorptions in the far-infrared region are very broad and could involve other low-energy vibrational modes (e.g., H_2O–M–H_2O bend, contribution from lattice modes, oscillation of polyatomic anion, acoustic modes (in neutron experiment only)) and other vibrations mixed with lattice modes.

The fundamental modes of the water molecule are found at 3600–3200 cm^{-1} (antisymmetric and symmetric O–H stretching mode) and at 1600–1650 cm^{-1} (HOH bending mode). In some cases these absorptions may show fine structure.

Figure 5-6 shows the characteristic low-frequency vibrations observed for metal hydrates.

Table 5-7A. Absorptions Found for the Three Types of Rare Earth Oxides (M_2O_3)[14]

Type	Absorptions, cm^{-1}
Hexagonal	644, 415, 346, 320
Monoclinic	630, 510, 365
Cubic	530–557, 400

Table 5-7B. Far-Infrared Frequencies of Metal Oxides[14]*

Metal	Formula	Region, cm^{-1}
Aluminum	α-Al_2O_3	575s, 432s, 375sh
Antimony	Sb_2O_3 (cubic)	740s, 685sh, 590m, 550sh, 482m, 383s, 345w, 322w, 265s
Arsenic	As_2O_3 (cubic)	500sh, 475s, 355sh, 340s
Barium	BaO	503sh, 483s, 305sh, 283m, 255m
	BaO_2	(No bands 700–250)
Beryllium	BeO	(No bands 700–250)
Bismuth	α-Bi_2O_3	645w, 505s, 345s
Boron	B_2O_3	765sb, 638m, 543m
Cadmium	CdO	(no bands 700–250)
Calcium	CaO	400sb, 290m
Cerium	CeO_2	525sh, 425s,vb
Chromium	CrO_3	(No bands 700–250)
	Cr_2O_3	625s, 555s, 435w, 407w
Cobalt	Co_3O_4	655s, 635sh, 562s, 460b,sh, 350w
Copper	CuO	610m, 500s, 410m
	Cu_2O	615
Dysprosium	Dy_2O_3 (cubic)	550s, 408s,vb, 320m, 307w, 284w, 273w
Erbium	Er_2O_3 (cubic)	563s, 465b,sh, 367s, 325m, 285m
Europium	Eu_2O_3 (monoclinic)	630w, 510sh, 365s,vb
Gadolinium	Gd_2O_3 (cubic)	535s, 465b,sh, 350s, 310m, 297w, 270m
Gallium	β-Ga_2O_3	663s, 450sb, 364m, 305mb
Germanium	GeO_2 (hexagonal)	585s, 550s, 512s, 332s, 255m
Gold	Au_2O_3	607
Hafnium	HfO_2 (monoclinic)	755m, 645m, 530s, 450sh, 425s, 375w, 350sh
Holmium	Ho_2O_3 (cubic)	559s, 470b,sh, 370s, 325m, 310sh, 285m
Indium	In_2O_3	(No bands 700–250)
Iron	α-Fe_2O_3	560sb, 468s, 370sh, 325s,
	γ-Fe_2O_3[a]	555, 468, 336
	Fe_3O_4	570sb, 385mb
Lanthanum	La_2O_3	644w, 415sb
Lead	PbO (orthorhombic)	500m, 377s, 300s
	PbO_2 (tetragonal)	(No bands 700–250)
	Pb_3O_4	650w, 525s, 445s, 380s, 320m
Lutetium	Lu_2O_3	570s, 485b,sh, 382s, 337m, 297m
Magnesium	MgO	445vb
Manganese	MnO_2 (tetragonal)	615s,vb, 400m, 335w
	Mn_3O_4[a]	600, 475, 393
Mercury	HgO (yellow)	588s, 480s
	HgO (red)	573s, 475s
Molybdenum	MoO_3[a]	660, 375, 300

Table 5-7B (Continued)

Metal	Formula	Region, cm^{-1}
Neodymium	Nd_2O_3	655
Nickel	NiO	650w, 465s,vb
Niobium	α-Nb_2O_5[a]	478, 295
	δ-Nb_2O_5[a]	575, 455sh, 357
Praseodymium	Pr_6O_{11}	655m, 500sh, 350s,vb
Ruthenium	RuO_2	(No bands 700–250)
Samarium	Sm_2O_3 (monoclinic)	640w, 530sh, 370s,vb
Scandium	Sc_2O_3	625s, 525sh, 425s, 382m, 365w, 343w
Silicon	SiO_2 (dehydrated silica gel)	460b
	SiO_2 (α-quartz)	775m, 693w, 510sh, 450s, 385sh, 362s, 257m
Silver	Ag_2O	645m, 540s
Strontium	SrO_2	590sb, 525sh
Tantalum	α-Ta_2O_5	612s,vb, 450b,sh, 300m
	β-Ta_2O_5	575, 455sh, 315
Thorium	ThO_2	645sh, 310s,vb
Thulium	Tm_2O_3	565s, 485sh, 380s, 335w, 295w
Tin	SnO[a] (tetragonal)	650, 480
	SnO_2	670m, 610sh, 312s
Titanium	TiO_2 (anatase)	700s,vb, 525s vb, 347m
	TiO_2 (rutile)	695sb, 608sb, 423w, 352w
	Ti_2O_3[a]	650 (No bands 600–250)
Tungsten	WO_2	(No bands 700–250)
	WO_3[a]	355, 315
Vanadium	V_2O_5	595s,vb, 395w, 288w
Ytterbium	Yb_2O_3	569s, 400sb, 330m, 322sh, 296m
Yttrium	Y_2O_3 (cubic)	561s, 423sb, 333m, 325sh, 300m
Zinc	ZnO	450vb
Zirconium	ZrO_2 (monoclinic)	745m, 620sh, 530sb, 450w, 420w, 375w, 360sh
	ZrO_2 (cubic CaO stabilized)	490vb

* Courtesy of Pergamon Press, New York.
[a] Bands weakly defined.
Legend: s=high intensity, m=medium intensity, w=low intensity, b=broad, v=very, sh=shoulder; (the intensity values should be compared only with bands in the same spectrum).

5.2. METALLIC OXIDES

McDevitt and Baun[14] have observed broad absorptions in various oxides from 700 to 240 cm^{-1}. In the case of rare earth oxides of the type R_2O_3, three types of oxides are found. Table 5-7a gives absorptions found for the three types of oxides. It is possible to distinguish between these types of oxides by observation of their low-frequency spectra. Table 5-7b gives absorptions for various oxides.

5.3. METAL–OXYGEN VIBRATIONS IN HYDROXYL-BRIDGED COMPLEXES

The hydroxyl stretching and bending vibrations for metal hydroxides, basic salts, and bridged hydroxyl complexes have been assigned.[15-21] The O–H stretching mode involving the hydroxyl bridge, observed at 3400–3600

Table 5-8. ν_{MO} Stretching Vibration in Several Hydroxyl-Bridged Complexes

Complex	ν_{MO}, cm^{-1}
L = 2-amino pyridine[19]	
$[LCu(OH)]_2[ClO_4]_2$	510
$[LCu(OH)]_2[NO_3]_2$	505
L′ = 2,2′-Bipyridyl[18,21]	
$[LCu(OH)]_2[SO_4]$	490
$[LCu(OH)]_2I_2$	478
$[LCu(OH)]_2Br_2$	490
$[LCu(OH)]^2{}_2[ClO_4]_2$	490
$[LCu(OH)]_2[PtCl_4]$	494
$[LCu(OH)]_2Cl_2$	489
$[LCu(OH)]_2[SCN]_2$	488
$[LCu(OH)]_2[PF_6]_2$	490
$[L_2Fe(OH)_2][SO_4]_2$	559, 542
$[L_2Cr(OH)_2][NO_3]_4$	559
$[L_2Cr(OH)_2]Br_4$	547
L″ = 1,10-phenanthroline[18,21]	
$[LCu(OH)_2][SO_4]$	480
$[LCu(OH)_2]I_2$	498
$[LCu(OH)_2]Br_2$	492
$[LCu(OH)_2][ClO_4]_2$	480
$[LCu(OH)_2]Cl_2$	480
$[LCu(OH)_2][SCN]_2$	482
$[L_2FeOH]_2Cl_4$	542
$[L_2FeOH]_2Br_4$	542
$[L_2FeOH]_2(ClO_4)_4$	542, 506
$[L_2FeOH]_2(SO_4)_2$	559, 542, 506
$[L_2Cr(OH)]_2(NO_3)_4$	563, 556
$[L_2Cr(OH)]_2Br_4$	567, 554
Raman Results[22]	
$[(NH_3)_5Cr]_2OH^{5+}$	569
$[(NH_3)_9H_2OCr_2(OH)]^{5+}$	578
$(NH_3)_9Cr_2(OH)_2{}^{4+}$	557

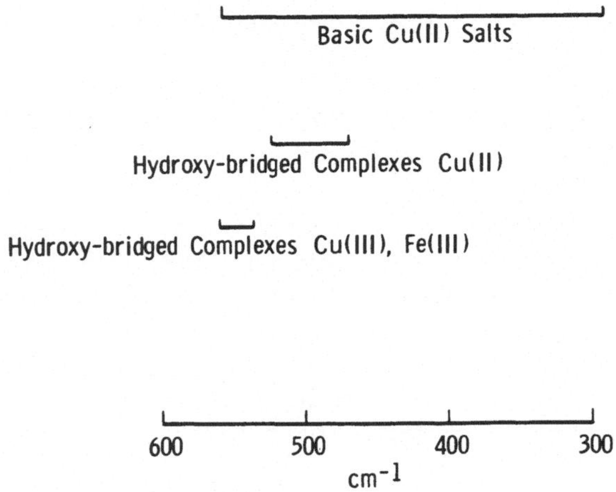

Fig. 5-7. Characteristic frequencies in hydrogen-bonded complexes and basic salts of copper.

cm^{-1}, is sharp and usually can be seen even in the presence of water absorption. However, very few investigations have been made in the frequency region below 650 cm^{-1}. Tarte[17] conducted a study on basic copper salts to 270 cm^{-1}. Absorption bands for these salts are found in the region of 300–550 cm^{-1} and were assigned to the Cu–O stretching vibration. Several hydroxyl-bridged copper(II) complexes have been studied by McWhinnie[19] and by Ferraro and Walker.[21] The copper–oxygen vibratión has been assigned at 480–515 cm^{-1}. Similar complexes involving Cr(III) and Fe(III) have been made and assignments made for ν_{CrO} at 547–567 cm^{-1} and ν_{FeO} at 506–550 cm^{-1}.[18] Table 5-8 lists the metal–oxygen stretching vibration in a number of hydroxyl-bridged complexes. Figure 5-7 lists the characteristic low frequencies in hydroxyl-bridged complexes.

5.4. METAL–OXYGEN VIBRATIONS IN SALTS INVOLVING OXYGENATED POLYATOMIC ANIONS

A number of inorganic salts of transition and rare earth metals involve a degree of covalency between the metal and the oxygen atoms of the polyatomic anion. As a result, an absorption, which can be assigned to a metal–oxygen vibration, can be observed in the low-frequency infrared spectrum. This possibility exists with many polyatomic anions such as those in a D_{3h} symmetry (e.g., nitrate, carbonate, borate), T_d symmetry (e.g., phosphate, sulfate, selenate, tungstate, permanganate, chromate, molybdate, perchlorate), and C_{3v} symmetry (e.g., chlorate, bromate, iodate, sulfite, selenite).

For polyatomic anions such as those mentioned, the ionic environment of the anions is one which is of rather high symmetry. For these molecules degenerate vibrations are involved. This degeneracy may be removed whenever the symmetry is lowered; then forbidden vibrations become active and additional vibrations are seen in the infrared spectrum. Lowering of the symmetry in salts involving polyatomic anions can be caused by a number of mechanisms. Some of these can be cited below:

Site Symmetry Lowering. The selection rules generally used for molecules are based on a point symmetry. These selection rules are based on isolated molecules, which probably exist only in the gaseous state. These selection rules may hold in the liquid state, but certainly do not hold in the crystalline state. For the solid state some of the degenerate vibrations can split and forbidden vibrations become active. New selection rules are thus necessary, and these are based on a site group symmetry.[23] A classical example of site symmetry lowering is that of the carbonate ion, which exists in both the calcite crystal (symmetry D_3) and aragonite (symmetry C_s). The magnitude of splitting caused by this method is generally very small.

Coordination of the Anion to the Metal. The lowering of symmetry caused by this method will not only cause the splitting of degenerate vibrations and the appearance of forbidden vibrations, but also results in a low-frequency metal–oxygen vibration. The splitting in this case may be greater than in the case of site symmetry lowering.

Hydrogen Bonding of Water to Anion. This type can be caused by hydrogen bonding of water molecules to the anions in the solid, where the anions may be uncoordinated or coordinated to the metal.

Other low-frequency vibrations in some polyatomic anions involve the bending vibrations, and these may be found below 650 cm^{-1}.

D_{3h} Symmetry—Nitrates, Carbonates, and Borates

The normal modes of vibration for a molecule in D_{3h} symmetry are shown in Fig. 5-8. The infrared active modes are ν_2, ν_3, and ν_4. All of these will be seen above 650 cm^{-1}. The ν_3 and ν_4 vibrations are doubly degenerate modes, and if the symmetry of the polyatomic anion is lowered these vibrations may split. Table 5-9 shows the relationship between D_{3h} symmetry and

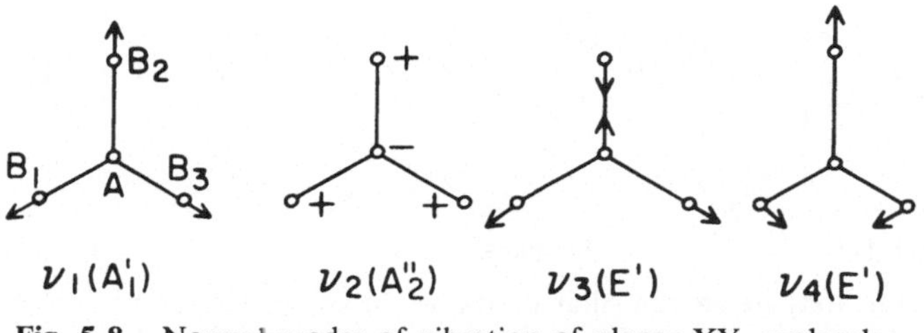

Fig. 5-8. Normal modes of vibration of planar XY_3 molecules.

Table 5-9. Selection Rules for D_{3h}, C_{2v}, and C_s Symmetry

Point group	ν_1	ν_2	ν_3	ν_4
D_{3h}	a_1'(R)	a_2''(I)	e'(I,R)	e'(I,R)
C_{2v}	a_1(I,R)	b_1(I,R)	a_1(I,R)+b_2(I,R)	a_1(I,R)+b_2(I,R)
C_s	a'(I,R)	a''(I,R)	a'(I,R)+a'(I,R)	a'(I,R)+a'(I,R)

lower symmetries. It is observed that ν_1 may become infrared active and thus six absorptions may now be found in the infrared spectrum above 650 cm^{-1}. Further, if the symmetry is lowered by coordination of the anion to the metal, bands attributed to metal–oxygen modes may also be found in the low-frequency region. There are several ways in which coordination may occur; Figure 5-9 illustrates this for the nitrate case. Evidence for this has increased in the case of nitrates, and coordination of the nitrate is now considered quite common. In fact, all three types of coordination have been observed. Table 5-10 lists the symbolism and types of vibrations involved for monodentate and bidentate nitrates, and Table 5-11 shows some examples of complexes with monodentate and bidentate nitrates with their infrared vibrations. The examples of bidentate nitrate complexes are increasing.[24–27] Bridging nitrates are found in $Be_4(NO_3)_2$[24] and in anhydrous $Cu(NO_3)_2$.[25] It is not possible to differentiate by infrared methods between the three types of coordinated nitrate. However, recent Raman studies have lent support to the belief that one may be able to distinguish between monodentate and bidentate or bridging nitrates.[26,27] The band in the 1400–1600 cm^{-1} region is polarized for bridging and bidentate nitrates and is depolarized for monodentate nitrates.[26,27]

Recent studies in the low-frequency region for anhydrous nitrates of transition metals and rare earth metals have been made. Absorptions in the 200–350 cm^{-1} region have been attributed to metal–oxygen stretching vibrations.[29,30] Table 5-12 lists the assignments made with ν_{MO} in these studies. Brintzinger and Hester have considered the perturbation of several polyatomic oxyanions (e.g., perturbation due to coordination of oxygen to metal) and have made a normal coordinate treatment for them.[31]

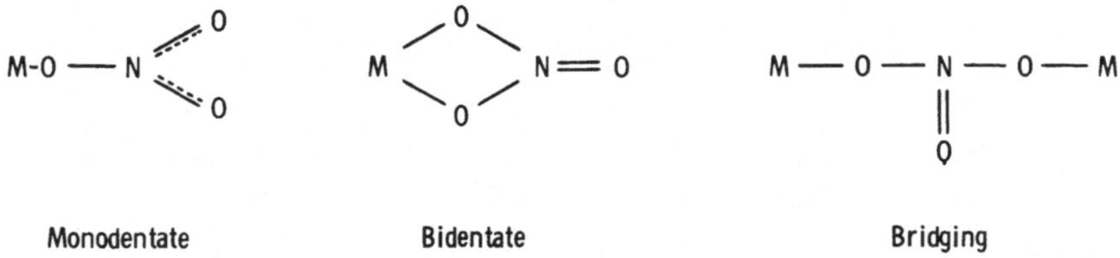

Fig. 5-9. Methods of coordination of nitrate in complexes. (Note: terdentate is a possibility, but no example of this type of coordination has been found.)

Table 5-10. Symbolism and Types of Vibrations for Monodentate* and Bidentate Nitrates

Type of vibration	Species	ν, cm^{-1} Monodentate nitrate	Bidentate nitrate
NO_2 sym. stretch	a_1	ν_1(1290)	ν_2(985)
N–O stretch	a_1	ν_2(1000)	ν_1(1630)
NO_2 sym. bend	a_1	ν_3(740)	ν_3(785)
NO_2 asym. stretch	b_2	ν_4(1410–1550)	ν_4(1250)
NO_2 asym. bend	b_2	ν_5(715)	ν_5(750)
Out-of-plane rocking	b_1	ν_6(800)	ν_6(700)

* Recent work has cast doubt on the data for monodentate-type metal nitrates [see D. W. James and G. M. Kimber, *Austral. J. Chem.* **22,** 2287 (1969)]. Clearly, further x-ray studies are necessary in this area.

Table 5-11. Mid-Infrared Frequencies for Several Monodentate and Bidentate Nitrate Complexes[27]

Monodentate complexes	Frequency, cm^{-1} ν_4	ν_1	ν_2
$Mn(CO)_5NO_3$	1486	1284	1010
$Pd(bipyridyl)(NO_3)_2$	1517	1274	979
Me_3SnNO_3	1488	1268	1031
$Me_2Sn(NO_3)_2$	1550	1270	1000
$Sn(Py)_2(NO_3)_4$	1556	1304	1008
$Mn(CO)_3(Py)_2NO_3$	1467	1287	1015
$Mn(CO)_3(bipyridyl)(NO_3)$	1457	1288	1018
$[Co(NH_3)_5NO_3]X_2$	1495	1268	1011

Bidentate complexes	Frequency, cm^{-1} ν_1	ν_2	ν_3	ν_4	ν_5	ν_6
$Ti(NO_3)_4$	1635vs 1615sh 1565m	993s	785s	1225s 1190s	773s	678w
$Sn(NO_3)_4$	1622vs 1610sh 1556w	978s	802s	1240s 1202m 1170m	783s	696m
$VO(NO_3)_3$	1640s,b 1567m 1556m	1015m 995m 962m	780s,b	1202s,b		693w 685w
$Co(NO_3)_3$	1649 1621	965s,sp	761s	1158s 1166sh		

Legend: s = high intensity, m = medium intensity, w = low intensity, sp = sharp, b = broad, v = very.

Table 5-12. Observed Low-Frequency Absorptions in Several Anhydrous Nitrate Complexes

Nitrates[29,30]	ν_{MO}, cm^{-1}
$Cu(NO_3)_2$	336, 299
$Zn(NO_3)_2$	317, 284
$Co(NO_3)_2$	316, 271
$Mn(NO_3)_2$	231, 199
$Cd(NO_3)_2$	199
$Pr(NO_3)_3$	237, 197
$Nd(NO_3)_3$	237, 195
$Sm(NO_3)_3$	237, 186
$Eu(NO_3)_3$	239, 191
$Gd(NO_3)_3$	241, 206
$Tb(NO_3)_3$	244, 215
$Dy(NO_3)_3$	261, 232, 184
$Ho(NO_3)_3$	265, 235, 188
$Er(NO_3)_3$	266, 235, 188
$Tm(NO_3)_3$	266, 235, 190
$Yb(NO_3)_3$	240
$Lu(NO_3)_3$	226

Nitrates[27]	ν_{MO} cm^{-1}	Raman data[27]
$Ti(NO_3)_4$	448	444, 310, 285, 98
$Sn(NO_3)_4$	344	302, 247, 98
$VO(NO_3)_4$	450, 379	358, 303, 285, 236, 157, 76
$Co(NO_3)_3$	499	

	ν_{MO} for complexes of the type $L_2M(NO_3)_2$[28]			
Ligand	Co	Ni	Cu	Zn
Pyridine			328,288	305,285
Quinoline	300,204	305,294	323,303	291,274
α-Picoline	306,280	312,286	326,282	280b
Triphenylphosphine oxide	303,256	325,260	356,300	303,256

Legend: b=broad.

Similar arguments can be made with the carbonates. Table 5-13 lists several low-frequency absorptions that were assigned as ν_{MO} to complexes involving coordinated carbonate. Rare earth carbonates have been prepared[32-34], and low-frequency data on these compounds have recently become available.[34]

Several borates have been examined in the infrared region.[35,36] The borates of the rare earths have been shown to have absorptions below 650 cm^{-1}. The bands in the 550–637 cm^{-1} region may be assigned to δ_{BO_2}. Lower-

Table 5-13. Observed Low-Frequency Absorptions in Several Carbonate and Borate Complexes

Carbonates[34]	ν_4(CoO), cm^{-1}		Considering C_{2v} symmetry
$[Co(NH_3)_5CO_3]Br$	362		ν_1 is at ~1370 cm^{-1}
$[Co(ND_3)_5CO_3]Br$	351		ν_2 is at ~1050–1070 cm^{-1}
$[Co(NH_3)_5CO_3]I$	360		ν_3 is at ~750 cm^{-1}
$[Co(ND_3)_5CO_3]I$	341		ν_5 is at ~1450–1470 cm^{-1}
$[Co(NH_3)_5CO_3]NO_3 \cdot \frac{1}{2} H_2O$	351		ν_6 is at ~680 cm^{-1}
$[Co(ND_3)_5CO_3]NO_3 \cdot \frac{1}{2} D_2O$	344		ν_8 is at 850 cm^{-1}
$[Co(NH_3)_5CO_3]ClO_4$	350		ν_7 not observed
$[Co(ND_3)_5CO_3]ClO_4$	338		
	ν_7[CoO(asym)], cm^{-1}	ν_4[CoO(sym)], cm^{-1}	
$[Co(NH_3)_4CO_3]Cl$	430	395	ν_1 is at ~1590–1650 cm^{-1}
$[Co(ND_3)_4CO_3]Cl$	418	378	ν_2 is at ~1030 cm^{-1}
$[Co(NH_3)_4CO_3]ClO_4$	428	396	ν_3 is at ~740–760 cm^{-1}
$[Co(ND_3)_4CO_3]ClO_4$	415	374	ν_5 is at ~1260–1275 cm^{-1}
$[Co(en)_2CO_3]Br$	399	353	ν_6 is at ~830 cm^{-1}
$[Co(enD)_2CO_3]Br$	382	347	ν_8 is at 830 cm^{-1}
$[Co(en)_2CO_3]ClO_4$	393	372	
$[Co(enD)_2CO_3]ClO_4$	382	366	
$K[Co(NH_3)_2(CO_3)_2]$	444, 441	366	
$K[Co(ND_3)_2(CO_3)_2]$	438, 400	366	
$[Co(NH_3)_6][Co(CO_3)_3]$	488, 465	394, 351	
Borates[35,36]	ν_4, cm^{-1}		Considering D_{3h} symmetry
$LaBO_3$	595, 614		ν_1 is at ~939 cm^{-1}
$InBO_3$	614		ν_2 is at ~718–765 cm^{-1}
$ScBO_3$	637, 422, 285, 262		ν_3 is at ~1260–1275 cm^{-1}
$PrBO_3$	610, 590, 300		
$ErBO_3$	565, 401, 365, 267		
$SmBO_3$	686, 558, 396, 358, 263		
$GdBO_3$	697, 561, 389, 368, 269		
$DyBO_3$	705, 562, 400, 362, 271		
$HoBO_3$	707, 563, 400, 368, 274		
$ErBO_3$	711, 565, 401, 356, 267		
$YbBO_3$	720, 570, 402, 363, 270		

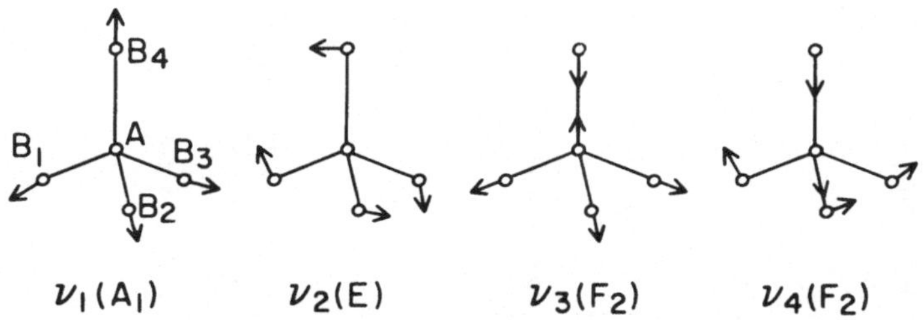

Fig. 5-10. Normal modes of vibration of a molecule in T_d symmetry.

frequency bands in the 250–400 cm^{-1} were assigned to metal–oxygen vibrations.[36] Table 5-13 lists the low-frequency absorptions for several borates.

T_d Symmetry—Sulfates, Selenates, Phosphates, Tungstates, Permanganates, Chromates, Molybdates, and Perchlorates

The normal modes of vibration for a tetrahedral molecule are illustrated in Fig. 5-10. All four modes are Raman active, but only ν_3 and ν_4 are infrared active. Only the ν_4 vibration, which is an YXY bending mode will appear in the low-frequency region. However, as shown in Table 5-14, if the tetrahedral anion is perturbed, the degenerate vibrations may split. For example, ν_4 is a triply degenerate vibration and may split into three bands if the symmetry is lowered. Furthermore, the ν_2 (*e*-type—doubly degenerate) vibration may become infrared active. The symmetry may be lowered to D_{2d}, C_{3v}, or C_{2v}; the additional absorptions that can occur are illustrated in Table 5-14. If the lowering of the symmetry is caused by coordination of the anion by the metal, a metal–oxygen vibration may be observed in the low-frequency region. Raman evidence[37] has been provided for an indium–oxygen symmetrical stretching vibration at 255 cm^{-1} in $In_2(SO_4)_3$. Infrared evidence is also available. The sulfate can be coordinated in three ways—unidentate, bidentate, and bridging, as illustrated in Fig. 5-11. Table 5-15 shows the vibrations of an ionic sulfate and compares them with those for a sulfate in a C_{3v} and C_{2v} symmetry.

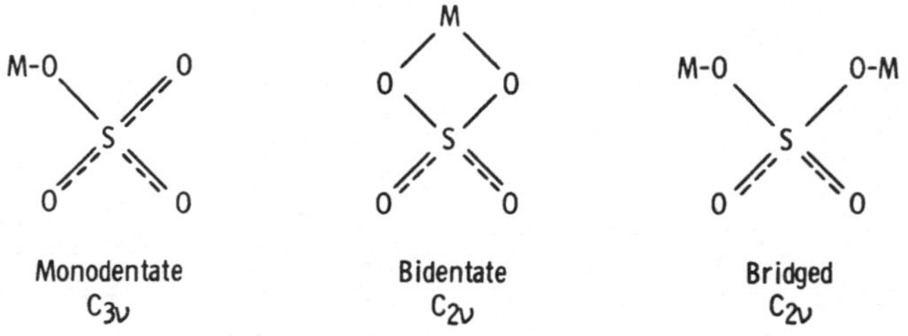

Fig. 5-11. Various ways in which a sulfate can be coordinated to a metal.

Table 5-14. Correlation Table for T_d, D_{2d}, C_{3v}, and C_{2v}

Point group	Number of infrared-active bands	ν_1	ν_2	ν_3	ν_4
T_d	2	a_1(R)	e(R)	f_2(I,R)	f_2(I,R)
D_{2d}	4	a_1(R)	a_1(R) + b_1(R)	b_2(I,R) + e(I,R)	b_2(I,R) + e(I,R)
C_{3v}	6	a_1(I,R)	e(I,R)	a_1(I,R) + e(I,R)	a_1(I,R) + e(I,R)
C_{2v}	8	a_1(I,R)	a_1(I,R) + a_2(R)	a_1(I,R) + b_1(I,R) + b_2(I,R)	a_1(I,R) + b_1(I,R) + b_2(I,R)

Table 5-15. Vibration Frequencies for Several Types of Sulfates

		Frequency, cm^{-1}			
Sulfate	Symmetry	ν_1	ν_2	ν_3	ν_4
Na_2SO_4	T_d			1104	613
$[Co(NH_3)_5SO_4]Br$	C_{3v}	970	438	1032–1044	645
				1117–1143	604
$[(NH_3)_4Co\langle^{NH_2}_{SO_4}\rangle Co(NH_3)_4][NO_3]_3$	C_{2v}	995	462	1050–1060	641
				1170	610
				1105	571

Although most studies with these salts have not been extended below 300 cm^{-1}, one recent investigation with rare earth salts has indicated that at 200–300 cm^{-1} a broad absorption involving metal–oxygen stretching can be observed.[12] This evidence together with the splitting of the ν_3 and ν_4 vibrations and the appearance of the forbidden ν_1 vibration is indicative of coordination to the metal.

Similar considerations are possible for other salts in a T_d symmetry. Table 5-16 summarizes data for various salts.[12,38–68]

C_{3v} Symmetry—Chlorates, Bromates, Iodates, Sulfites, and Selenites

The normal modes of vibration for a molecule in a C_{3v} symmetry are illustrated in Fig. 5-12. All vibrations are Raman and infrared active as determined by the selection rules based on a point group symmetry. The correlation Table 5-17 shows the relationship between C_{3v} and C_s symmetry. Only the ν_2 and ν_4 vibrations, which are the YXY bending modes, will appear below 650 cm^{-1}. The ν_4 vibration is doubly degenerate and can split into two bands if the symmetry of the anion is lowered. Dasent and Waddington[59] studied several metal iodates and found extra bands at 420–480 cm^{-1}. They ascribed these bands to metal–oxygen vibrations and considered these iodates to be in a lower symmetry (e.g., C_s). Six bands, active in the Raman and infrared, are allowed for a C_s symmetry. Further study of these salts

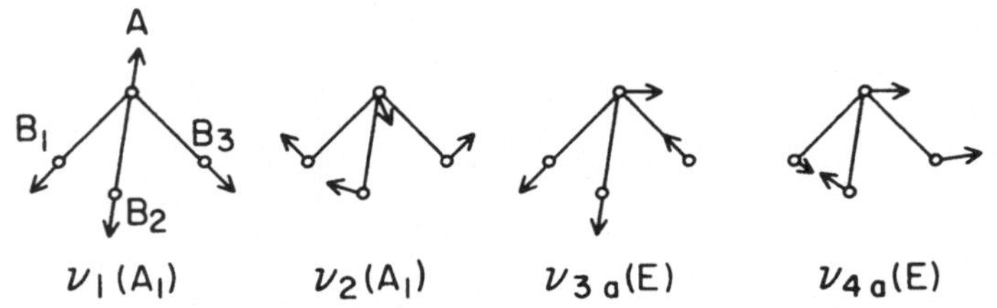

Fig. 5-12. Normal modes of vibrations of pyramidal XY_3-type molecules.

Table 5-16. Low-Frequency Absorptions for Several Salts with Polyatomic Anions in a T_d Symmetry or Lower

Sulfates[12,38,40]	Frequency, cm^{-1}		Remarks
	ν_4	ν_2	
$MnSO_4 \cdot xH_2O$	655, 627, 607, 510		ν_3 at 1100 cm^{-1}
$FeSO_4 \cdot 7H_2O$	655	422	
$CuSO_4 \cdot xH_2O$	680, 630, 594	472,400	ν_1 and ν_2 are normally forbidden in T_d symmetry
$Fe_2(SO_4)_3 \cdot xH_2O$	641, 635, 581	483, 426	
$CoSO_4 \cdot 7H_2O$	623		
$Cr_2(SO_4)_3 \cdot K_2SO_4 \cdot 24H_2O$	615, 562	325	
$Co(NH_3)_5SO_4 \cdot Br$	645, 604	438	
$[(NH_3)_4Co]_2NH_2SO_4(NO_3)_3$	641, 610, 571	462	
$LiSO_4 \cdot H_2O$	646, 577	480, 437, 367, 322	
Na_2SO_4	641, 620		
K_2SO_4	620		
$ZnSO_4 \cdot 4H_2O$	657, 625	445, 328	
$MgSO_4 \cdot 7H_2O$	588	305	
$PbSO_4$	623, 592		
$SrSO_4$	641, 606		
$Bi_2(SO_4)_3$	633, 606	405, 302	
$BaSO_4$	633, 606		
$Al_2(SO_4)_3 \cdot xH_2O$	641, 602	493, 408	
Ag_2SO_4	588		
$CdSO_4 \cdot xH_2O$	654, 617, 588, 513	417	
Hg_2SO_4	643, 593, 568, 508		
$Sc_2(SO_4)_3$	667vs*, 631s, 613sh, 575vw, 495sh	451m, 431sh	
$Y_2(SO_4)_3$	658s, 628s, 615s, 579vw, 468m	451m, 433s	
$La_2(SO_4)_3$	667s, 651s, 621s, 603vs, 587s	469m, 444vw, 420m	ν_3 at 1050–1270 cm^{-1}
$Ce_2(SO_4)_3$	668s, 650s, 618w, 598s, 583s	468m, 418m	ν_1 at 970–1020 cm^{-1}
$Pr_2(SO_4)_3$	671s, 658s, 625s, 606s, 598s	474m, 427m, 422m	
$Nd_2(SO_4)_3$	670s, 658vs, 621s, 602s, 584s, 497vw	471m, 424m, 419m	

Table 5-16 (Continued)

Sulfates[12,38,40]	Frequency, cm^{-1} ν_4	ν_2	Remarks
$Sm_2(SO_4)_3$	688m, 658s, 624s, 601s, 583m	475m, 455vw, 425m, 420m	
$Eu_2(SO_4)_3$	670m, 657s, 637w, 624s, 606sh, 597s	473m, 445s, 442w, 429w	
$Gd_2(SO_4)_3$	658s, 627sh, 615vs	452sh, 440vs, 435sh	
$Tb_2(SO_4)_3$	649s, 620s, 611vs	442sh, 437s, 432sh	
$Dy_2(SO_4)_3$	653s, 622sh, 612vs	437sh, 433s, 402vw	
$Ho_2(SO_4)_3$	657s, 622sh, 612vs, 568vvw	432s	
$Er_2(SO_4)_3$	661s, 627s, 615vs	435s	
$Tm_2(SO_4)_3$	656s, 618s, 607vs	431s	
$Yb_2(SO_4)_3$	666s, 627s, 615vs, 466vw	435s	

Perchlorates[40–44]	ν_3	ν_7	ν_9	ν_4	Remarks
$Mn(ClO_4)_2 \cdot 2H_2O$	666	635, 613	467	453	C_{2v} symmetry
$Co(ClO_4)_2 \cdot 2H_2O$	647	628, 611	497		"
$Ni(ClO_4)_2 \cdot 2H_2O$	645	625, 612	490	454	"
$Cu(ClO_4)_2$(anhydrous)	665, 647	624, 600	497	466	"
$Cu(ClO_4)_2 \cdot 2H_2O$	648	620, 605	480, 460		C_{3v} symmetry

Chromates[38,46–48]	ν_4	Remarks
$ZnCrO_4 \cdot 7H_2O$	408, 383	
K_2CrO_4	397–384	ν_3 at 839, 915 cm^{-1}
$PbCrO_4$	388	
$(NH_4)_2CrO_4$	369	
Perrhennates[49–52]		
$KReO_4$	332	ν_3 at 898–915 cm^{-1}
NH_4ReO_4	318, 303	
Pertechnetates[52]		
NH_4TcO_4	329, 317	ν_3 at 900–925 cm^{-1}
$TlTcO_4$	327, 322	

Table 5-16 (Continued)

Selenates[53–54]	Frequency, cm^{-1} ν_4	ν_2	Remarks
$CdSeO_4$	502, 481, 459, 420	389	ν_3 at ~895 cm^{-1}
$CaSeO_4$	455, 436	379	
Ag_2SeO_4	417	388	
Na_2SeO_4	425	392	
Phosphates[38,40,45]			
$Na_3PO_4 \cdot 12H_2O$	572, 553, 422, 402		ν_3 at ~1080 cm^{-1}
K_3PO_4	545, 516		
$Mg_3(PO_4)_2 \cdot 4H_2O$	603, 570, 489, 460, 446, 430, 400	341, 325	
$Ca_3(PO_4)_2$	635, 606, 568	353	
$CrPO_4 \cdot H_2O$	570		
$Co(NH_3)PO_4$	643, 576, 531		ν_1 at 894, ν_3 at 1106, 1055, 1040, 917 cm^{-1}
$Mn_3(PO_4)_2 \cdot 7H_2O$	602, 586, 528	372, 334, 317	
$Ni_3(PO_4)_2 \cdot 7H_2O$	625, 572		
$Cu_3(PO_4)_2 \cdot 3H_2O$	643, 620, 562, 508, 432	398, 323	
$Pb_3(PO_4)_2$	581, 545		
Permanganates[46–55]			
$KMnO_4$	400, 383		ν_3 found at 800–901 cm^{-1}
K_2MnO_4	320		
Na_3MnO_4	348		
Ruthenates[55,56]			
$KRuO_4$	282		ν_3 found at 806–856 cm^{-1}
$BaRuO_4$	330		
Ferrates[48,55]			
K_2FeO_4	339, 320		ν_3 at 800 cm^{-1}
Molybdates[40,51,57]	ν_4 and ν_2 (overlapping)		
Na_2MoO_4	313, 290		ν_3 at 838 cm^{-1}
Ag_2MoO_4	420, 350, 285		
Li_2MoO_4	440, 370		
$CaMoO_4$	430, 325		
$SrMoO_4$	400, 325, 270		
$BaMoO_4$	372, 320, 291, 285		

Table 5-16 (Continued)

	Frequency, cm^{-1}			
Molybdates[40,51,57]	ν_3	ν_4 and ν_2 (overlapping)		Remarks
$CdMoO_4$		430, 305		
$PbMoO_4$		370, 350, 305, 265		
$CoMoO_4$	640	470	295, 285, 265, 258	
$NiMoO_4$	605	425, 345, 305	260	
K_2MoO_4		485	328, 310	
Rb_2MoO_4			335, 315	
Cs_2MoO_4			335, 316, 303	
$Zr(MoO_4)_2$			345, 290, 280, 250	
$Hf(MoO_4)_2$	610, 515		325, 290, 270	
$Cr_2(MoO_4)_3$	610, 555		320, 310, 250	
$CuMoO_4$		485	290, 260	
$ZnMoO_4$		435	335, 310, 270	
Tungstates[49,51,57,58]	ν_4 and ν_2 (overlapping)			
Na_2WO_4	320			ν_3 at 835 cm^{-1}
$CaWO_4$	440, 325, 285			
$SrWO_4$	410, 315, 265			
$BaWO_4$	380, 310, 280			
$PbWO_4$	375, 355, 290			
Li_2WO_4	415, 330, 290			
$MnWO_4$	450, 420, 330			
$CoWO_4$	340			
$NiWO_4$	450, 350			
$MgWO_4$	608, 525, 480, 445, 380, 320, 290, 250			
$ZnWO_4$	590, 430, 320			
K_2WO_4	315, 290			
Rb_2WO_4	290			
Cs_2WO_4	340, 290			
$CuWO_4$	605, 550, 415, 375, 340, 290, 270			
Ag_2WO_4	590, 380, 335, 270			
$CdWO_4$	560, 505, 455, 410, 355, 310			
Vanadates[68]				
Aqueous solution of vanadates	340 (Raman)			ν_3 is at 780 cm^{-1} ν^1 at 827 cm^{-1}

*Legend: s=high intensity, m=medium intensity, v=very, w=low intensity, b=broad, sh=shoulder.

appears necessary, for these measurements were made only to 400 cm^{-1} and the asymmetric and symmetric IO_2 bending vibrations, at ~330 and ~390 cm^{-1} respectively, were not observed. Rocchiccioli[60] investigated several metal chlorates as well as several metal sulfites.[65-67] The infrared spectra of

Table 5-17. Relationship Between C_{3v} and C_s Symmetry

C_{3v} (XY_3^{n-})	$\nu_1(a_1)$ $\nu_s(XY)$ (R,I)	$\nu_2(a_1)$ $\delta_s(YXY)$ (R,I)	$\nu_3(e)$ $\nu_d(XY)$ (R,I)		$\nu_4(e)$ $\delta_d(YXY)$ (R,I)	
	↓	↓	↙	↘	↙	↘
$C_s(Y^*\text{-}XY_2^{n-})$	$\nu_1(a')$ $\nu_s(XY^*)$ (R,I)	$\nu_3(a')$ $\delta_s(YXY^*)$ (R,I)	$\nu_2(a')$ $\nu_s(XY)$ (R,I)	$\nu_5(a'')$ $\nu_{asym}(XY)$ (R,I)	$\nu_4(a')$ $\delta_s(YXY)$ (R,I)	$\nu_6(a'')$ $\delta_{asym}(YXY^*)$ (R,I)

Y*—coordinated atom.

several metal selenites have also been measured.[53] Table 5-18 shows a compilation of low-frequency data for these salts. Figure 5-13 shows the characteristic low-frequency vibrations in salts of various polyatomic anions.

5.5. β-DIKETONE COMPLEXES

Metal chelate complexes of the β-diketones have been extensively studied in the mid-infrared region. Normal coordinate analysis (NCT) has been made on Cu(acac*)$_2$[69] and on several transition metal acetylacetonates.[70] This has aided in making assignments for these compounds. X-ray

*acac = acetylacetone.

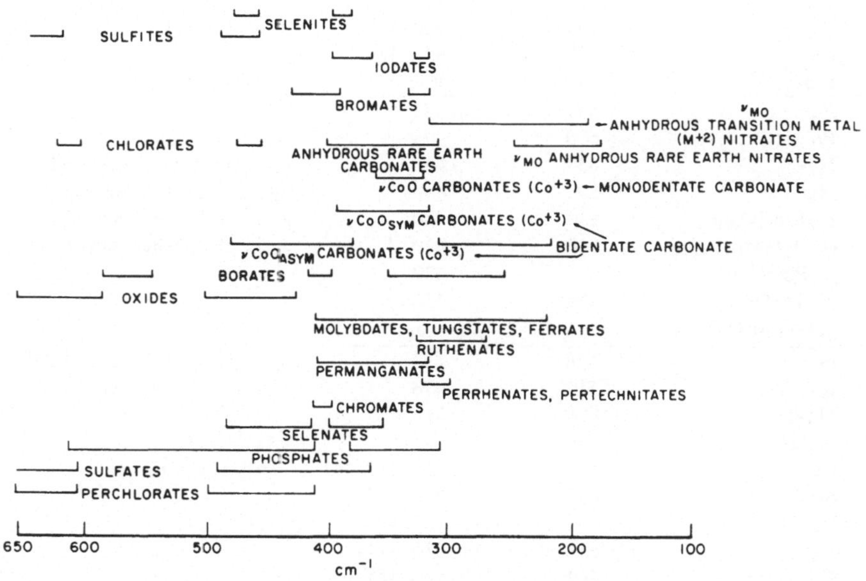

Fig. 5-13. Characteristic low-frequency vibrations in salts of polyatomic anions.

Table 5-18. Tabulation of Low-Frequency Data for Several Salts in C_{3v} Symmetry or Lower

	Frequency, cm^{-1}		
Chlorates[38,61,63]	ν_2	ν_4	Remarks
Na	619	478	ν_1 is at ~910 cm^{-1}
K	619	479	ν_3 is at ~960 cm^{-1}
Mg	609	499, 478	
Ca	626	493, 478	
Sr	610	505, 471	
Ba	605	503, 483	
Cr	610	495, 481	
Nd	607	498, 478	
Co	610	498, 479	
Ni	609	478	
Cu	609	480, 478	
Ag	612	493, 470	
Zn	607	498, 478	
Cd	607	501, 478	
Pb	606	499, 467	
Bromates[38]			
Na	421	356	ν_1 is at ~806 cm^{-1}
K	431	362	ν_2 is at ~836 cm^{-1}
Ag	440	358, 388	
Iodates[38,59,64]			
Na	385	334	
K	359	330, 311	
Ca	398, 367	335, 324	
Iodates[59]	ν_1	ν_{MO}	
Fe	697	451	C_s symmetry considered; ν_3, ν_4, ν_6 not observed; ν_2 is at 719–758 cm^{-1}; ν_5 is at 751–808 cm^{-1}
Pb	690	423	
Hg^{2+}	633	436	
Hg^{+}	650	448	
$K_2Mn(IO_3)_6$	630	480	
$(NH_4)_2Mn(IO_3)_6$	640	479	
$K_2Pb(IO_3)_6$	695	420	
$K_2Tl(IO_3)_6$	656	443	
Sulfites[38,60,65-67]	ν_2	ν_4	
Na	626	493	ν_1 is at ~1000–1138 cm^{-1}
K	619	478	ν_3 is at 892–965 cm^{-1}
NaK	616	512, 485	
NH_4	616		
Ca	649	515, 487, 445	
Sr	640	519, 497	
Ba	630	508, 493	
Ag	631	476, 474	
Cd	649	527, 475	
Pb	625	483, 472	

Table 5-18 (Continued)

Sulfites[38,60,65–67]	Frequency, cm^{-1} ν_2	ν_4	Remarks
UO_2	633	413	
$[Co(en)_2SO_3]Cl$	647, 625		No measurements made
$[Co(en)_2SO_3]Br$	649, 621		at lower frequencies
$[Co(en)_2SO_3]I$	649, 621		ν_1 is at 1070–1119 cm^{-1}
			ν_2 is at 983–1042 cm^{-1}
Selenites[53]			
Ba	463	418	ν_1 is at ~800 cm^{-1}
Cd	487	384, 403, 420	ν_3 is at ~740 cm^{-1}
Cu	436	379, 398	
Na	462	383	
Zn	470		

analysis is available for $Fe(acac)_3$, and conclusions from this work have indicated that the chelate ring is planar and symmetrical.[71] Thus, the two carbonyl and C=C bonds in the ring are equivalent. Certain questions existed concerning the assignment of the $\nu_{C=O}$ and $\nu_{C=C}$ stretching vibrations. In a series of metal acetylacetonates Nakamoto[72] assigned the $\nu_{C=C}$ vibration at about 1570–1601 cm^{-1} and the $\nu_{C=O}$ vibration at 1524–1601 cm^{-1}. Because of a recent isotopic study[73] and a normal coordinate treatment of $CH_3COCD_2COCH_3$[74] doubt has been cast on these two assignments and the above assignments have now been reversed.

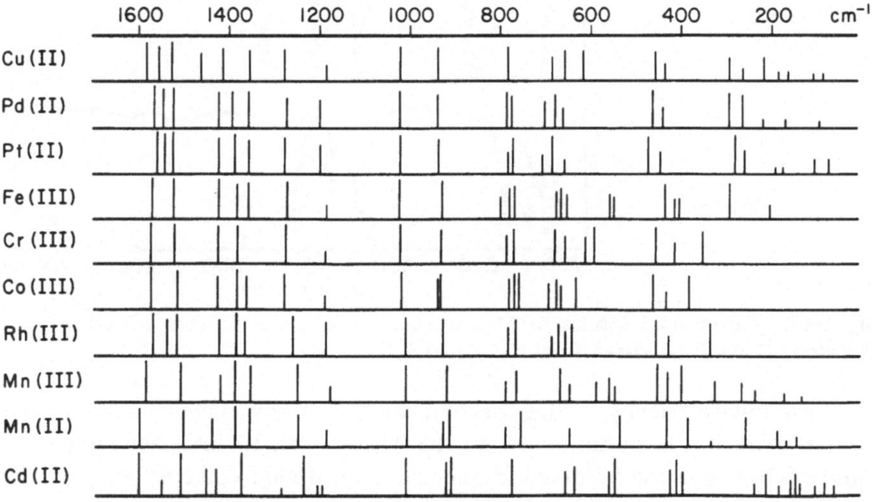

Fig. 5-14. Infrared frequencies of various acetylacetonates of transition metals (1600–60 cm^{-1}). (Courtesy of Pergamon Press, New York. From Mikami *et al.*[70])

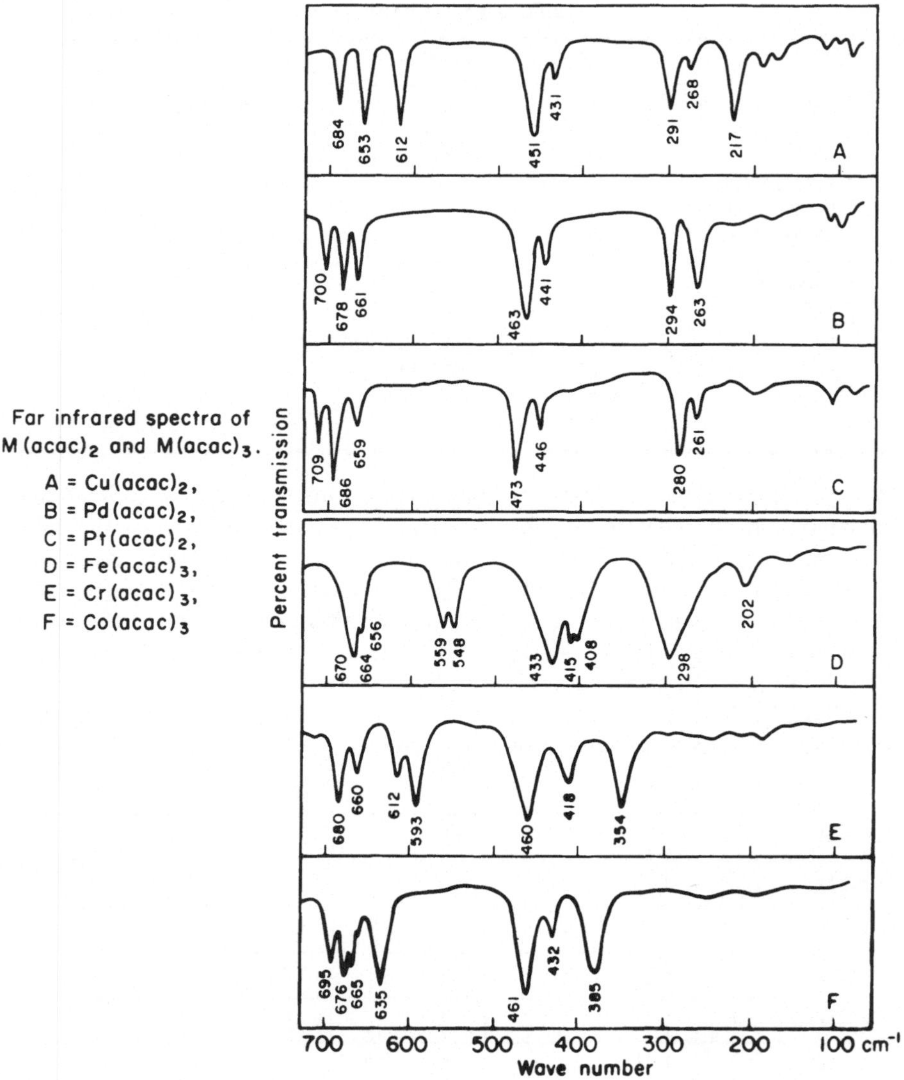

Fig. 5-15. Far-infrared spectra of M(acac)$_2$ and M(acac)$_3$ complexes. (Courtesy of Pergamon Press, New York. From Mikami *et al.*[80])

Few low-frequency studies beyond 400 cm^{-1} have been made.[70,75–78] Raman work has been done only on the Al, Ga, and In acetylacetonates.[79] Figure 5-14 shows the observed frequencies from 1600 to 60 cm^{-1} of several acetylacetonates,[70] and Fig. 5-15 shows the far-infrared spectra of several M(acac)$_2$ and M(acac)$_3$ complexes.[80] Table 5-19 records the low-frequency

Table 5-19. Low-Frequency Infrared Absorption Found in Several β-Diketone Complexes, cm^{-1}

	β-Diketone[72,81]		
Metal ion	(acac)	(TFTF)	(PP)
Cu(II)	455	415	462
	431		
	291, 268		
	217		
Ni(II)	452	399	458
Co(II)	422		
Zn(II)	422		
Be(II)	500		
Pt(II)	473		
	446		
	280, 261		
Pd(II)	464		
	441		
	294, 233		
Co(III)	466		
	432		
	395		
Cr(III)	459		
	418		
	354		
Fe(III)	434		
	415, 408		
	298		
	202		
Al(III)	490		

Out-of-plane vibrations also observed at 549–720 cm^{-1} and 411–442 cm^{-1} for acetylacetonates. Abbreviations: acac = acetylacetone; TFTF = $CF_3COCH_2COCF_3$; PP = $\varphi COCH_2CO\varphi$.

vibrations observed for various β-diketones.[72,81] Recent studies by Nakamoto *et al.*[82] have indicated metal–olefinic and metal–carbon-bonded chelates in acetylacetonates of Pt(II). Acetylacetonates of Zr(IV) and Hf(IV),[75–77] Sn(IV),[84] Ti(IV),[84] and the rare earth metals[78] have been studied. Table 5-20 shows some low frequencies observed in mixed acetylacetonates.[83–85] Table 5-21 shows a comparison with Raman data for $Al(acac)_2$[79] and Table 5-22 tabulates the low-frequency data and assignments made by Mikami *et al.*[80] based on a NCT.

Some of the actual vibrational modes corresponding to the assignments are illustrated in Fig. 5-16. In $Cu(acac)_2$, which has a square planar structure, two ν_{MO} stretching vibrations and one ring out-of-plane bending vibration are predicted by the selection rules. The three vibrations at 451, 291, and 217

Table 5-20. Low-Frequency Bands (400–500 cm^{-1}) in Mixed Acetylacetonates[83-85] Complexes

$Cu(acac)_2$	456, 432
$Cu(TFM)_2$	445
$Cu(TFTF)_2$	415
Cu(acac) (TFM)	457, 449, 433
Cu(acac) (TFTF)	466, 420, 403
Cu(TFM) (TFTF)	442, 437, 406
$SnCl_2(acac)_2$[71a]	461
$SnBr_2(acac)_2$	453
$SnI_2(acac)_2$	445
$GeCl_2(acac)_2$	478
$TiCl_2(acac)_2$	459
$WO_2(acac)_2$[71b]	578,551
$MO_2(acac)_2$	576, 550

Abbreviations: TFM = trifluoroacetylacetone; TFTF = $CF_3COCH_2COCF_3$; acac = acetylacetone.

Table 5-21. Typical Assignments for an Acetylacetonate $[Al(acac)_3]$, cm^{-1}

Infrared[72]	Raman[79]	
1590 $\nu_{C=C}$?	1600w	$\nu_{C=O}$
1545, 1530 $\nu_{C=O}$?		
1466 $\nu_{C=O} + \delta_{CH}$	1450w	
1387 δ_{CH_3} sym and δ_{CH_3} asym	1373s, sh, d	δ_{CH_3}
1288 $\nu_{C=C} + \nu_{CCH_3}$	1298vvs, sh, p	ν_{CC} sym
1191 δ_{CH}	1190ms, sh, d	δ_{CCH}
1028 CH_3 rock	1033mw, sh, p	CH_3 rock
	956mw, sh, p	$\nu_{CCH} + \nu_{CO_3}$
935 $\nu_{CCH_3} + \nu_{C=O}$	938mw, sh, d	ν_{CCH_3}
738 π_{CH}		
685 ring def. $+\nu_{MO}$	690m, sh, sl. p	$\nu_{MO} + \delta$ (ring)
658 $\delta_{CH_3} + \nu_{MO}$		
594, 577 π	570mw, sh, d	$C{-}C\langle^{C}_{C}$ wag
490 ν_{MO}	495mw, sh, d	ν_{MO} asym
	465s, sh, p	ν_{MO} sym
425, 416 π	420w, d	δ ring, out-of-plane
	400 m, sh, d	δ ring, in plane
	340vw	δ_{CCH}
	220w, d	δ ring, in plane

Legend: π = out-of-plane vibration; s = strong; m = medium; w = weak; v = very; sh = shoulder; sp = sharp; d = depolarized; p = polarized; acac = acetylacetone.

Table 5-22. Observed Low Frequencies and Assignments of $Cu(acac)_2$[70]

Frequency, cm^{-1}	b_{1u}	b_{2u}	b_{3u}
684	ν_{10}		
653			ν_{31}
612		ν_{23}	
451	ν_{11}		
431		ν_{24}	
291		ν_{25}	
268	ν_{12}		
217			ν_{32}
187	ν_{13}		
167			ν_{33}
108			ν_{34}
84		ν_{26}	

*For some of the actual vibration modes see Fig. 5-16.

cm^{-1} are assigned to these vibrations. For the $Fe(acac)_2$ complex, these vibrations are assigned at 433 and 291 cm^{-1}. The original assignment of Nakamoto of a metal–oxygen stretching vibration, occurring at near 450 cm^{-1}, appears to be correct. The force constants of the metal–oxygen band obtained for the various acetylacetonates indicate the following sequence:

1:3 complex Rh(III) > Co(III) > Cr(III) > Fe(III) > Mn(III)
1:2 complex Pt(II) > Pd(II) > Cu(II) > Mn(II).

5.6. COMPLEXES OF ORGANIC ACIDS

Although a number of complexes of organic acids have been prepared and studied in the mid-infrared region, very few far-infrared and Raman studies have been made. Thus a great many of the assignments made must be considered tentative, and further work is indicated.

Oxalato Complexes

Mid-infrared data and a normal coordinate treatment of oxalato complexes are available.[86] However, there is a lack of infrared data below 300 cm^{-1} and of Raman data. As in the case with the β-diketones, additional doubt is cast on the low-frequency assignments because of the suspected nonpurity of the vibrations. Table 5-23 lists the assignments made for the Al oxalate complex.[86,87] Table 5-24 lists some assignments made for the symmetric ν_{MO} vibration from Raman data.[88]

Amino Acid Complexes

Several metal–leucine complexes have been prepared and Table 5-25 summarizes the assignments made on the basis of a normal coordinate

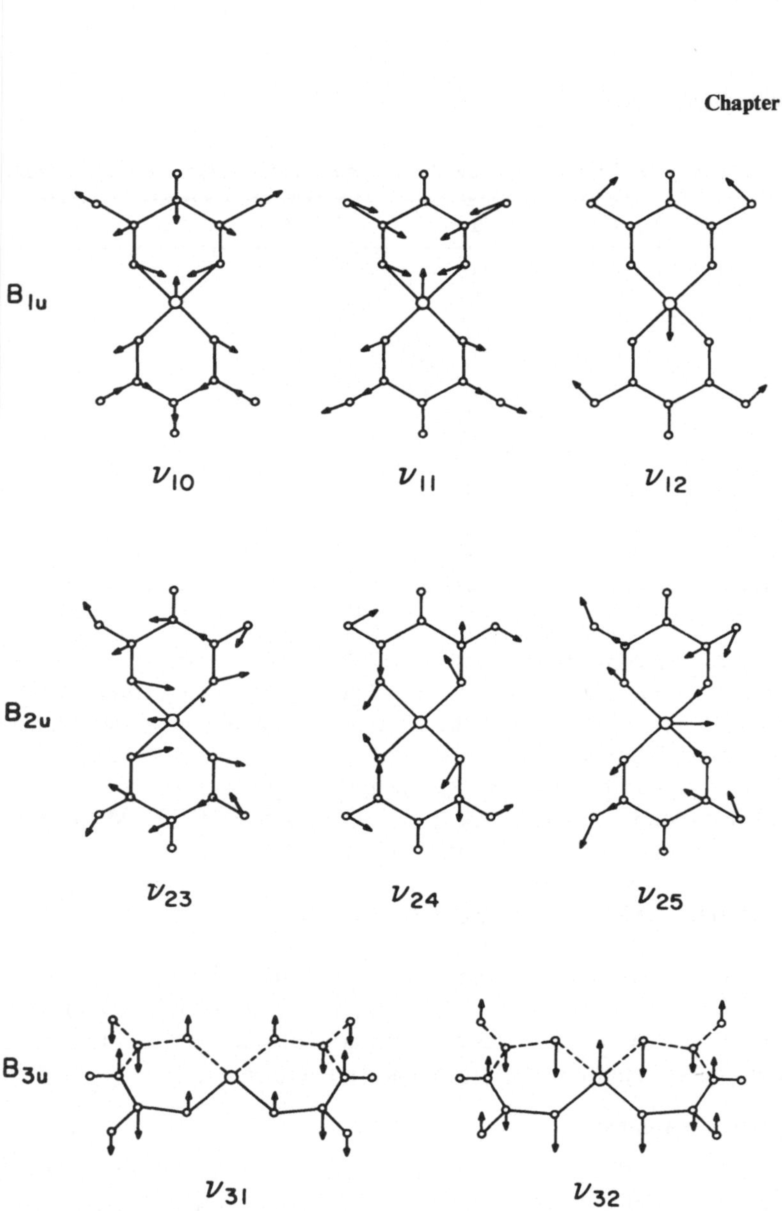

Fig. 5-16. Schematics of vibrational modes for $Cu(acac)_2$.

treatment.[89] The ν_{MN} stretching vibration for transition metal complexes has been assigned from 312–409 cm^{-1}; the ν_{MO} stretching vibration at 140–200 cm^{-1}. Metal–valine complexes have been similarly studied.[90] Table 5-26 summarizes these data. The assignment for ν_{MN} in transition metal complexes has been made in the region of 365–399 cm^{-1}, whereas the ν_{MO} vibration was not assigned since the lower frequency limit studied was 270 cm^{-1}.

Table 5-23. Typical Assignments in $K_3[Al(C_2O_2)_3]$, cm^{-1}

Infrared[86]	Raman[87]	
1722 $\nu_{C=O}$ asym	1750vs, sh, p	$\nu_{C=O}$ sym
1700, 1683 $\nu_{C=O}$ sym	1685s, b, d	$\nu_{C=O}$ asym
1405 ν_{CO} sym + ν_{CC}	1420vs, sh, p	ν_{CO} sym
1292, 1269 ν_{CO} sym + $\delta_{O\text{-}C=O}$	1410s, sh, d	ν_{CO} asym
	920m, sh, p	ν_{CC}
904 ν_{CO} sym + $\delta_{O\text{-}C=O}$	865vw, d	δ_{OCO}
	600w, b, d	ν_{MO} asym
820, 803 $\delta_{O\text{-}C=O}$ + ν_{MO}	585m, sh, p	ν_{MO} sym
	470w, b, d	ring deformation
587 ν_{MO} + ν_{CC}	370w, d	CCO, OMO
	275w, d	COM, deformations
436 ring deformation + $\delta_{O\text{-}C=O}$		
485 ν_{MO} + ring deformation		
354 $\delta_{O\text{-}C=O}$		

For oxalato complexes of type $K_2M(C_2O_4)_2{\cdot}2H_2O$, $K_3M(C_2O_4)_3{\cdot}3H_2O$:*

Assignments[86]	Frequency range
ν_{MO} + ν_{CC}	519–587
$\delta_{O\text{-}C=O}$ + ring deformation	436–519
ν_{MO} + ring deformation	366–485
$\delta_{O\text{-}C=O}$	340–382
π	291–350

Legend: s = strong, m = medium, w = weak, v = very, sh = shoulder, sp = sharp, p = polarized, d = depolarized.
*M = Transition metal.

Glycolato Complexes

Nakamoto and co-workers[91] have prepared and made assignments for several metal–glycolato complexes. Table 5-27 summarizes the assignments made. In these complexes no pure ν_{MO} stretching vibration has been found, since the vibrations are mixed with ring deformation modes.

Table 5-24. Assignments from Raman Data for Several Oxalato Complexes[88] (Measured in Aqueous Solution)

Complex	ν_{MO} sym, cm^{-1}
$Zn(C_2O_4)_2^{2-}$	515p
$Mg(C_2O_4)_3^{4-}$	510p
$Al(C_2O_4)_3^{3-}$	583p
$Ga(C_2O_4)_3^{3-}$	573p

Legend: p = polarized.

Table 5-25. Observed Frequencies Related to the Metal–Leucine Complexes,[89] cm^{-1}

Vibrational mode	Pt(II)	Pd(II)	Cu(II)	Ni(II)	Zn(II)	Cd(II)	Co(II)
NH_2 asym. stretch	3217	3280	3304	3368	3345	3359	3368
NH_2 sym. stretch	3120	3224	3275	3276	3253	3257	3257
CO_2 asym. stretch	1640	1636	1640	1595	1598	1570	1583
NH_2 scissors	1600	1601	1573	—	1555	—	—
CO_2 sym. stretch	1379	1365	1398	1395	1412	1401	1400
CO_2 scissors	723	727	727	831	830	830	830
NH_2 rocking	801	786	655	625	645	576	579
CO_2 wagging	658	663	655	656	655	645	655
CO_2 rocking	606	587	580	570	580	576	579
M–N stretch	409	385	400	323	314	290	312
M–O stretch	200	195	185	140	160	130	160

Biuret Complexes

Nakamoto and co-workers[92] have prepared several biuret complexes of transition metals. They have made a NCT for these complexes and have presented evidence for linkage isomers. The bonding to biuret (H_2N–CO–NH–CO–NH_2) may be to the nitrogen or to the oxygens. The isomers may be distinguished by their infrared spectra. If the bonding is to the oxygen, then a spectrum showing free amino groups, bonded carbonyl groups, and a metal–oxygen absorption should be obtained. If the bonding is to the nitrogen, a metal–nitrogen absorption and a free carbonyl band should be seen. The ν_{MO} in the Cu(biuret)$_2$Cl$_2$ and Ni(biuret)$_2$Cl$_2$ complexes is assigned at 284 and 268 cm^{-1} respectively. In the $K_2M(biuret)_2 \cdot 2H_2O$ complexes where

Table 5-26. Observed Frequencies Related to the Metal–Ligand Bonds in Valine Complexes,[90] cm^{-1}

Vibrational mode	Pt(II)	Pd(II)	Cu(II)	Ni(II)
NH_2 asym. stretch	3227	3246	3297	3326
NH_2 sym. stretch	3137	3115	3250	3270
CO_2 asym. stretch	1645	1640	1610	1591
NH_2 scissors	1604	1630	1572	1578
CO_2 sym. stretch	1360	1363	1381	1405
NH_2 rocking	833	749	640	638
CO_2 scissors	711	799	737	790
CO_2 wagging	594	600	578	553
CO_2 rocking	522	572	494	539
MN stretch	399	385	394	365
MO stretch	<270 (not assigned in this paper)			

Table 5-27. Observed Frequencies of Various Metal-Glycolato Complexes,[91] cm^{-1}

$Cu(CH_2OHCOO)_2$	$Ni(CH_2OHCOO)_2$	$Co(CH_2OHCOO)_2$	$Zn(CH_2OHCOO)_2$	$Mn(CH_2OHCOO)_2$	Band assignment
3000	3070	3090	3040	3100	ν_{OH}
—	2990	2990	2980	2960	ν_{CH_2}
—	2940	2935	2920	2920	ν_{CH_2}
1575	1590	1595	1590	1605	$\nu_{C=O}$
1485	1480	1482	1475	1480	δ_{OH}
1440	1440	1438	1440	1440	δ_{CH_2}
1400	1403	1402	1400	1390	δ_{CO_2}
1305	1303	1300	1300	1300	$\rho_w CH_2$
1225	1233	1237	1238	1240	$\rho_t CH_2$
1065	1065	1062	1062	1075, 1050	ν_{CO_3}
940	950	944	940	935	ν_{CC}
920	—	920	—	—	$\rho_r CH_2$
750	780	762	770	770	π_{OH}
710	700	689	685	710	$\delta_{C=O}$
601, 554	583, 568	580, 564	583, 567	578, 561	$\pi_{C=O}$
515	543	540	541	530	ring def.
483	474, 442	459, 427	468, 435	445, 418	ν_{MO} + ring def.
408, 378	370, 327sh, 317	343, 307	327, 284sh, 270	304, 250	ν_{MO} + ring def.

Legend: sh=shoulder band; π denotes out-of-plane bending.

M is Cu, Ni, or Co, a metal-sensitive band appears at 405, 476, and 458 cm^{-1}, respectively. This is assigned to a ν_{MN} coupled to a $\delta_{C=O}$ vibration.

5.7. COMPLEXES OF ORGANIC OXIDE LIGANDS

Complexes of pyridine-N-oxides, substituted pyridine-N-oxides, arsine oxide, and phosphine oxide are known. A good number of these have been studied in the mid-infrared region, but only few have been examined below 650 cm^{-1}.[93,95] Some low-frequency bands have been assigned (see Table 5-28). However, since the ligands show rich far-infrared spectra and the complexes themselves show complicated spectra, the assignments are difficult to make. The absorption at 400–450 cm^{-1} in trivalent transition metals appears to be the ν_{MO} in pyridine-N-oxide. For divalent metals, it appears to occur at 311–368 cm^{-1}. In copper complexes the absorption varies, depending

Table 5-28. Low-Frequency Vibrations in Organic Oxide Complexes[93-100]

	Frequency, cm^{-1}		
	C_6H_5NO	$\varphi_3As{=}O$	$\varphi_2MeAs{=}O$
Mn^{2+}	311	370	395
Fe^{2+}	320	384	
Co^{2+}	331	382	418
Ni^{2+}	342	432	425, 414
Cu^{2+}	368		
Zn^{2+}	319	382	408
Al^{3+}	442,325		
Cr^{3+}	438		
Fe^{3+}	385		

Complex	ν_{MO}, cm^{-1}
$CuCl_2$·2(4-methoxypyridine)	347
$CuBr_2$·2(4-methoxypyridine)	330
$CuCl_2$·2(4-picoline-N-oxide)	426
$CuBr_2$·2(4-picoline-N-oxide)	417, 408
$CuBr_2$·2(pyridine-N-oxide)	367
$CuCl_2$·2(4-chloropyridine-N-oxide)	420
$CuCl_2$·2(4-nitropyridine-N-oxide)	370
$CuCl_2$·2(3-picoline-N-oxide)	373
$CuBr_2$·2(3-picoline-N-oxide)	380, 359
$CuCl_2$·2(2-picoline-N-oxide)	380
$CuCl_2$·2(2,6-lutidine-N-oxide)	408, 370
$CuBr_2$·2(2,6-lutidine-N-oxide)	418, 368
$CuCl_2$·2(2,4,6-collidine-N-oxide)	453, 435
$CuBr_2$·2(2,4,6-collidine-N-oxide)	450, 436
$(CH_3)_3PO{=}L$[96]	
SnL_2F_4	472, 448
SnL_2Cl_4	428
SnL_2Br_4	418
SnL_2I_4	405
GeL_2Cl_4	450
ZrL_2Cl_4	417
CoL_2Cl_2	444
CoL_2Br_2	436
CoL_2I_2	423
ZnL_2Cl_2	439
ZnL_2Br_2	433
$\varphi_3PO{=}L'$[97-100]	
$UO_2X_2L'_2$ (X = Cl, Br, NO_3, NCS)	408–417
$ZnCl_2L'_2$	410
$ZnI_2L'_2$	415
$\varphi_3AsO{=}L''$	
$UO_2X_2L''_2$ (X = Cl, NCS)	435, 440

on the particular oxide, and is at its highest frequency for the 2,4,6-collidine-N-oxide.

5.8. MISCELLANEOUS COMPLEXES

Peroxy Complexes

Several peroxy complexes of transition metals have been prepared. Very few spectroscopic investigations have been conducted.[101–106] These materials are highly colored and Raman studies have been practically neglected. With the advent of laser Raman instrumentation, they may now receive the attention they deserve.

Table 5-29 lists some results of Griffiths[104,105] with permolybdate and pertungstate complexes and hydroperoxide complexes. Mid-infrared results

Table 5-29. Assignments Made in Several Peroxy Complexes,[102-104] cm^{-1}

$K_2[Mo_2O_{11}]$	$K_2[W_2O_{11}]$		
IR	IR	R	Assignment
971	963		M=O stretch
961	966	961	" "
953	940		" "
861	860	845	O–O stretch
847	836		" "
700	750	750	M–O–M asym. stretch
580	530	620, 556	M–O–M sym. stretch
353	357		deformation modes
	330		" "
292	294	326	" "
245	250		" "
$K_3[CO(CN)_5]OOH$			
1264			OOH deformation
820			O–O stretch
545			CoO stretch
Other	$\nu_{O\text{-}O}$[107]		
Li_2O_2	1093		
Na_2O_2	1081		
K_2O_2	1062		
Rb_2O_2	1054		
MgO_2	1088		
CaO_2	1086		
SrO_2	1073		
BaO_2	1061		
KO_2	1146		
RbO_2	1141		

Table 5-30. Assignments Made for ν_{MO} Vibration in Alkoxides[108]

Alkoxide		ν_{MO}, cm^{-1}	
$Ti(OCHMe_2)_4$		619	
$Zr(OCHMe_2)_4$		559, 548	
$Al(OCHMe_2)_3$		699, 678, 610, 566, 535	
$Ta(OCHMe_2)_5$		540	
$Zr(OCMe_3)_4$		557, 540	
$Hf(OCMe_3)_4$		567, 526	
$Ti(OCEt_2Me)_4$		615, 596	
$Zr(OCEt_2Me)_4$		586, 559, 521	
$Ti(OEt)_4$		625, 500	
$Ta(OEt)_5$		556	
$Nb(OEt)_5$		571	
	IR[109,110]	R[109,110]	
$Ti(OMe)_4$		588, 553	ν_{TiO} sym, ν_{TiO} asym
$Ti(OEt)_4$	618, 578	616, 575	" "
$Ti(O\textit{i}Pr)_4$	621, 560	620, 566	" "
$Ti(O\textit{n}Pr)_4$	641, 600	657, 610	" "
$Ti(O\textit{n}Bu)_4$	633, 591	633, 590	" "

with perchromates have been interpreted and a strong band between 922 and 984 cm^{-1} has been assigned as the M=O stretch. Even with limited infrared data it appears that the absorptions of interest in peroxy complexes are the following:

(1) M=O stretch at about 900–980 cm^{-1}.
(2) The O–O stretch at about 850 cm^{-1}.
(3) The M–O–H asymmetric stretch at about 700–750 cm^{-1} and the M–O–H symmetric stretch at about 530–580 cm^{-1}.

In peroxides and superoxides of Group IA and IIA the ν_{OO} vibration appears to be of higher bond order than the peroxy complexes of molybdenum and tungsten.[108]

Alkoxides

A few low-frequency assignments have been made with alkoxides. Some of these compounds show complicated spectra and the assignments are therefore difficult. Barraclough[108] has conducted infrared studies on a series of metal alkoxides and Table 5-30 tabulates some of these results along with those of Kriegsmann and Licht.[109,110] The assignment for a ν_{MO} vibration in these compounds has been made at 500–650 cm^{-1}. Recently, the ν_{CrO} vibration was assigned at 500–505 cm^{-1} and the δ_{OCrO} vibration at 250–253 cm^{-1} in $Cr(OMe)_3$.[111]

Table 5-31. Infrared Absorption Maxima in the Spectra of 3:1 Metal Aluminates[112]

Maxima, cm^{-1}	
$3CaO \cdot Al_2O_3$	$3SrO \cdot Al_2O_3$
898	877
861	850
841	826
818	809
803	
787	788
760	775
742	742
707	724
645	690
617	
536	515
520	501
509	478
400–450	400–450

Aluminates

Aluminates result from the heating at 1000°C of Al_2O_3 and the appropriate metal carbonate. For $3CaO \cdot Al_2O_3$, which is a body-centered cubic structure, the aluminum atoms are in a position of six-fold coordination, but the six Al–O distances are not equal. Four of the oxygens are at the corners of a square with aluminum at the center (Al–O distance, 1.9 Å). The other two oxygens are above and below the aluminum (Al–O distance, 2.6 Å). Each oxygen is associated with only one aluminum and four calcium

Table 5-32. Low-Frequency Assignments for Several Sulfoxide Complexes[113,114]

Complex	ν_{MO}, cm^{-1}
$[M(DMSO)_6](ClO_4^-)_n$	
Fe(II)	438, 415
Mn(II)	418
Zn(II)	431
Co(II)	436
Ni(II)	444
Cr(III)	529
$M(TMSO)_6(ClO_4^-)_n$	
Mn(II)	388
Al(III)	499

atoms. Two types of Al–O bonds are therefore apparent. Table 5-31 tabulates the results of Schroeder and Lyons[112] for $3CaO \cdot Al_2O_3$ and $3SrO \cdot Al_2O_3$. No attempts at assignment have been made.

Sulfoxide Complexes

The recent interest in dimethlysulfoxide as a solvent has lead to investigations of its donor properties. Recently, a series of complexes of the type $[M(DMSO)_6](ClO_4)_n$ were studied at low frequencies and the ν_{MO} vibration was assigned.[113] Related complexes of the type $M(TMSO)_6(ClO_4)_n$ with the metals Mn(II), Zn(II), Cu(II), Ni(II), Fe(III), Cr(III), and Al(III) have also been reported, where TMSO is tetramethylsulfoxide.[114] Table 5-32 illustrates the results of these studies.

5.9. NATURE OF THE METAL–OXYGEN VIBRATION

In general, the metal–oxygen stretching vibration is found to be intense and broad in nature. The vibration is more intense than the ν_{MN} vibration. This may be due to the possibility that a larger dipole moment change occurs with the metal–oxygen band.

The metal–oxygen bond appears to be a strong one. The metal–water librational modes for transition metals (M^{2+}) overlap the region of the transition metal (M^{3+})–nitrogen stretching vibrations in amines. The absorption is sensitive to oxidation state and the (M^{3+}) transition metal vibrations are found at higher frequencies than (M^{2+}) transition metals. The weaker complexing nature of the rare earth metals is also reflected, as the rare earth–oxygen stretching vibrations are found to occur at lower frequencies than the transition metal (M^{2+} and M^{3+})–oxygen stretches. All metal–oxygen stretching vibrations are observed to occur above 200 cm^{-1}, regardless of the type of compound (see Fig. 5-13). The metal–oxygen bending vibrations, which occur at frequencies less than 200 cm^{-1}, have as yet not been assigned.

It should be noted that metal–oxygen vibrations may occur anywhere from 1100 to 200 cm^{-1}, depending on the mass of the metal and the double-bond character in the bond. Assignment of $\nu_{M=O}$ in compounds containing V=O, U=O, and Mo=O (monoxo system) from 1050 cm^{-1} to 950 cm^{-1} have been made by Barraclough.[115] In dioxo systems, the lower limit has been extended from 750 to 780 cm^{-1} depending on whether the vibration is asymmetric or symmetric.

5.10. RAMAN SPECTROSCOPIC STUDIES

Most of the Raman studies with inorganic and coordination complexes have been made in aqueous solution with noncolored materials. This has been due to (1) the difficulty in obtaining good solid Raman spectra; (2) the diffuse Raman spectrum of water (by contrast the difficulty in working with aqueous solutions in the infrared); and (3) the absorption of ultraviolet-exciting radiation by colored materials.

It is hoped that some of these problems can be ameliorated with the advent of laser Raman instrumentation. Considerable potential research is now possible with the colored transition metal complexes and their solutions.

Some of the early studies pertaining to low-frequency vibrations involving metal–oxygen were those of Mathieu,[116] who characterized a zinc–water vibration in crystalline $Zn(H_2O)_6^{2+}$ and in aqueous solution at 370–390 cm^{-1}, and a similar vibration at 390–435 cm^{-1} in copper salts. LaFont[117] characterized an absorption in magnesium and aluminum salts at ~360 cm^{-1} as due to magnesium–water and at 525 cm^{-1} as due to aluminum–water. Recently, Hester and Plane[118] have examined a series of inorganic salts and aqueous solutions and characterized metal–water vibrational modes. These are tabulated in Table 5-33. Low-frequency assignments in $Al(OH)_4^-$ and $Zn(OH)_4^{2-}$ have also been made.[119] Maroni *et al.*[120,121] have recently studied the vibrational spectrum of $Pb_4(OH)_4^{4+}$ and $Bi_6(OH)_{12}^{6+}$.

Studies of aqueous solutions of salts of polyatomic anion have also been made. Compounds of NO_3^-, SO_4^{2-}, PO_4^{3-}, and ClO_4^- have been researched. Hester and Plane[118] have indicated that the tendency toward metal complexation decreases, with $NO_3^- > SO_4^{2-} > ClO_4^-$. $In(NO_3)_3$ was studied in aqueous solution by Raman techniques, and an absorption at 270

Table 5-33. Low-Frequency Raman Results Involving Metal–Water Vibrations[118]

Salt*	Frequency, cm^{-1}
$Cu(NO_3)_2$	440
$Zn(NO_3)_2$	390
$Hg(NO_3)_2$	380
$Mg(NO_3)_2$	370
$In(NO_3)_3$	410, 460
$CuSO_4$	440
$MgSO_4$	360
$ZnSO_4$	400
$Ga_2(SO_4)_3$	475
$In_2(SO_4)_3$	350–550
Tl_2SO_4	470
$Cu(ClO_4)_2$	440
$Hg(ClO_4)_2$	380
$In(ClO_4)_3$	420
$Mg(ClO_4)_2$	360

* From aqueous solutions.

cm^{-1} was assigned to a ν_{InO} stretching vibration.[116,122] Likewise in $In_2(SO_4)_3$ the band at 250 cm^{-1} was attributed to the ν_{InOSO_3} vibration.[122] Evidence for a $HgONO_2$ vibration in CH_3ONO_2 at 292 cm^{-1} has also been reported,[123,124] although a recent study of $Hg(NO_3)_2$ in water failed to reveal such a vibration.[125] Evidence for a MO vibration in $Bi(NO_3)_3$ (235 cm^{-1}), Ce(IV) (238 cm^{-1}), and Sm(III) (280–290 cm^{-1}) has been reported.[126–128] A recent study of aqueous $Cd(NO_3)_2$ solutions indicated the presence of nitrato cadmium species, but no evidence for a cadmium–oxygen stretching vibration was found.[129]

Much work remains to be done, as only the surface has been scratched in systems involving possible metal–oxygen vibrations in the low-frequency region. Greater use of the Raman technique in this area in the future is a virtual certainty.

BIBLIOGRAPHY

1. J. C. Taylor, M. H. Mueller, and R.C.H. Hitterman, *Acta Cryst.* **20,** 842 (1966).
2. J. C. Taylor and M. H. Mueller, *Acta Cryst.* **19,** 536 (1965).
3. R. E. Richards and J.A.S. Smith, *Trans. Faraday Soc.* **47,** 1261 (1951).
4. Y. Kakiuchi, H. Shono, K. Komatsu, and J. Kigoshi, *J. Phys. Soc. Japan* **7,** 102 (1956).
5. F. S. Lee and G. B. Carpenter, *J. Phys. Chem.* **63,** 279 (1959).
6. Y. K. Yoon and G. B. Carpenter, *Acta Cryst.* **12,** 17 (1959).
7. C. C. Ferriso and D. F. Hornig, *J. Chem. Phys.* **23,** 1464 (1955).
8. J. M. Williams, *Inorg. Nucl. Chem. Letters* **3,** 297 (1967).

8a. R. Chidambaram, A. Sequera, and S. K. Sikka, *J. Chem. Phys.* **41,** 3616 (1964).

8b. W. C. Hamilton and J. A. Ibers, *Hydrogen Bonding in Solids,* W. A. Benjamin, Inc., New York (1968).

9. G. Sartori, C. Furlani, and A. Damiani, *J. Inorg. Nucl. Chem.* **8,** 119 (1958).
10. J. Van der Elsken and D. W. Robinson, *Spectrochim. Acta* **17,** 1249 (1961).
11. I. Nakagawa and T. Shimanouchi, *Spectrochim. Acta* **20,** 429 (1964).
12. C. Postmus and J. R. Ferraro, *J. Chem. Phys.* **48,** 3605 (1968).
13. L. J. Basile, D. Gronert, and J. R. Ferraro, in *Solvent Extraction Chemistry* (A. S. Kertes and Y. Marcus, Eds.), J. Wiley and Sons Inc., New York (1969).
14. N. T. McDevitt and N. L. Baun, *J. Am. Ceramic Soc.* **46:** 294 (1963); *Spectrochim. Acta* **20,** 799 (1964).
15. L. H. Jones, *J. Chem. Phys.* **22,** 217 (1954).
16. D. Scargill, *J. Chem. Soc.,* 4440 (1961).
17. P. Tarte, *Spectrochim. Acta* **13,** 107 (1958).
18. J. R. Ferraro, R. Driver, W. R. Walker, and W. Wozniak, *Inorg. Chem.* **6,** 1586 (1967).
19. W. R. McWhinnie, *J. Inorg. Nucl. Chem.* **27,** 1063 (1965).
20. W. R. McWhinnie, *J. Chem. Soc.,* 2959 (1964).
21. J. R. Ferraro and W. R. Walker, *Inorg. Chem.* **4,** 1382 (1965).
22. W. P. Griffith and D. J. Hewkin, *J. Chem. Soc.* (A), 472 (1966).
23. R. S. Halford, *J. Chem. Phys.* **14,** 8 (1946).
24. C. C. Addison and A. Walker, *Proc. Chem. Soc.,* 242 (1961).

25. S. C. Wallwork, *Proc. Chem. Soc.*, 311 (1959).
26. J. R. Ferraro, C. Cristallini, and I. Fox, *J. Inorg. Nucl. Chem.* **29,** 139 (1967).
27. C. C. Addison, D. W. Amos, D. Sutton, and W.H.H. Hoyle, *J. Chem. Soc.* (A), 808 (1967).
27a. R. J. Fereday and N. Logan, *Chem. Commun.*, 271 (1968).
27b. J. Milton and S. C. Wallwork, *Chem. Commun.*, 871 (1968).
28. R. H. Nuttall and D. W. Taylor, *Chem. Commun.*, 1417 (1968).
29. J. R. Ferraro and A. Walker, *J. Chem. Phys.* **42,** 1273 (1965).
30. A. Walker and J. R. Ferraro, *J. Chem. Phys.* **43,** 2689 (1965).
31. H. Brintzinger and R. E. Hester, *Inorg. Chem.* **5,** 980 (1966).
32. J. A. Goldsmith and S. D. Ross, *Spectrochim. Acta* **22,** 1069 (1966).
33. J. A. Goldsmith & S. D. Ross, *Spectrochim. Acta* **23A,** 1909 (1967).
34. J. R. Ferraro, A. Quattrochi, K. C. Patel, and C.N.R. Rao, *J. Inorg. Nucl. Chem.* **31,** 3667 (1969).
35. J. P. Laperches and P. Tarte, *Spectrochim. Acta* **22,** 1201 (1966).
36. C. E. Weir and E. R. Lippincott, *J. Res. Natl. Bur. Std.* **65A,** 173 (1961); **68A,** 465 (1964).
37. R. E. Hester, R. A. Plane, and G. E. Walrafen, *J. Chem. Phys.* **38,** 249 (1963).
38. F. A. Miller, G. L. Carlson, F. F. Bentley, and W. H. Jones, *Spectrochim. Acta* **16,** 135 (1960).
39. K. Nakamoto, J. Fujita, S. Tanaka, and M. Kobayashi, *J. Am. Chem. Soc.* **79,** 4904 (1957).
40. Landolt-Börnstein, *Phys.-Chem. Tabellen,* **2** (1951).
41. B. J. Hathaway and A. E. Underhill, *J. Chem. Soc.*, 3091 (1961).
42. H. Colm, *J. Chem. Soc.*, 4282 (1952).
43. B. J. Hathaway, O. G. Holah, and M. Hudson, *J. Chem. Soc.*, 4586 (1963).
44. S. D. Ross, *Spectrochim. Acta* **18,** 225 (1962).
45. A. Hezel and S. D. Ross, *Spectrochim. Acta* **24A,** 985 (1968).
46. J. E. Guerchais, *Compt. Rend* **261,** 3628 (1965).
47. D. Bassi and O. Sala, *Spectrochim. Acta* **12,** 403 (1958).
48. P. Tarte and G. Nizet, *Spectrochim. Acta* **20,** 503 (1964).
49. L. A. Woodward and H. G. Roberts, *Trans. Faraday Soc.* **52,** 615 (1956).
50. H. H. Claassen and A. Zielen, *J. Chem. Phys.* **22,** 701 (1954).
51. R. H. Busey and O. L. Keller, *J. Chem. Phys.* **41,** 215 (1965).
52. A. Müller and B. Krebs, *Z. Naturforsch.* **20A,** 967 (1965).
53. K. Sathiandan, L. D. McCory, and J. L. Margrave, *Spectrochim. Acta* **20,** 957 (1964).
54. H. Siebert, *Z. Anorg. Allgem. Chem.* **275,** 225 (1954).
55. W. P. Griffiths, *J. Chem. Soc.* (A), 1467 (1966).
56. R. K. Dodd, *Trans. Faraday Soc.* **55,** 1480 (1959).
57. G. M. Clark and W. P. Doyle, *Spectrochim. Acta* **22,** 1441 (1966).
58. H. Siebert, *Z. Anorg. Allgem. Chem.* **275,** 225 (1954).
59. W. E. Dasent and T. C. Waddington, *J. Chem. Soc.*, 2429, 3350 (1960).
60. C. Rocchiccioli, *Compt. Rend.* **242,** 2922 (1956).
61. A. K. Ramdas, *Proc. Indian Acad. Sci.* **37A,** 451 (1953); **36A,** 55 (1952).
62. N. Duveau, *Bull. Soc. Chem. France* **10,** 374 (1943).
63. J. L. Hollenberg and D. A. Dows, *Spectrochim. Acta* **16,** 1155 (1960).
64. N. R. Rao, *Indian. J. Phys.* **16,** 17 (1942).
65. J. C. Evans and H. J. Bernstein, *Can. J. Chem.* **33,** 1270 (1955).
66. A. Simon and K. Waldmann, *Z. Phys. Chem. Leipzig* **204,** 235 (1955).

67. M. E. Baldwin, *J. Chem. Soc.*, 3123 (1961).
68. W. P. Griffiths and T. D. Wickens, *J. Chem. Soc.*, 1087 (1966).
69. K. Nakamoto and A. E. Martell, *J. Chem. Phys.* **32,** 588 (1960).
70. H. Mikami, I. Nakagawa, and T. Shimanouchi, *Spectrochim. Acta* **23A,** 1037 (1967).
71. R. B. Roof, *Acta Cryst.* **9,** 781 (1956).
72. K. Nakamoto, P. J. McCarthy, A. Ruby, and A. E. Martell, *J. Am. Chem. Soc.* **83,** 1066, 1272 (1961).
73. S. Pinchas, B. L. Silver, and I. Laulicht, *J. Chem. Phys.* **46,** 1506 (1967).
74. H. Ogoshi and K. Nakamoto, *J. Chem. Phys.* **45,** 3113 (1966).
75. R. D. Gillard, H. G. Silver, and J. L. Wood, *Spectrochim. Acta* **20,** 63 (1964).
76. K. Lawson, *Spectrochim. Acta* **17,** 248 (1961).
77. R. C. Fay and T. J. Pinnavaia, *Inorg. Chem.* **7,** 508 (1968).
78. M. F. Richardson, W. F. Wagner, and D. E. Sands, *Inorg. Chem.* **7,** 2495 (1968).
79. R. E. Hester and R. A. Plane, *Inorg. Chem.* **3,** 513 (1964).
80. H. Mikami, I. Nakagawa, and T. Shimanouchi, *Spectrochim. Acta* **23A,** 1037(1967).
81. K. Nakamoto, Y. Morimoto, and A. E. Martell, *J. Phys. Chem.* **66,** 346 (1962).
82. G. T. Behnke and K. Nakamoto *Inorg. Chem.* **7,** 330 (1968); **7,** 2030 (1968).
83. M. F. Farona, D. C. Perry, and H. A. Kuska, *Inorg. Chem.* **7,** 2415 (1968).
84. Y. Kawasaka, T. Tanaka, and R. Okawara, *Spectrochim. Acta* **22,** 1571 (1966).
85. B. Soptrasonov, A. Nikolovski, and I. Petrov, *Spectrochim. Acta* **24A,** 1617 (1968).
86. K. Nakamoto, *Infrared Spectra of Inorganic and Coordination Compounds*, J. Wiley and Sons, New York (1963).
87. R. E. Hester and R. A. Plane, *Inorg. Chem.* **3,** 513 (1964).
88. E. C. Gruen and R. A. Plane, *Inorg. Chem.* **6,** 1123 (1967).
89. J. F. Jakovitz and J. L. Walter, *Spectrochim. Acta* **22,** 1393 (1966).
90. I. Nakagawa, R. J. Hooper, J. W. Walter, and T. J. Lane, *Spectrochim. Acta* **21,** 1 (1965).
91. K. Nakamoto, P. J. McCarthy, and B. Miniatas, *Spectrochim. Acta* **21,** 379 (1965).
92. B. B. Kedzin, P. X. Armendarez, and K. Nakamoto, *J. Inorg. Nucl. Chem.* **30,** 849 (1968).
93. Y. Kakiuti, S. Kida, and J. V. Quagliano, *Spectrochim. Acta* **19,** 201 (1963).
94. G. A. Rodley, D.M.L. Goodgame, and F. A. Cotton, *J. Chem. Soc.*, 1499 (1965).
95. R. Whyman and W. E. Hatfield, *Inorg. Chem.* **6,** 1859 (1967).
96. S. H. Hunter, V. M. Langford, G. A. Rodley, and C. J. Wilkins, *J. Chem. Soc.* (A), 305 (1968).
97. F. A. Hart and J. E. Newbury, *J. Inorg. Nucl. Chem.* **30,** 318 (1968).
98. G. B. Deacon and J.H.S. Green, *Spectrochim. Acta* **24A,** 845 (1968).
99. G. A. Rodley, D.M.L. Goodgame, and F. A. Cotton, *J. Chem. Soc.*, 1499 (1965).
100. J. P. Clark, V. M. Langford, and C.J. Wilkins, *J. Chem. Soc.* (A), 792 (1967).
101. W. P. Griffith, *J. Chem. Soc.*, 3948 (1962).
102. W. P. Griffith, *J. Chem. Soc.*, 5345 (1963).
103. W. P. Griffith and T. D. Wilkins, *J. Chem. Soc.*, 397 (1968).
104. W. P. Griffith, *J. Chem. Soc.*, 5248 (1964).
105. W. P. Griffith and T. D. Wilkins, *J. Chem. Soc.*, 400 (1968).
106. J. E. Guerchais and R. Rohmer, *Compt. Rend.* **259,** 1135 (1964).
107. F. J. Blunt, P. J. Hendra, and J. R. Mackenzie, *Chem. Commun.*, 278 (1969).
108. C. C. Barraclough, D. C. Bradley, J. Lewis, and I. M. Thomas, *J. Chem. Soc.*, 2601 (1961).
109. H. Kriegsmann and K. Licht, *Z. Elektrochem.* **62,** 1163 (1958).
110. H. Kriegsmann and K. Licht, *Z. Elektrochem.* **68,** 617 (1964).

111. D. A. Brown, D. Cunningham, and W. K. Glass, *J. Chem. Soc.* (A), 1563 (1968).
112. R. A. Schroeder and L. L. Lyons, *J. Inorg. Nucl. Chem.* **28,** 1155 (1966).
113. C. V. Berney, J. H. Weber, *Inorg. Chem.* **7,** 283 (1968).
114. C. V. Berney and J. H. Weber, Abstracts, Pittsburgh Conference (1969).
115. C. C. Barraclough, J. Lewis, and R. S. Nyholm, *J. Chem. Soc.*, 3552 (1959).
116. J. P. Mathieu, *Compt. Rend.* **231,** 896 (1950).
117. R. L. LaFont, *Compt. Rend.* **244,** 1481 (1957).
118. R. E. Hester and R. A Plane, *Inorg. Chem.* **3,** 768 (1964).
119. E. R. Lippincott, J. A. Psellas, and M. C. Tobin, *J. Chem. Phys.* **43,** 843 (1965).
120. V. A. Maroni and T. G. Spiro, *Inorg. Chem.* **7,** 183 (1968).
121. V. A. Maroni and T. G. Spiro, *Inorg. Chem.* **7,** 188 (1968).
122. R. E. Hester, R. A. Plane, and G. E. Walrafen, *J. Chem. Phys.* **38,** 249 (1963).
123. D. L. Goggin and L. A. Woodward, *Trans. Faraday Soc.* **58,** 1495 (1962).
124. J.H.R. Clark and L. A. Woodward, *Trans. Faraday Soc.* **62,** 3022 (1968).
125. A. R. Davis and D. E. Irish, *Inorg. Chem.* **7,** 1699 (1968).
126. R. P. Oertel and R. A. Plane, *Inorg. Chem.* **7,** 1192 (1968).
127. J. T. Miller and D. E. Irish, *Can. J. Chem.* **45,** 147 (1967).
128. B. Strauch and L. N. Komissarova, *Z. Chem.* **6,** 4748 (1966).
129. A. R. Davis and R. A. Plane, *Inorg. Chem.* **7,** 2565 (1968).

Chapter 6

METAL HALIDE VIBRATIONS*

6.1. INTRODUCTION

The assignments for metal–halogen vibrations have been, for the most part, well established. In most cases it is the investigation of the metal chloride vibration that has received the most attention. Usually, the metal halide stretching vibration can be assigned unequivocally, for the frequency of the vibration will decrease as the mass of the halogen atom increases. By noting the disappearance of the stretching vibration (e.g., metal–chlorine) and the appearance of a new vibration (e.g., metal–bromine and/or metal–iodine), as one goes from the chloride to the bromide and to the iodide, accurate assignments can be made. The method is useful except when there is a change in structure in the series. Examples are known in which the central metal atom may be in an octahedral environment in the chloride and a tetrahedral environment in the bromide or iodide. This causes the coordination-number effect to counteract the mass effect, and little difference in frequency position may be noted for the metal halide vibration.[1]

The infrared intensity of the metal halide stretching vibration (ν_{MX}) in inorganic and coordination compounds appears to be of moderate-strong intensity. There has been a lack of Raman data available because many of the materials are colored. There has also been a lack of normal coordinate treatment and, as a result, hardly any force constants have been calculated. As a consequence, the discussion in this chapter will deal mainly with frequencies rather than force constants. In addition, the discussion will deal with inorganic metal halides (e.g., MX_m, MX_m^{n-}) and coordination compounds (e.g., L_nMX_m). The discussion will be in terms of coordination number (CN) and the symmetry of the molecule.

It should be mentioned that in some cases it is incorrect to speak of a metal–halogen vibration in the far infrared since many of these vibrations are coupled and are therefore not "pure" vibrations. Despite this, certain

* With Louis J. Basile.

Table 6-1. Factors Determining Position of Metal–Ligand Stretching Vibration

Oxidation state of metal	⟶	Higher oxidation state—Higher frequency
Mass of metal and ligand	⟶	Larger mass—Lower frequency
Coordination number (CN) of metal	⟶	Higher CN—Lower frequency
Stereochemistry of complex	⟶	Frequency decreases from T_d—O_h structure
Basicity of ligand	⟶	Higher basicity—Higher frequency[a]
Counter-ion effect	⟶	Increased size of counter-ion—Decreased frequency
Bridging or nonbridging	⟶	Nonbridging ligand—Vibration at higher frequency
High ligand field stabilization energy	⟶	Higher energy—Higher frequency

[a] Only when sigma bonding is involved.

very valuable and useful correlations and trends are possible. Table 6-1 cites several of these. Most of these trends have been cited by Clark.[1]

6.2. HEPTA- AND OCTACOORDINATED MOLECULES—MX_7, MX_8 (CN = 7, 8)

Thus far only a few molecules having a coordination of seven or eight have been studied by spectroscopic techniques. These molecules may possess

Table 6-2A. Frequency Assignments for 7- and 8-Coordinated Halides

Compound	ν_{MX} (IR), cm^{-1}	Reference
K_2NbF_7	524	5
K_2MoF_7	645	5
K_2TaF_7	526, 518	5
KWF_7	620	5
$CsReF_7$	598	5
K_2PaF_7	430, 356	8
$(NH_4)_2PaF_7$	434, 357	8
Rb_2PaF_7	438, 356	8
Cs_2PaF_7	438, 356	8
Li_3PaF_8	404	8
Na_3PaF_8	468, 422	8
K_3PaF_8	401	8
Cs_3PaF_8	395	8
$(Me_4N)_3UCl_8$	310	6, 7
$(Me_4N)_3PaCl_8$	290	6, 7
$TiCl_4\cdot[o\text{-}(Me_2As)_2\varphi]_2$	317	9
$TiCl_4\cdot[o\text{-}(Me_2P)_2\varphi]_2$	312	9

Table 6-2B. Vibrational Frequencies of Some D_{5h} Point Group Molecules, cm^{-1}

Compound	a_1' R	a_1' R	a_2'' IR	a_2'' IR	e_1' IR	e_1' IR	e_1' IR	e_2' R	e_2' R	e_1'' R	e_2'' IA	Reference
ReF_7 (R) (gas)	736(p)	645						489(d)	352	597	ia	3, 3a
(IR) (gas)			703	299	703	353	217					
OsF_7 (IR)[a]	(715, 550, 483, 366, 336(?), 282—all unassigned)											4
IF_7 (R) (gas)	676(p)	635(p)						352	310	510(d)	ia	2–3a
(IR) (gas)			670	365	746	425	257					

[a] IR results only to 280 cm^{-1} on solid OsF_7 at −180°C.
Legend: d = depolarized, p = polarized.

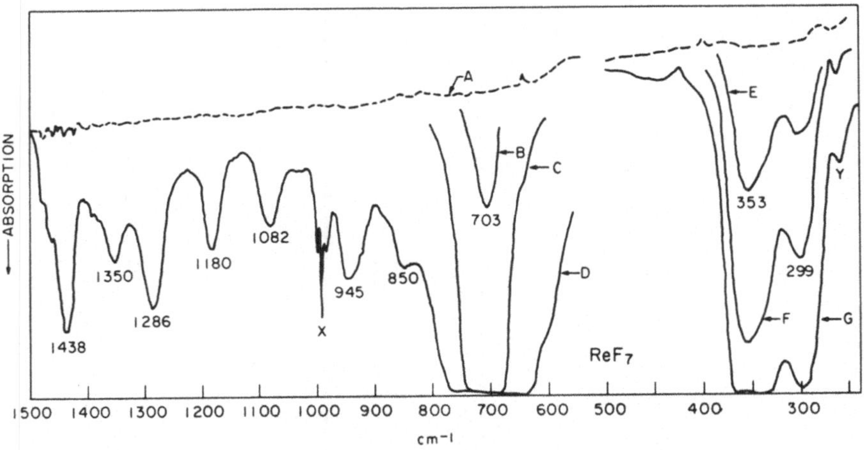

Fig. 6-1. Infrared spectrum of ReF_7 vapor. *A* is background, *B* (unknown, <1 mm, path length 60 cm), *C* (3 mm, path length 60 cm), *D* (87 mm, path length 60 cm), *E* (9 mm, path length 10 cm), *F* (25 mm, path length 10 cm), *G* (97 mm, path length 10 cm). (Courtesy of American Institute of Physics, New York. From Claassen and Selig.[3])

several types of structures. One of the first of the heptacoordinated molecules to be studied extensively was IF_7.[2] It was found to possess a pentagonal bipyramidal (D_{5h}) structure. Five infrared and Raman fundamentals were found to be active, and the results were consistent with the above structure. Recent results with ReF_7 and OsF_7 are also suggestive of a D_{5h} symmetry.[3,4]

A tabulation of the spectroscopic data available for seven- and eight-coordinated compounds is presented in Tables 6-2A and 6-2B. Figure 6-1 shows the infrared spectrum of ReF_7 vapor. Some of the newly developed "cluster" compounds fall in the octacoordinated group of molecules.

6.3. HEXACOORDINATED MOLECULES—O_h SYMMETRY (CN = 6)

Molecules in an octahedral environment of point group O_h will show six normal vibrational modes. These vibrations are depicted in Fig. 6-2. The selection rules governing a molecule of seven atoms (AB_6) are illustrated in Table 6-3. Figure 6-3 shows the Raman spectrum of NpF_6 vapor. Molecules of this type have a center of symmetry. Because of the rule of mutual exclusion and the lack of coincidences it is absolutely necessary to use the Raman experiment to complement the infrared measurements. Many of the molecules falling into this symmetry class are colored materials, and until recently

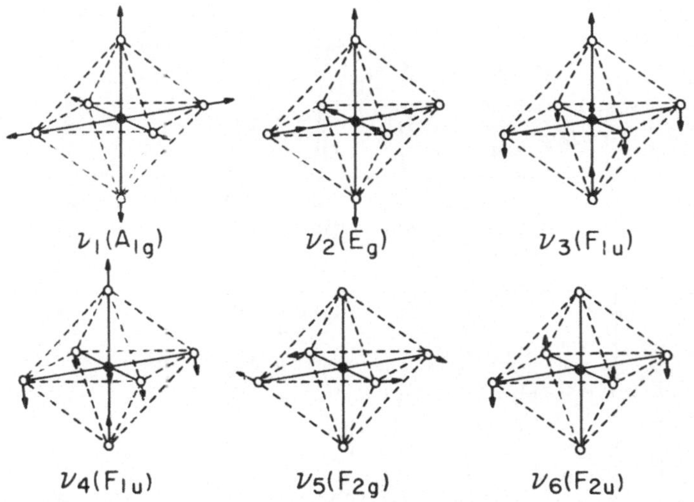

Fig. 6-2. Normal modes of vibrations of octahedral AB_6 molecule point group (O_h).

Raman spectra were unavailable. Recent advancements in Raman instrumentation using laser sources makes experiments with colored materials easier to accomplish.

The Hexafluorides—MF_6 (Gaseous)

An excellent review on the vibrational properties of the gaseous hexafluoride molecules was written by Weinstock and Goodman.[10] For a detailed

Table 6-3. Selection Rules for the Fundamental Vibrations in O_h Symmetry

Type	Fundamentals*	Activity: Raman	Activity: Infrared
a_{1g}	1	a	ia
a_{1u}	0	ia	ia
a_{2g}	0	ia	ia
a_{2u}	0	ia	ia
e_g	1	a	ia
e_u	0	ia	ia
f_{1g}	0	ia	ia
f_{1u}	2	ia	a
f_{2g}	1	a	ia
f_{2u}	1	ia	ia

* For a molecule of type AB_6

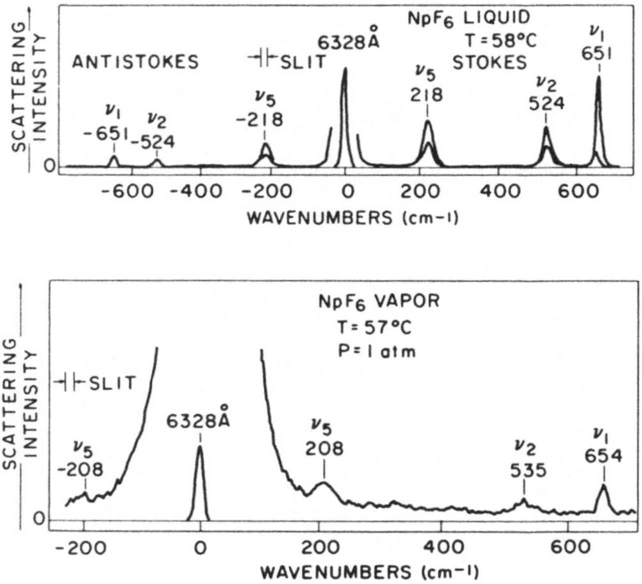

Fig. 6-3. Raman spectrum of NpF_6 vapor and liquid. (Courtesy of American Institute of Physics, New York. From Gasner and Frlec. [9a])

discussion concerning these molecules this paper should be consulted.

Table 6-4 collects the spectroscopic results for 12 gaseous metal hexafluorides and three gaseous nonmetal hexafluorides. The Raman vibrations change in the order $\nu_1 > \nu_2 > \nu_5$ and the infrared vibrations in the order $\nu_3 > \nu_4$. Since the primary interest of this book is frequencies below 650 cm^{-1}, it can be noted that, generally speaking, ν_2, ν_5 (Raman active), and ν_4 (infrared active) are low-frequency vibrations. The table demonstrates the need to obtain direct experimental data for practically all of the compounds, particularly for the Raman results. To a large extent, the lack of such data has been due to experimental difficulties. In some cases, the compounds are colored, decompose photochemically, are radioactive, or the instrumentation available did not have an extended low-frequency range. As a result, the spectra of PoF_6 and AmF_6 are unknown. The hexafluoride of palladium has, as yet, not been prepared. The compound CrF_6 has been reported, but no spectral data appears to be available.[11] The discussion of XeF_6 will be deferred until later in this chapter in the section on xenon compounds.

Of the fifteen hexafluoride molecules, eleven have either nondegenerate electronic ground states or degenerate ground states with spin character. For these compounds the spectra appear normal. Four molecules [$TcF_6(4d^1)$,

Table 6-4. Fundamental Vibration Frequencies of Hexafluoride Molecules MF_6 (Gaseous)[10]

Compound	$\nu_1(a_{1g})$ R	$\nu_2(e_g)$ R	$\nu_3(f_{1u})$ IR	$\nu_4(f_{1u})$ IR	$\nu_5(f_{2g})$ R	$\nu_6(f_{2u})$ IA	f, mdyn/Å
	Frequency, cm^{-1}						
Metal hexafluorides[a]							
MoF_6	741	643	741	262	(312)	(122)[c]	4.84
TcF_6	(712)	(639)[b]	748	265	(297)[b]	(174)	4.81
RuF_6	(675)	(624)	735	275	(283)[b]	(186)	4.57
RhF_6	(634)	(592)	724	283	(269)	(189)	4.27
WF_6	(771)	(673)	711	258	(315)	(134)	5.20
ReF_6	755	(671)[b]	715	257	(295)[b]	(193)	5.17
OsF_6	(733)	(668)[b]	720	272	(276)[b]	(205)	5.14
IrF_6	(701)	(646)	719	276	(258)	(206)	4.94
PtF_6	(655)	(600)	705	273	(242)	(211)	4.52
UF_6	667	535	624	(184)	(201)	(140)	3.81
NpF_6	(648)	(528)	624	(198)	(205)	(165)	3.73
PuF_6	(628)	(523)	616	(203)	(211)	(173)	3.62
Nonmetal hexafluorides							
SF_6	770	640	939	614	622	(349)	5.53
SeF_6	708	(661)	780	437	(403)	(262)	5.04
TeF_6	701	674	752	325	313	(185)	5.11

[a] Frequencies that have not been observed directly, but are derived from combination bands, are placed in parentheses.
[b] Values obtained by interpolation of the corresponding values for the non-Jahn-Teller active hexafluorides.
[c] Discrepancy exists for this vibration. Claassen *et al.*[12] gave a value of 190 cm^{-1} for this vibration.

$ReF_6(5d^1)$, $RuF_6(4d^2)$, $OsF_6(5d^2)$]* show some anomalies in their vibrational spectra. These have been interpreted in terms of a Jahn–Teller effect.[10] Corresponding anomalies have not been observed for d^3 hexafluorides.

Generally speaking, the stretching modes, ν_1 and ν_3, of the metal hexafluorides lie in the range 600–770 cm^{-1}. It is interesting to note that for the third-row transition metal hexafluorides, ν_1 and ν_2 lie about 30–60 cm^{-1} higher than the corresponding second-row transition metal compound, while ν_3 is found at lower frequency by about 30 cm^{-1}. The ν_4 vibration is observed to increase as the mass increases and to decrease as one goes from the second-row transition metal hexafluorides to the third-row transition metal hexafluorides. Until more meaningful data become available for ν_5, no attempts are made to indicate any trends. There appears to be a correlation between the vibrational spectra and the force constants in both the metal and nonmetal hexafluorides. This indicates that there may be similarities in the nature of the bonding between these two classes of hexafluorides.

* Nonbonding valence electrons on *d* orbitals.

Anionic Hexahalides (MX_6^{n-})

Hexafluorides (MF_6^{n-})

The majority of data available on metal–fluorine vibrations for hexafluoride anions are due to the work of Peacock and Sharp.[5] Much of the Raman data for these compounds are as yet unavailable. The results for some compounds with $n = 2, 1$ are summarized in Table 6-5. Table 6-6 relates the ν_3 frequencies for several fluoroanions where $n = 3, 2, 1$. The symmetry of these anions may be considered to be octahedral, particularly where the cation is an alkali metal and where cation–anion interaction is considered to be minimal.

The mass of the central metal atom is sufficiently large for all vibrations to be observed at frequencies less than 650 cm^{-1}. The dependence of ν_3 on the oxidation state is apparent when one compares the results of Table 6-6 with Table 6-4. For example, ReF_6^{2-}, 541 cm^{-1}; ReF_6^-, 627 cm^{-1}; ReF_6, 715 cm^{-1}; and RuF_6^{2-}, 588 cm^{-1}; RuF_6^-, 640 cm^{-1}; RuF_6, 735 cm^{-1}. A decrease in the frequency of the ν_3 and ν_1 vibrations occurs as the central metal atom increases in mass within a particular group in the periodic table. From Table 6-5 the ν_3 and ν_1 frequencies change in the order $SiF_6^{2-} > GeF_6^{2-} > SnF_6^{2-} > PbF_6^{2-}$. For the group IV hexafluorides, the relationship for the ν_1 vibration follows the trend $TiF_6^{2-} > ZrF_6^{2-} \simeq HfF_6^{2-}$; the lack of difference

Table 6-5. Vibrational Frequencies for Anionic Hexafluorides

	Frequency, cm^{-1}						
Compound	$\nu_1(a_{1g})$ R	$\nu_2(e_g)$ R	$\nu_3(f_{1u})$ IR	$\nu_4(f_{1u})$ IR	$\nu_5(f_{2g})$ R	$\nu_6(f_{2u})$ IA	Reference
TiF_6^{2-}	613		560		275		13
ZrF_6^{2-}	581				228		13
HfF_6^{2-}	589				230		13
SiF_6^{2-}	655	474	740	485	395		14
GeF_6^{2-}	627	454	600	350	318		14
SnF_6^{2-}	585	470	556		241		13
PbF_6^{2-}	543		502				13
NiF_6^{2-}	562	520	654	345	310		15
PdF_6^{2-}			602				12
PtF_6^{2-}	600	576	571	281	210	(143)	16
PF_6^-	741						17
AsF_6^-	682	583	706	402, 384	372		18
NbF_6^-	683	562	692sh, 585	256, 232	280		19
TaF_6^-	692	581	560	240	272		20
UF_6^{-a}	506		503	150	145	100	21

( ) Estimated from combination bands.
^{a} Deduced from splittings observed in the near-infrared and visible regions.
Legend: sh = shoulder.

Table 6-6. Infrared Absorption Frequencies (M–F) $\nu_3(f_{1u})$ of Complex Fluoroions,[5] cm^{-1}

Compound	Transition series		Number of *d* electrons: 0	1	2	3	4	5	6	7	8	9	10
KMF_3	1	M=	Mg 462 Ca <400				Cr 481	Mn 407	Fe 431	Co 439	Ni 445	Cu 489	Zn 437 Cd <400
K_3MF_6	1	M=	Al 570 Sc 479	Ti 452	V 511	Cr 535, 522	Mn 560, 617	Fe 465	Co 480				Ga 464
	2	M=	Y <400 La <400					Ru 514, 497, 479	Rh 530, 512, 500				In 446 Tl <400
K_2MF_6	1	M=	Ti 560	V 583	Cr 556	Mn 622			Ni 654				Si 726, 480 Ge 600
	2	M=	Zr <400				Ru 588	Rh 589	Pd 602				Sn 552
	3	M=	Th 400			Re 541	Os 548	Ir 568	Pt 583				Pb 502

Table 6-6 (Continued)

Compound	Transition series	Number of d electrons: 0	1	2	3	4	5	6	7	8	9	10
KMF_6												P 845, 559
	1 M=	V 715										As 700, 400
	2 M=	Nb 580	Mo 623		Ru 640							Sb 660
	3 M=	Ta 580	W 594	Re 627	Os 616	Ir 667						

between Zr^{4+} and Hf^{4+} may be due to the lanthanide contraction effect. From Table 6-6 it can be observed that the ν_3 frequency for MF_6^{2-} anions decreases as one proceeds from the first transition series to the second and third (e.g., NiF_6^{2-}, 654 cm^{-1}; PdF_6^{2-}, 602 cm^{-1}; PtF_6^{2-}, 583 cm^{-1}) and correspondingly increases as the number of *d* electrons increases (ReF_6^{2-}, 541 cm^{-1}; OsF_6^{2-}, 548 cm^{-1}; IrF_6^{2-}, 568 cm^{-1}; PtF_6^{2-}, 583 cm^{-1}). Although the data are incomplete, reverse trends for ν_3 apparently exist for the MF_6^{3-} anions. For the MF_6^- anions, the ν_3 vibration appears to increase as the number of *d* electrons increases.

For the KMF_3 salts (perovskites) no discrete MF_6^{4-} ions exist. Potassium and fluorine atoms are in an infinite-chained packed-cubic structure with the central atom in an octahedral hole. No attempts at defining trends for these molecules will be made, since insufficient data are available.

Table 6-7. Vibrational Frequencies for the MCl_6^{2-} Anions

	Frequency, cm^{-1}						
Compound	$\nu_1(a_{1g})$ R	$\nu_2(e_g)$ R	$\nu_3(f_{1u})$ IR	$\nu_4(f_{1u})$ IR	$\nu_5(f_{2g})$ R	$\nu_6(f_{2u})$ IA	References
$TiCl_6^{2-}$	321–331	284, 271, 236	302–330	188–193	186–194	(142)	22–24
$ZrCl_6^{2-}$	323–333	275, 237	276–297	145–156	157–159	(90)	22–24
$HfCl_6^{2-}$	328–333	264, 237	273–288	138–147	153–163	(80)	22–24
$GeCl_6^{2-}$			293–312	205			25, 26
$SnCl_6^{2-}$	311–318	229–235	295–318	161–177	157–169		17, 25–30
$PbCl_6^{2-}$	288	215	265		137		26, 30, 31
$SeCl_6^{2-}$	346	273	294	186	166		27, 28
$TeCl_6^{2-}$	301	250–261	226–249	94–127	140–144		27, 30
$MoCl_6^{2-}$			325–340	170–174			32
$RuCl_6^{2-}$			332–346	188			26, 33
$PdCl_6^{2-}$	317	292	322–358	175	164		34, 26, 28, 33, 35
WCl_6^{2-}			306–324	160–166			32
$ReCl_6^{2-}$	346	(275)	300–332	173–177	159		16, 26, 33, 35
$OsCl_6^{2-}$	346	(274)	305–240	176	165		16, 26, 33, 35
$IrCl_6^{2-}$			316–333	184–192			26, 33, 36
$PtCl_6^{2-}$	341–345	319–324	330–345	182–200	161–163		13, 26, 28–29, 33–37
$CeCl_6^{2-}$	295	205	265–294	114–120	120		38
$ThCl_6^{2-}$			263				26
UCl_6^{2-}			260				26

( ) Calculated from combination bands.

Table 6-8. Vibrational Frequencies for MBr_6^{2-} and MI_6^{2-} Anions

Compound	$\nu_1(a_{1g})$ R	$\nu_2(e_g)$ R	$\nu_3(f_{1u})$ IR	$\nu_4(f_{1u})$ IR	$\nu_5(f_{2g})$ R	$\nu_6(f_{2u})$ IA	References
	Frequency, cm^{-1}						
$TiBr_6^{2-}$	192	141	250–256 268	121	110		22
$ZrBr_6^{2-}$	198		276	104, 114	116		22
$HfBr_6^{2-}$	201	157	193	105, 112	116		22
$SnBr_6^{2-}$	183–185	138–144	206–220	94–122	69, 95		17, 27, 28, 30
$SeBr_6^{2-}$			215–222	94–98			27
$TeBr_6^{2-}$	166–174	151	181–198	79–127	70–85		27, 30
WBr_6^{2-}			214–229	60–78			32
$ReBr_6^{2-}$	213	174	217	118	104		16
$PtBr_6^{2-}$	207	190	240–244	78–90	97		28, 33, 34
SnI_6^{2-}			165	48			27
SeI_6^{2-}			142	45			27
TeI_6^{2-}			142–146 160	57–66			27
PtI_6^{2-}			186	46			33

Other Hexahalides (MX_6^{n-}, $X = Cl, Br, I$)

The environment of the anions MX_6^{3-}, MX_6^{2-}, and MX_6^{-}, where X can be Cl, Br, or I, may also be considered to involve octahedral symmetry. Table 6-7 presents the available data on the MCl_6^{2-} anions. Since Raman data are not complete, correlations are only available for the ν_3 and ν_4 vibrations. Both ν_3 and ν_4 have been reported to be cation dependent, and for this reason some of the data are given in a range of frequencies to save space. Table 6-8 presents data on the MBr_6^{2-} and MI_6^{2-} anions. Fewer data are available at present for these anions. Tables 6-9 and 6-10 record data for the MCl_6^{-} and MCl_6^{3-} anions.

Table 6-9. Vibrational Frequencies (ν_3) of MCl_6^- Anions

Anion	ν_3, cm^{-1} IR	References
$NbCl_6^-$	333–336	26, 39
$SbCl_6^-$	336	40
$TaCl_6^-$	319–330	26, 39
WCl_6^-	305–329	26, 41
$PaCl_6^-$	305–310	39
UCl_6^-	303–310	41
PCl_6^-	449[a]	42

[a] Note: PCl_6^- also has frequencies ν_1, 360; ν_2, 282; ν_4, 262; ν_5, 150 cm^{-1}.

Table 6-10. Vibrational Frequencies MCl_6^{3-} Anions

Anion	$\nu_1(a_{1g})$ R	$\nu_2(e_g)$ R	$\nu_3(f_{1u})$ IR	$\nu_4(f_{1u})$ IR	$\nu_5(f_{2g})$ R	$\nu_6(f_{2u})$ IA	References
	Frequency, cm^{-1}						
$CrCl_6^{3-}$			315	200			37
$MnCl_6^{3-}$			342	183			37
$FeCl_6^{3-}$	283		227–248	166–181			37
$InCl_6^{3-}$	275	175	248	161	130		37, 43
$IrCl_6^{3-}$			290–300	170–200			33, 36, 37

Relatively little change in ν_3 and ν_4 in the MCl_6^{2-} anions is observed for the second and third transition series as one increases the number of *d* electrons. Original data for ν_1 in $TiCl_6^{2-}$ appears to be in error in light of new data. The ν_1 vibration in $TiCl_6^{2-}$, $ZrCl_6^{2-}$, and $HfCl_6^{2-}$ shows little difference and this has been attributed to the lanthanide contraction. Some controversy exists for the Raman vibration ν_2 in $TiCl_6^{2-}$. Adams and Newton[24] and Clark[23] have assigned this vibration at 284 cm^{-1} and 271 cm^{-1} respectively, while Brisdon *et al.*[22] have assigned it at 236 cm^{-1}. Similar controversy exists for ν_2 in $ZrCl_6^{2-}$ and $HfCl_6^{2-}$.

6.4. PENTACOORDINATED MOLECULES—MX_5, MX_5^{n-} (CN = 5)

A molecule with the stoichiometric ratio MX_5 may possess a D_{3h} trigonal bipyramid or C_{4v} tetragonal pyramid structure. The selection rules for these structures are shown in Table 6-11. Figure 6-4 depicts the normal

Table 6-11. Selection Rules for the D_{3h} and C_{4v} Point Groups

Species	Fundamentals*	R	IR
	D_{3h} point group		
a_1'	2	a	ia
a_1''	0	ia	ia
a_2'	0	ia	ia
a_2''	2	ia	a
e'	3	a	a
e''	1	a	ia
	C_{4v} point group		
a_1	3	a	a
a_2	0	ia	ia
b_1	2	a	ia
b_2	1	a	ia
e	3	a	a

*For an AB_5-type molecule.

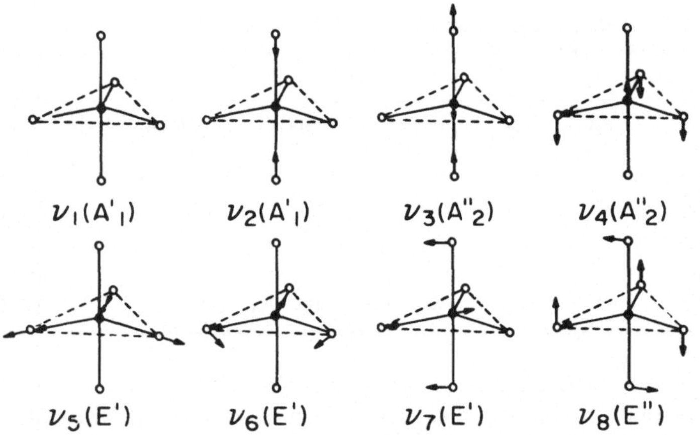

Fig. 6-4. Normal modes of vibrations of trigonal bipyramidal AB_5 molecule point group (D_{3h}).

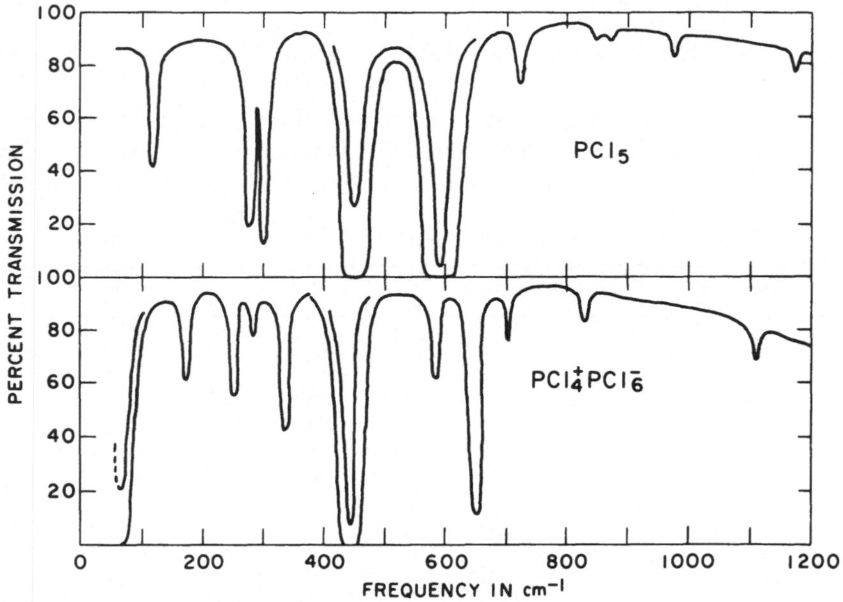

Fig. 6-5. Infrared spectrum of PCl_5 vapor and solid. Top: Spectrum of solid obtained by condensing PCl_5 vapor onto cold window (—90°K). Bottom: Infrared spectrum of solid obtained as Nujol mull or by allowing deposited film in cell to warm up. (Courtesy of Pergamon Press, New York. From Carlson.[42])

Table 6-12. Vibrational Frequency Assignments for Several MX_5 Molecules (Neutral and Anionic)

Compound		Frequency, cm^{-1}									References
		$\nu_1(a_1')$ R	$\nu_2(a_1')$ R	$\nu_3(a_2'')$ IR	$\nu_4(a_2'')$ IR	$\nu_5(e')$ R,IR	$\nu_6(e')$ R,IR	$\nu_7(e')$ R,IR	$\nu_8(e'')$ R		
					D_{3h} point group						
PF_5		817	640	945	576	1026	534	301	514		44
PF_2Cl_3		633	387	867	328	625	404	122	357		45
SbF_5		667	264			716	90	491	228		46, 47
SbF_3Cl_2		610	392	399		655	442	147	292		48
VF_5		719	608	784	331	810	282	~200	350		44
NbF_5		(750)		684	510	748, 732					49
PCl_5		393	282	448	300	581	273	100	261		42, 50
$SbCl_5$		353	303	372	156	395	178	68	164		42, 50
$TaCl_5$		490	410	402	146	365	170	108	194		42
$NbCl_5$		500	412	420	153	355	170	106	(200)		42
$GeCl_5^-$		348	232	319	~200	398	~200				50
$SnCl_5^-$		331	252	326	~160	359	~160				50
$InCl_5^{2-}$		295		267		281					43
	Activity	$\nu_1(a_1')$ R,IR	$\nu_2(a_1')$ R,IR	$\nu_3(a_1)$ R,IR	$\nu_4(b_1)$ R	$\nu_5(b_1)$ R	$\nu_6(b_2)$ R	$\nu_7(e)$ R,IR	$\nu_8(e)$ R,IR	$\nu_9(e)$ R,IR	
					C_{4v} point group						
ClF_5	R	709(p)	538(p)	480(p)	480	346	375			296	
	IR	(712)	541	486	(488)			732		302	51
BrF_5	R	682(p)	570(p)	365(p)	535	281	312		414	237	51–54
	IR	683	587	369	(547)			644	415		51–54
IF_5	R	698(p)	593(p)	315(p)	575	257	273	?	374	189	51
	IR	710	(595)	318				640	372		2, 51
$SbCl_5^{2-}$	R	445	290		420		117(?)	300	255	90	55
	IR	445	285	180				300	255, 230	90	55
TeF_5^-	R	616	511	294	570		247	484	347	132	
	IR	616	521	293				472	347	140–130, 119	56

( ) Calculated from combination bands. [a] Results for potassium salts reported. Legend: p = polarized.

modes of vibration for a trigonal bipyramid molecule. Table 6-12 shows the infrared and Raman assignments made for several MX_5-type molecules (neutral and anionic). PCl_5 is a classical compound which exists in the pentacoordinated state (PCl_5) in the gaseous and liquid phases whereas it exists as the ionic structure $[PCl_4]^+[PCl_6]^-$ in the crystalline state. Figure 6-5 illustrates the infrared spectrum of PCl_5 vapor and solid. The data on $[PCl_4]^+$ will be discussed in the section on tetracoordinated compounds (T_d), while the data on $[PCl_6]^-$ have already been discussed in the section on hexacoordinated compounds (O_h). The phosphorus chlorofluorides PCl_aF_{5-a} also appear to exist in a molecular form at low temperature and in an ionic form at room temperature. At -40°C PF_2Cl_3 has a D_{3h} symmetry, indicating that the two fluorines are axial.

Only a few examples are known for the C_{4v} symmetry. Probably the first compounds of this class, which were characterized by spectroscopic techniques and found to exist in a tetragonal pyramid (C_{4v}) structure, were BrF_5 and IF_5. Later ClF_5 was found to have a similar structure.

It may be observed from Table 6-12 that if the halogen is chlorine all of the vibrations are found below 650 cm^{-1}. Thus, one would expect that the bromides and iodides, when prepared, would be also low-frequency vibrations. On the other hand, if the halogen is fluorine only some of the vibrations are found in the low-frequency range.

6.5. TETRACOORDINATED MOLECULES—MX_4, MX_4^{n-} (CN = 4)

Tetracoordinated molecules or ions may have a tetrahedral (T_d) or a square planar structure (D_{4h}). The discussion of the assignments of the vibrational modes of these molecules will be made for each structure separately.

Tetrahedral Molecules and Ions

The selection rules for a tetrahedral molecule are listed in Table 6-13. Figure 5-10 depicts the normal fundamental modes for a MX_4 (T_d) molecule. All four fundamentals are Raman active while only two modes are infrared active, and thus more information is obtainable from the Raman experiment. Further, in many instances the tetrahedral ions are water soluble and are more easily studied by Raman methods.

Tables 6-14 to 6-16 tabulate the available Raman and infrared data for tetrahalides. The largest research effort has been made with the tetrachlorides and the smallest with the tetrafluorides. Because of the lighter mass of the fluorine atom, only the bending modes (ν_2, ν_4) are found in the low-frequency range for the tetrafluorides. As soon as the halogen becomes chlorine, all four modes can be expected in the low-frequency region. The general frequency

Table 6-13. Selection Rules for the T_d and D_{4h} Point Groups

Species	Fundamentals*	R	IR
	T_d *point group*		
a_1	1	a	ia
a_2	0	ia	ia
e	1	a	ia
f_1	0	ia	ia
f_2	2	a	a
	D_{4h} *point group*		
a_{1g}	1	a	ia
a_{1u}	0	ia	ia
a_{2g}	0	ia	ia
a_{2u}	1	ia	a
b_{1g}	1	a	ia
b_{1u}	0	ia	ia
b_{2g}	1	a	ia
b_{2u}	1	ia	ia
e_g	0	ia	ia
e_u	2	ia	a

*For an AB_4-type molecule.

position trend for the vibrations is $\nu_3 > \nu_1$ and $\nu_4 > \nu_2$. Some gap in the infrared frequencies appears for the ν_4 vibration in the tetraiodides, since this mode should occur at a frequency <80 cm^{-1}.

Table 6-15 shows that only a small variation occurs for ν_3 in the MX_4^{2-} ions, where M can be Mn, Fe, Co, Ni, or Zn and X can be Cl, Br, or I. The maximum frequency is demonstrated by CoX_4^{2-}, which corresponds to the maximum ligand field stabilization energy for Co(II) ($e_g^4 + t_{2g}^3$). The ν_3(MX) frequencies follow the trend Mn$\leq$Fe$<$Co$>$Ni$>$Zn. The ν_3(MCl) vibration occurs in the region of ~290 cm^{-1}, ν_3(MBr) at ~220 cm^{-1}, and ν_3(MI) at ~180 cm^{-1}. The other (MX) vibrations follow similar trends. Clark[1] has reported that the ratio ν_3(MBr)/ν_3(MCl) is ~0.76 and ν_3(MI)/ν_3(MCl)

Table 6-14. Vibrational Frequencies of Tetrafluoride Molecules

		Frequency, cm^{-1}				
Compound	Activity	$\nu_1(a_1)$ R	$\nu_2(e)$ R	$\nu_3(f_2)$ R,IR	$\nu_4(f_2)$ R, IR	References
GeF_4	R	738	205			57
	IR			800	260	57, 58
ZrF_4	IR			668		59
HfF_4	IR			645		59
ThF_4	IR			520		59

Table 6-15. Vibrational Frequencies of Tetrachloride Molecules

Compound	Activity	Frequency, cm^{-1} $\nu_1(a_1)$ R	$\nu_2(e)$ R	$\nu_3(f_2)$ R, IR	$\nu_4(f_2)$ R, IR	References
$TiCl_4$	R	389	120	490, 506	140	60
	IR			495		60
$ZrCl_4$	R	383	117–123		117–123	61, 61a
	IR			421–423	(112)	59
$HfCl_4$	IR			393		59
$ThCl_4$	IR			335		59
$GeCl_4$	R	397	132	451	171	62, 43
	IR			461		63
$SnCl_4$	R	368	105	403	132	62, 43
	IR			407	127	64
$PbCl_4$	R	327	90	348	90	65
VCl_4	R	383	128	475	128	66
	IR			461		60
PCl_4^+	R	458	171	658	251	42
	IR		171	651	251	42
$AsCl_4^+$	R	422	156	500	187	18
$SbCl_4^+$	R	353	143	399	153	67
$AlCl_4^-$	R	349	146	575	180	68
	IR			495		42
$GaCl_4^-$	R	346	114	386	149	69, 43
	IR			373		26
$InCl_4^-$	R	321	89	337	112	43
$FeCl_4^-$	R	330	106(114)[a]	385(378)[a]	133(136)[a]	43, 70
	IR			377		26
$ZnCl_4^{2-}$	R	282(288)[b]	82(166)[b]	298[b]	116(130)[b]	71, 43
	IR			277	130	72
$CdCl_4^{2-}$	IR			260		26, 43
$HgCl_4^{2-}$	R	267	180	276	192	73
	IR			228		26
$MnCl_4^{2-}$	IR	(258)[a]		284	118(116)[a]	72
$FeCl_4^{2-}$	IR	(266)[a]		286	119	72
$CoCl_4^{2-}$	IR			297	130	72
$NiCl_4^{2-}$	IR			289	112	72

[a] Recent Raman data from Avery *et al.*[78a]
[b] From Quicksall and Spiro.[78b]

is ~0.62, for a given metal and counter-ion. A large counter-ion effect is noted in the MX_4^{2-} ions. The larger cation causes a shift of the ν_{MX} frequency to lower regions.

Wherever comparisons are possible, the effect of the oxidation state in the tetrahalide anions may be demonstrated. For example, ν_3 is 377 cm^{-1} for the $FeCl_4^-$ anion and 286 cm^{-1} for the $FeCl_4^{2-}$ anion.

Table 6-16. Vibrational Frequencies of Tetrabromides and Tetraiodides

Compound	Activity	Frequency, cm^{-1} $\nu_1(a_1)$ R	$\nu_2(e)$ R	$\nu_3(f_2)$ R, IR	$\nu_4(f_2)$ R, IR	References
$TiBr_4$	R	230	74	384	73	74
	IR			383		74, 75
$GeBr_4$	R	234	78	328	111	62, 76
$SnBr_4$	R	222	65	281	87	76, 77
	IR			280	86	64
$GaBr_4^-$	R	210	71	278	102	69, 43
$InBr_4^-$	R	197	55	239	79	78, 43
$TlBr_4^-$	R	190	51	209	64	79
$FeBr_4^-$	IR	(200)[a]		289(290)[a]		72
$MnBr_4^{2-}$	IR	(157)[a]		221	81	72
$FeBr_4^{2-}$	IR	(162)[a]		219	84	72
$CoBr_4^{2-}$	IR			231	91	72
$NiBr_4^{2-}$	IR			224, 231	83	72
$CuBr_4^{2-}$	IR			174, 216	85	72
$ZnBr_4^{2-}$	R	172(178)[b]	61(80)[b]	210(212)[b]	82(89)[b]	71
	IR			215, 223		80
$CdBr_4^{2-}$	R	167	53	185	66	81
$HgBr_4^{2-}$	R	166				81
GeI_4	R	159	60	264	80	82
SnI_4	R	149	47	216	63	82
	IR			219	71	64
GaI_4^-	R	145	52	222	73	83, 43
InI_4^-	R	139	42	185	58	83, 43
MnI_4^{2-}	IR	(111)[a]		185		72
FeI_4^{2-}	IR			186		72
CoI_4^{2-}	IR			192, 197		72
NiI_4^{2-}	IR			189		72
ZnI_4^{2-}	R	122(130)[b]	44(60)[b]	170(172)[b]	62(70)[b]	71
	IR			165		72
CdI_4^{2-}	R	117	37	145	44	81
HgI_4^{2-}	R	126	35		41	84

[a] Recent Raman data from Avery *et al.*[78a]
[b] From Quicksall and Spiro.[78b]

Most of the tetrahalide compounds described thus far are in a T_d environment. However, it is possible for distortion of this symmetry to occur. The $CuCl_4^{2-}$ ion is interesting for it demonstrates a D_{2d} symmetry.[72] Table 6-17 demonstrates the correlation table for T_d, D_{2d}, the C_s site symmetry, and the factor group symmetry. For the D_{2d} symmetry, the two f_2 vibrations split into two b_2 and two e vibrations, which are both Raman and infrared active. For $CuBr_4^{2-}$ the symmetry has lowered to C_s, all the degeneracy has been removed, and the spectrum becomes quite complex, since all

Table 6-17. Correlation Table for T_d, D_{2d}, C_s, and D_{2h} Symmetries[72]

Nondistorted ion symmetry T_d	Distorted ion symmetry D_{2d}	Site symmetry C_s	Factor group symmetry D_{2h}

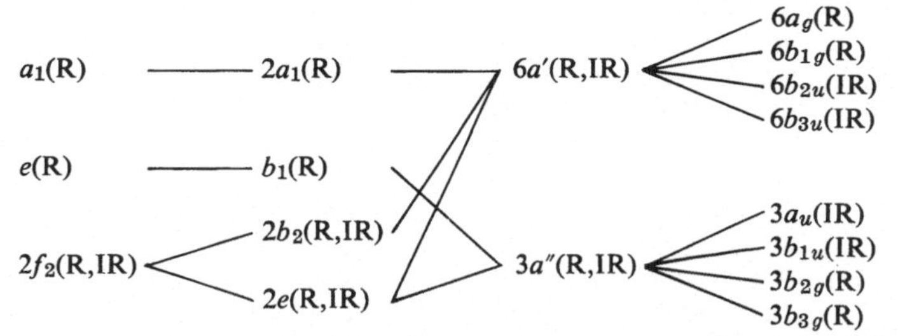
a_1(R)
$2a_1$(R)
$6a'$(R,IR)
$6a_g$(R)
$6b_{1g}$(R)
$6b_{2u}$(IR)
$6b_{3u}$(IR)
e(R)
b_1(R)
$2b_2$(R,IR)
$2f_2$(R,IR)
$2e$(R,IR)
$3a''$(R,IR)
$3a_u$(IR)
$3b_{1u}$(IR)
$3b_{2g}$(R)
$3b_{3g}$(R)

nine modes are observed in the Raman and infrared spectrum. Table 6-18 tabulates data for several vibrations for the ν_{CuX} and δ_{CuX} vibrations in compounds with D_{2d} symmetry. Effects of the cation on these vibrations can be easily observed.

Square Planar Ions—MX_4^{n-}

The selection rules for the square planar (D_{4h}) ions MX_4^{n-} are shown in Table 6-13, and Fig. 6-6 illustrates the normal fundamental modes. The center-of-symmetry present in such molecules illustrates why the use of both

Table 6-18. ν_{MX} and δ_{MX} Vibrations of R_2CuX_4 Compounds in D_{2d} Symmetry

Compound	Frequency, cm^{-1}		Reference
	ν_{CuX}	δ_{XCuX}	
	Infrared		
Cs_2CuCl_4	292, 257		72
$(Me_4N)_2CuCl_4$	281, 237	145, 128	"
$(Et_4N)_2CuCl_4$	267, 248	136, 118	"
Cs_2CuBr_4	224, 189		"
$(Et_4N)_2CuBr_4$	216, 174	85	"
	Raman		
Cs_2CuCl_4	295, 250	134, 116	78a
$(Me_4N)_2CuCl_4$	276, 232	133, 114	"
$(Et_4N)_2CuCl_4$	277		"

Note:
The structure of the $CuCl_5^{3-}$ has recently been the subject of debate. In molecules of the type $[M(NH_3)_6][CuCl_5]$, where M is Co or Cr, the $CuCl_5^{3-}$ ion has been reported to involve a trigonal bipyramidal structure haying 32 molecules per unit cell with all bond lengths Cu–Cl being equal.[90a] Recent far-infrared data[260] and electronic and magnetic results[90b] are not explainable on the basis a of pentacoordinate molecule. Far-infrared results show only one ν_{CuCl} stretching vibration, whereas two are expected for a D_{3h} structure. Adams and Lock[260] have concluded that the results can be explained on the basis of a T_d or D_{4h} $CuCl_4^{2-}$ ion and a free Cl^-.

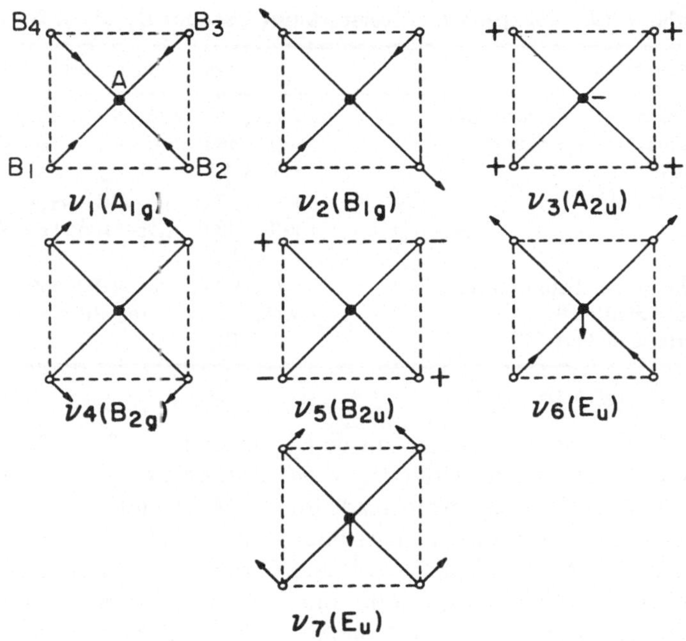

Fig. 6-6. Normal modes of vibrations of square planar AB_4 molecule point group (D_{4h}).

the Raman and infrared methods is necessary. Three Raman and three infrared active fundamentals with no coincidences are found in a molecule with D_{4h} symmetry. Thus, none of the information obtained from either experiment is redundant.

Most of the ions are highly colored and Raman spectra were unobtainable until recently. The advent of laser Raman excitation has spurred new interest in these molecules.

There is confusion in the literature as to the nomenclature and description of vibrations used for the D_{4h} point group. Apparently three different systems of classification are used: that used by Nakamoto,[85] Ferraro and Ziomek,[86] and James and Nolan[87]; that used by Adams,[88] Fertel and Perry,[89] and Sabatini, Sacconi, and Schettino[90]; and that used by Hendra.[91] The basis for confusion is the choice of axes. The system used in this book is that of Nakamoto and Ferraro and Ziomek. Because of the confusion, Table 6-19 has been compiled to attempt to help the reader interested in the literature on D_{4h} molecules move between the different nomenclatures. It is the hope of the authors that this problem will be rectified and one system of nomenclature be used in the future.

Table 6-19. Comparison of Nomenclature Used for D_{4h} Point Group

	Nomenclature	Vibrational species
1.	Nakamoto[85]; Ferraro and Ziomek[86]; James and Nolan[87]	$\nu_1(a_{1g})$(R); $\nu_2(b_{1g})$(R); $\nu_3(a_{2u})$(IR); $\nu_4(b_{2g})$(R); $\nu_5(b_{2u})$(IA); $\nu_6(e_u)$(IR); $\nu_7(e_u)$(IR).
2.	Hendra[91]	$\nu_1(a_{1g})$(R); $\nu_2(b_{2g})$(R); $\nu_3(a_{2u})$(IR); $\nu_4(b_{1g})$(R); $\nu_5(b_{2u})$(IA); $\nu_6(e_u)$(IR); $\nu_7(e_u)$(IR).
3.	Adams[89]; Sabatini, Sacconi, and Schettino[90]; Fertel and Perry[88]	$\nu_1(a_{1g})$(R); $\nu_2(a_{2u})$(IR); $\nu_3(b_{1g})$(R); $\nu_4(b_{1u})$;(IA); $\nu_5(b_{2g})$(R); $\nu_6(e_u)$(IR); $\nu_7(e_u)$(IR).

The MX_4^{n-} ions, where X can be Cl, Br, or I, are formed by Pd(II), Pt(II), and Au(III) metals with $4d^8$ and $5d^8$ electronic configurations. Table 6-20 summarizes the assignments made for the MX_4^{n-} ions.

Some slight cation effects on the internal modes of MX_4^{n-} ions may be observed and Table 6-21 tabulates these. Certain differences may also be observed in the spectra of the solid and in solution. Raman studies with $PtCl_4^{2-}$ indicate that ν_1 and ν_4 are similar in the solid state and in solution, but ν_2 shifts toward lower frequency by $\sim$30 cm^{-1}.[91,96] Comparison of the spectra of $PdCl_4^{2-}$ and $PtCl_4^{2-}$ indicates only minor differences in the infrared frequencies (ν_3, ν_6, ν_7) which involve the movement of the central atom,

Table 6-20. Frequency Assignments for Several MX_4^{2-} Molecules (D_{4h} Point Group)

Compound	Frequency, cm^{-1}						References
	$\nu_1(a_{1g})$ R	$\nu_2(b_{1g})$ R	$\nu_3(a_{2u})$ IR	$\nu_4(b_{2g})$ R	$\nu_6(e_u)$ IR	$\nu_7(e_u)$ IR	
$K_2[PdCl_4]$	310	198	170	275	336	193	36, 88, 91–95
$K_2[PtCl_4]$	333	196	168	306	321	191	88, 90–100
$K_2[PtCl_4]^{2-}$ R = solution	335	164		304	316	185	"
$K[AuCl_4]$			151		350	179	96, 90, 101, 102
$K[AuCl_4]^-$ solution	347	171		324			"
$K[AuCl_4]^-$ R = solution	347	171		324	356	173	"
$K_2[PdBr_4]$	192	125	130	165	260	140	91–93, 95, 97
$K_2[PtBr_4]$	205	125	135	190	232	135	88, 91, 97
$K[AuBr_4]$	214	102		196	260, 249	134	90, 96, 97, 102
$K[AuBr_4]^-$ R = solution	212	102		196	252	100	"
$K_2[PtI_4]$	142		105	126	180	127	91, 93
$K[AuI_4]$	148	75	113	110	192	113	"

Table 6-21. Cation Effects on Infrared Vibrations of M'_2MX_4 Compounds[103]

Complex	MX_4^{2-} frequencies, cm^{-1}				
	Internal modes			Lattice vibrations	
	$\nu_6(e_u)$	$\nu_7(e_u)$	$\nu_3(a_{2u})$	$\nu_8(a_{2u})$	$\nu_9(e_u)$
K_2PtCl_4	321	191	168	116	103, 90
Rb_2PtCl_4	320	186	166	79	64
Cs_2PtCl_4	313	177	157	65	50
K_2PtBr_4	232	135	125	104	76
Rb_2PtBr_4	231	126	117	72	61
Cs_2PtBr_4	229	120	108	60	51
K_2PdCl_4	334	190	170	120	111, 95
Rb_2PdCl_4	331	188	166	88	70
Cs_2PdCl_4	328	183	160	75	50
$(NH_4)_2PdCl_4$	327	205	175	120	
K_2PdBr_4	260	140	130	100	85
Rb_2PdBr_4	258	135	125	76	68
Cs_2PdBr_4	249	130	114	67	38

despite the large differences in mass. Larger differences are noted for the Raman frequencies (ν_1 and ν_4). These results are similar to those noted for the metal hexachlorides.

Several far-infrared studies have recently been made together with space group selection rules and a normal coordinate treatment to determine the lattice modes in MX_4^{2-} solids. Table 6-22 tabulates these results for K_2PtCl_4, which has been extensively studied. Certain original assignments are observed to be in error. Verification of Hiraishi and Shimanouchi's[94] lattice mode assignments have been obtained recently by high-pressure studies by Ferraro.[100] Figure 6-7 shows the low-frequency spectra of K_2PdCl_4 and K_4PtCl_2.

Table 6-22. Lattice Mode Assignments in K_2PtCl_4

Species	Frequency, cm^{-1}					
	Fertel and Perry[88]	Poulet *et al.*[99]	Sabatini and Sacconi[90]	Adams and Gebbie[97]	Hiraishi and Shimanouchi[94]	Ferraro[100]
$\nu_6(e_u)$	321	322, 195	325	320	325	325
$\nu_7(e_u)$	191	172, 160	193	183	190	195
$\nu_3(a_{2u})$	168	111	175	93	170	173
$\nu_8(a_{2u})$	111		106		103	106
$\nu_9(e_u)$	89	39			116, 90	116, 89

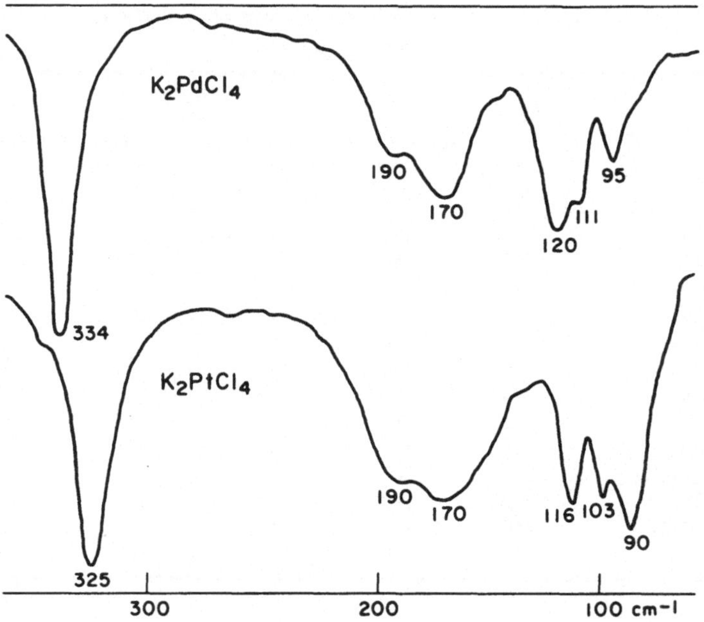

Fig. 6-7. Low-frequency spectra of K_2PdCl_4 and K_2PtCl_4. (Courtesy of Pergamon Press, New York. From Hiraishi and Shimanouchi.[94])

6.6. TRICOORDINATED MOLECULES—MX_3 (CN = 3)

Most of the halogen-containing molecules of the type MX_3 possess a pyramidal structure. The normal modes of vibration for these molecules of

Table 6-23. Selections Rule for C_{3v} and D_{3h} Point Groups

Species	Fundamentals*	R	IR
	C_{3v} point group		
a_1	2	a	a
a_2	0	ia	ia
e	2	a	a
	D_{3h} point group		
a_1'	1	a	ia
a_1''	0	ia	ia
a_2'	0	ia	ia
a_2''	1	ia	a
e'	2	a	a
e''	0	a	ia

*For an AB_3-type molecule.

Table 6-24. Vibrational Frequency Assignments for Several MX_3 Molecules (Neutral, anionic, and cationic species)

		Frequency, cm^{-1}				
		C_{3v} point group				
Molecule	Activity	$\nu_1(a_1)$	$\nu_2(a_1)$	$\nu_3(e)$	$\nu_4(e)$	References
NF_3	IR	1031–1032	642–647	902–907	492–497	104–106
	R	1050	667	905	515	105
PF_3	IR	892	487	860	344	104, 107
	R	890	531	840	486	76
AsF_3	IR	740	336	702	262	104
	R	707	341	644	274	76
PCl_3	IR	507	260	494	189	108
	R	510	257	480	190	76
$AsCl_3$	IR	412	194	307	155	108
	R	410	193	370	159	76
$SbCl_3$	IR	377	164	356	128	108
	R	360	165	320	134	76
$BiCl_3$	R	288	130	242	96	76
PBr_3	IR	392	161	342	116	108
	R	380	162	400	116	76
$AsBr_3$	R	(284)	128	275	98	109
	IR	(284)	128	274	~100	109
	IR	279	128	272	98	110
$SbBr_3$	R	227	110	236	92	111
	R	254	101	245	81	112
	IR	222–248[a]	109	222–248[a]	91	108
$BiBr_3$	IR	196	104	169	90	110
PI_3	R	303	111	325	79	113
AsI_3	IR	226	102	201	74	110
	R	216	94	221	70	113
SbI_3	IR	177	89	147	71	110
BiI_3	IR	145	90	115	71	110
$GeCl_3^-$	R	320	162	253	139	114
SnF_3^-[b]	IR	458	188	394	150	115
SnF_3^-[c]	IR	429	152	382	129	115
$SnCl_3^-$	R	297	128	256	103	112
$SnBr_3^-$	R	211	83	181	65	112
SCl_3^-	R	519	284	543	214	116
$SeCl_3^+$	R	437	200	390	168	116
$TeCl_3^+$	R	412	170	385	150	116
$TeCl_4$[d] ($TeCl_3^+$)	IR	363	186	344, 352	154	117
$TeBr_4$[d] ($TeBr_3^+$)	IR	240	110	222	87	117
TeI_4[d] (TeI_3^+)	IR	172	88	140	62	117
$GaCl_3$ in ($LGaCl_3$)[g]	IR(6)[e]	396 ± 7	153 ± 3	359 ± 2	134 ± 3	118
$GaBr_3$ in ($LGaBr_3$)[g]	IR(5)	299 ± 5	109 ± 4	226 ± 7	84 ± 2	118
GaI_3 in ($LGaI_3$)[g]	IR(4)	247 ± 3	87 ± 2	162 ± 3	66 ± 2	118

Table 6-24 (Continued)

Molecule	Activity	Frequency, cm^{-1}						References
		D_{3h} point group						
		$\nu_1(a_1')$ R	$\nu_2(a_2'')$ IR	$\nu_3(e')$ R, IR	$\nu_4(e')$ R, IR			
$HgCl_3^-$	R	282–294		287	210[f]			119, 119a
$HgBr_3^-$	R	179						120
HgI_3^-	R	125						120
AlF_3(gas)	IR			945				121
$AlCl_3$(gas)	IR			610				122
		C_{2v} point group						
		$\nu_1(a_1)$	$\nu_2(a_1)$	$\nu_3(a_1)$	$\nu_4(a_1)$	$\nu_5(b_1)$	$\nu_6(b_2)$	
ClF_3		741, 761	518, 535	319, 332	694, 703, 713	434	364	123
BrF_3		674	[528]	[300]	613	[384]	[289]	123

( ) Obtained from combination band at 657 cm^{-1}: $657 - 275 - 98 = 284$.
[a]Five bands in this region
[b]Univalent cations
[c]Bivalent cations
[d]Molecules which are considered to exist as ionic $TeX_3^+X^-$ molecules.
[e]Number of compounds used.
[f]Infrared results of melt of $HgCl_2$ in KCl.
[g]L = Me_3N, Me_2S, Et_3N, Et_2O, Et_2S, py.
[ ] Calculated from NCT.

C_{3v} symmetry are shown in Fig. 5-12. Table 6-23 tabulates the selection rules for a C_{3v} environment. Table 6-24 lists the assignments made for a series of molecules in a C_{3v} symmetry. As may be observed, all four vibrations are Raman and infrared allowed. The usual mass effects and trends are seen in the neutral, anionic, and cationic species of the MX_3 type, as one increases the mass of M or X. In general, the frequency positions vary $\nu_1 > \nu_3$ and $\nu_2 > \nu_4$; all four modes are found in the energy region < 650 cm^{-1}, except for the lighter molecules. Several $LGaX_3$ complexes[118] are listed, with L being an organic ligand. Halogen sensitive bands were found in the low-frequency range in $LGaX_3$ compounds, as indicated in Table 6–24.

Table 6-25. The Low-Frequency Vibrations of Se(IV) and Te(IV) Halides[124]

Compound	Activity	Frequency, cm^{-1}
$SeCl_4$	R	590, 375, 361, 348, 249, 163, 132
	IR	376, 351
$TeCl_4$	R	376, 351, 343
	IR	365, 350, 154
$TeBr_4$	R	250, 226, 220, 133, 125
	IR	242, 224, 111

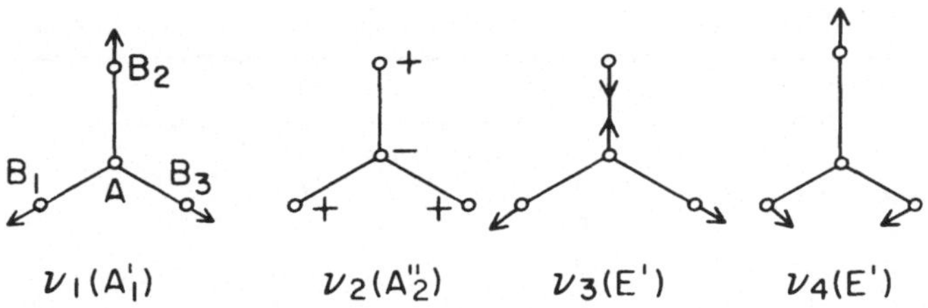

Fig. 6-8. Normal modes of vibration of planar AB_3 molecule point group (D_{3h}).

Greenwood *et al.*[117] assigned the far-infrared spectra of the solid Te(IV) chloride, bromide, and iodide on the basis of a C_{3v} pyramidal structure made up of ionic species $TeX_3^+X^-$. A more recent vibrational study of solid selenium and tellurium tetrachlorides by Hendra *et al.*[124] has contradicted this and a C_{2v} covalent structure has been suggested. Electron diffraction data on $TeCl_4$ have also suggested a C_{2v} structure. The low-frequency data of Hendra appear in Table 6-25.

Only a few MX_3-type molecules have been found to possess the D_{3h} structure. These are the HgX_3^- anions[119-120] and gaseous AlF_3[121] and $AlCl_3$.[122] Limited spectroscopic data are available for these anions, and these are tabulated in Table 6-24. Table 6-23 shows the selection rules for an MX_3 molecule of D_{3h} symmetry. Three vibrations are Raman and infrared active, with two coincidences occurring. Figure 6-8 shows the modes of vibration of a D_{3h} molecule.

A few molecules of the MX_3 type have been found to possess a *T* structure (C_{2v}). ClF_3 and BrF_3 are examples of this symmetry (see Table 6-24).[123]

The infrared spectra of rare earth fluorides have been studied. The main band in the 330–387 cm^{-1} region has been assigned as the ν_{MX} vibration.[123a]

6.7. DICOORDINATED MOLECULES—MX_2 (CN = 2)

A contrast in the infrared spectra may be observed in MX_2 molecules when comparing the gaseous and solid state. The bond type may vary from ionic to covalent in solids and may be considered predominantly covalent in the vapor.

Gaseous MX_2 Compounds

Considerable work has now been done with the gaseous metal halides of the type MX_2. Although the gaseous MX molecules may be explained on the basis of an ionic model, such is not the case for the gaseous MX_2 molecules. Thermodynamic properties indicate that polymerization of the

Table 6-26. Selection Rules for C_{2v} and $D_{\infty h}$ Point Groups

Type	Fundamentals*	Raman	Infrared
	C_{2v} point group		
a_1 (ν_1, ν_3)	2	a	a
a_2	0	a	ia
b_1	0	a	a
b_2 (ν_2)	1	a	a
	$D_{\infty h}$ point group		
a_{1g} (ν_1)	1	a	ia
a_{1u} (ν_3)	1	ia	a
e_u (ν_2)	1	ia	a

*For AB_2-type molecules.

dihalide vapors is possible, although it is only of minor importance. Four possible environments are possible for a MX_2-type molecule, the C_s, $C_{\infty v}$, C_{2v}, and $D_{\infty v}$ symmetries. The C_s and $C_{\infty h}$ symmetries involve bent and linear asymmetric M–X–X structures respectively. Thermodynamically, these structures would be very unstable and can be eliminated. Thus, one must choose between a bent C_{2v} structure and a linear $D_{\infty h}$ structure. Table 6-26 shows the selection rules and Figs. 6-9 and 6-10 show the motions of all the normal modes for molecules in these environments.

In the linear structure ν_1 is Raman active, ν_2 and ν_3 are infrared active, and no coincidences are present because of the center of symmetry existing in such a structure. For the bent structure all vibrations are allowed in the infrared and Raman spectra. Table 6-27 shows the spectra of several gaseous halides possessing a $D_{\infty h}$ symmetry and a few which have the C_{2v} symmetry. The difficulty in obtaining gas spectra by the conventional Raman techniques

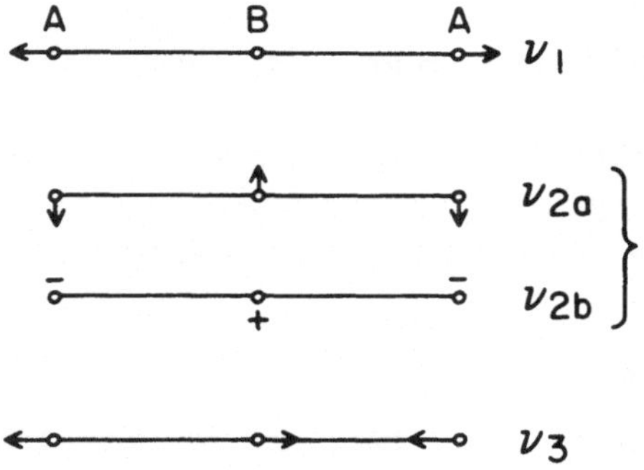

Fig. 6-9. Normal modes of vibration of linear A_3 and ABA molecule point group ($D_{\infty h}$).

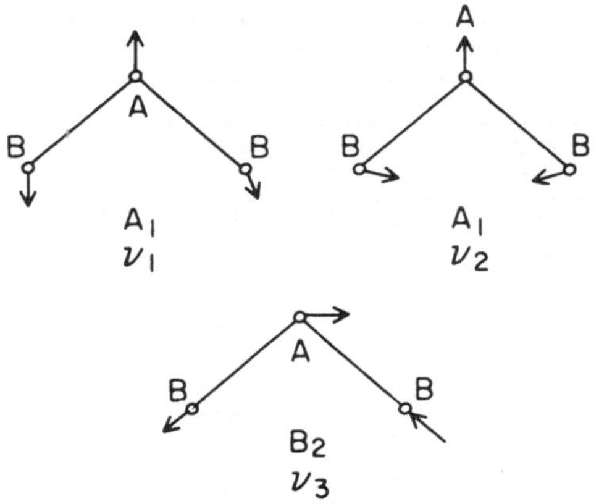

Fig. 6-10. Normal modes of vibration of bent AB_2 molecule point group (C_{2v}).

(mercury-arc excitation) and the fact that some of these gases are colored are the reasons for the lack of data for ν_1 of the linear molecules. Recent developments in the laser Raman technique should ameliorate this situation. For the gaseous transition metal dihalides, there has been a lack of data on the ν_2 vibration, which is of very low energy (<100 cm^{-1}). Thus, only inferences about structures were originally made. Recent matrix isolation studies by Thompson and Carlson[131] have provided values for the ν_2 vibration for the first time, and additional support has become available for $D_{\infty h}$ structure. The vibrational assignments of several interhalogen ions[136-140] and the KrF_2[141] molecule have been made. For the few bent molecules reported in Table 6-27, ν_2 was not observed because of the instrumental cutoff at 200 cm^{-1} in the work reported.

Thompson and Carlson[131] also reported on the evaluated and observed frequencies for the transition metal dihalide dimers of the type M_2X_4. These were considered to have a bridged structure of D_{2h} symmetry involving MX bridging and MX terminal vibrations. Observed frequencies for several of these compounds are seen in Table 6-28.

Solid MX_2 Compounds

These solids may involve ionic or covalent bonding. The first-row transition metal fluorides are ionic and have a fluorite or rutile structure. The other halides of this group of the metals Mn(II), Fe(II), Co(II), and Ni(II) are also ionic. Some of the other MX_2 compounds are more complex, since they involve higher states of aggregation. The Pt(II) and Pd(II) halides possess infinite-chain structures involving halide bridging. These structures have square planar (MX_4) units, sharing common edges. The Cu(II) halides

Table 6-27. Vibrational Frequency Assignments for MX_2 Gaseous Compounds

Molecule or ion	Point group	Frequency, cm^{-1} $\nu_1(a_{1g})$ R	$\nu_2(e_{1u})$ IR	$\nu_3(a_{1u})$ IR		References
$BeCl_2$	$D_{\infty h}$	(380)	482 [250]	1113 [1135]		125, 125a
$MgCl_2$	"		(297)	588		126
$MnCl_2$	"		83	477		127, 131
$FeCl_2$	"		88	499		127, 131
$CoCl_2$	"		95	493		127, 131
$NiCl_2$	"	341	85	521		126, 127
$CuCl_2$	"			492		127
$ZnCl_2$	"		295	516[a]		128, 132
$CdCl_2$	"			409[a]		128, 132
$HgCl_2$	"	360[c]	70[b]	413[a]		129, 130, 132, 133
$BeBr_2$	"		220	1010		125a
$MgBr_2$	"		(178)	490		126
$CoBr_2$	"			396		127
$NiBr_2$	"		69	415		131
$ZnBr_2$	"		225	400[a]		128, 132
$CdBr_2$	"			315[a]		132
$HgBr_2$	"	225[c]	41[b]	293		129, 130, 132, 133
BeI_2	"			873		125a
ZnI_2	"			340[a]		132
CdI_2	"			265[a]		132
HgI_2	"	156[c]	33[b]	237[a]		129, 130, 132
$CuCl_2^{2-}$	"	296			ether solution of HCl + CuX_2	134
$CuBr_2^{2-}$	"	190				134
$AuCl_2^{2-}$	"		340		mull $Cs_2Au_2Cl_6$	135
ICl_2^-	"	254		218		136
IBr_2^-	"	160		171		137, 138
$BrCl_2^-$	"	272		305		136
Cl_3^-	"	268		142		139
Br_3^-	"	162				136
I_3^-	"			193		138
ClF_2^-	"			137		140
KrF_2	"	449	233	588, 636		141
BeF_2	"	(680)	345	1555		142
MgF_2	"	(540)	270	875		142
CaF_2	C_{2v}	520[d](a_1)		595[d](b_2)		142
SrF_2	"	485[d](a_1)		490[d](b_2)		142
BaF_2	"	450[d](a_1)		430[d](b_2)		142

[a]Emission maxima.
[b]From electronic spectra.
[c]From Raman and electronic spectra.
[d]Corrected frequencies obtained by matrix isolation technique.
( ) Calculated.
[ ] Values from two sources; discrepancy exists for the ν_2 vibration.

Table 6-28. Observed Frequencies in M_2X_4 Dimers[131]

	Frequency, cm^{-1}			
Molecule	ν_9* $\nu_b(MX)$	ν_{10} in-plane bend	ν_{11} $\nu_b(MX)$	ν_{12}† $\nu_t(MX)$
Fe_2Cl_4		110		
Co_2Cl_4	323		289	436.2
				432.5
Ni_2Cl_4				440
Cu_2Cl_4				416.5
Zn_2Cl_4			297	435

* $\nu_b(MX)$ = bridged MX vibration.
† $\nu_t(MX)$ = terminal MX vibration.

have similar packing as the PdX_2 compound, but the copper atom is surrounded by a tetragonally distorted octahedron of chlorine atoms. The low-frequency infrared assignments for these solids are listed in Table 6-29.

Solid mercuric chloride and bromide involve a distorted octahedral environment. Solid $HgCl_2$ is orthorhombic and has a space group D_{2h}^{16}. The unit cell consists of four molecules. Solid $HgBr_2$ has a space group of C_{2v}^{12} with four molecules per unit cell. Both $HgCl_2$ and $HgBr_2$ may be considered to be molecular crystals. Red HgI_2 has a cubic close-packed structure, with a D_{4h}^{15} space group. Each mercury atom is surrounded by four equally distant iodine atoms arranged tetrahedrally and is considered to be a nonmolecular lattice with layers of HgI_4 units. At temperatures of 126°C the red form of HgI_2 undergoes a transition to yellow HgI_2. Yellow HgI_2 is orthorhombic having a C_{2v}^{12} space group with four molecules per unit cell. The cadmium halides, on the other hand, are considerably more ionic. The Cd^{2+} ion is in an octahedral environment; the CdX vibrations are essentially ionic and, indeed, are found at lower frequency than those for the mercury salts (see Table 6-29). Results for mercuric halide solids are

Table 6-29. Vibrational Frequencies for Some MX_2 Solids[143]

	Frequency, cm^{-1}		
Solid	ν_{MX}	δ_{MXM}	Other bands
$PdCl_2$	340	174	348, 297, 187
$PtCl_2$	318	200	
$PtBr_2$	242, 230		
$CuCl_2$	329, 277		
$CuBr_2$	254, 223		
$CdCl_2$	326		265, 220
$CdBr_2$	231		<170

Table 6-30. Vibrational Assignments for HgX_2, Hg_2X_2, cm^{-1}

Molecule	Point group	Species	Gas	Solid		Melt	Solution	References
$HgCl_2$	D_2	ν_1(R)	355–360	310–360		314	331(EtOH)	132, 133, 144
		ν_2(IR)	70	100–126		(100)		
		ν_3(IR)	413	368–379		376		
$HgBr_2$	$D_{\infty h}$	ν_1(R)	220–225	187		195	205(EtOH)	132, 133, 144
		ν_2(IR)	41	251		(90)		
		ν_3(IR)	293			271		
				Yellow[a]	Red[b]			
HgI_2	$D_{\infty h}$	ν_1(R)	156	138(R)	92, 102, 110(IR) 114(R)	138	150(EtOH)	130, 133, 144 145–149
		ν_2(IR)	53	50(IR) 37, 41(R)	25(IR) 17.5, 29(R)	41		
		ν_3(IR)	237	200(IR)	226(R)			
			Hg_2Cl_2	Hg_2Br_2	Hg_2I_2			
Hg_2X_2	$D_{\infty h}$	ν_1(R)	279	220	194			
		ν_2(R)	169	136	113			
		ν_3(IR)	~250	184	138			150, 151a, 151b, 151c
		ν_4(R)	42	37	31			
		ν_5(IR)	107	71	49			
		ν_{HgCl}	ν_{HgBr}	ν_{HgI}	δ_{HgXY}			
HgClBr		335	236		111			152
HgClBr		347	232		139			153
HgBrI			233	168				154

[a]High-temperature or high-pressure form.
[b]Room-temperature and ambient-pressure form.

Table 6-31. Correlation Diagrams

Mode	Point group	Site group	Factor group
	Yellow HgI$_2$ and HgBr$_2$ crystals		
	$D_{\infty h}$	C_s	$C_{2v}{}^{12}$
ν_1	a_{1g}(R)	a'(R,IR)	a_1(R,IR)
ν_3	a_{1u}(IR)		a_2(R)
ν_2	e_{1u}(IR)	a''(R,IR)	b_1(R,IR)
			b_2(R,IR)
	HgCl$_2$ crystals		
	D_2	C_s	$D_{2h}{}^{16}$
ν_1	Σ^+(R)		a_g(R)
		a'(R,IR)	b_{1g}(R)
ν_3	Σ^-(R)		b_{2g}(R)
		a''(R,IR)	b_{3g}(R)
ν_2	π(IR)		a_u(IR)
			b_{1u}(IR)
			b_{2u}(IR)
			b_{3u}(IR)

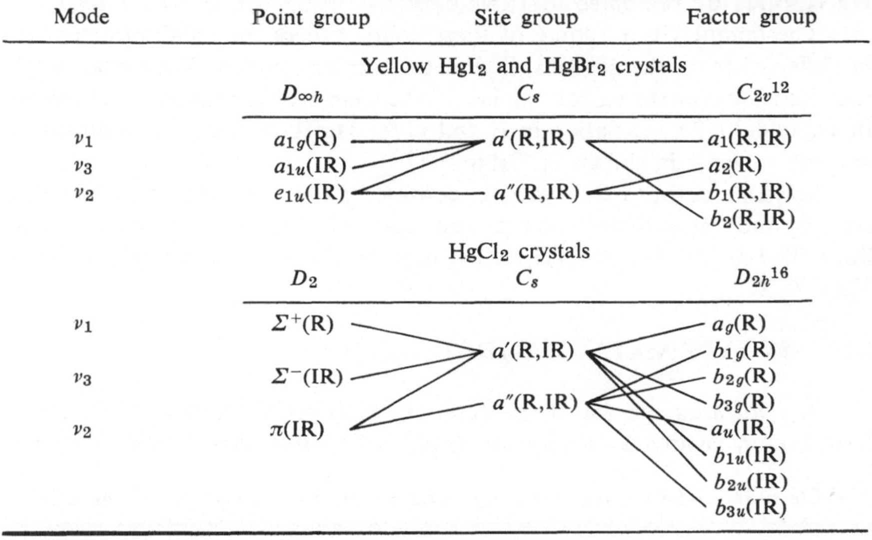

Table 6-32. Variation of ν_1 and ν_3 of HgX$_2$ in Various Solvents

Compound	Solvent	Frequency, cm^{-1}		Reference
		ν_1 (R)	ν_3 (IR)	
HgCl$_2$	Benzene	350	392	155
	Ethyl acetate	334	382	155
	Methyl formate	331	380	155
	Acetonitrile	331	380	155
	Acetone	331	372	155
	Ethyl alcohol	324		155
	Ethyl alcohol	320		153
	Ethyl alcohol	331		156
	Methyl alcohol	323		155
	Benzyl alcohol	323		155
	Isopropyl alcohol	323		155
	Water	320	372	155
	Tributyl phosphate	320		120
	Dioxane	302		155
	Pyridine	280		155
HgBr$_2$	Tributyl phosphate	207		120
	Ether	205		156
HgI$_2$	Ethanol	150		156

presented in Table 6-30. The site and factor group correlation charts for HgX_2 solids are presented in Table 6-31.

The complexity of some of these solids causes the solid spectrum to be different from the gaseous, liquid, or solution spectra. The variation of the vibrations for the various phases in the solid HgX_2 compounds is shown in Table 6-30. The variation in ν_1 and ν_3 for Hg(II) halides for solutions in various solvents is shown in Table 6-32.

Solid mercurous halides may be considered to be covalent molecules and recently their vibrational spectra were reported. They have been treated as linear X–Hg–Hg–X molecules, $D_{\infty h}$ symmetry. These results are presented in Table 6-30.

6.8. MX-TYPE METAL HALIDES

The infrared spectra of the gaseous alkali metal halides to 200 cm^{-1} have been measured by Klemperer *et al.*[159,159a] Vibrations below 200 cm^{-1}

Table 6-33. Vibrational Frequency Assignments for Several MX Molecules

Compound	Frequency, cm^{-1}		References
	Gaseous	Solid (ν_{TO})[a]	
LiF	906		159, 159*a*
^{6}LiF	917[b]		160
^{7}LiF	867[b]		160
LiCl	641		159, 159*a*
NaCl	366	164	157
KCl	281	146	157
RbCl	228	116	157
CsCl	209	99	157
LiBr	563		159, 159a
NaBr	302	134	157
KBr	213	113	157
RbBr	(166)	88	157
CsBr	(139)	73	157
LiI	498		159, 159a
NaI	258	117	157
KI	(173)	101	157
RbI	(128)	75	157
CsI	(101)	62	157
^{24}MgF	738.2		158
^{25}MgF	732.8		158
^{26}MgF	726.5		158
AlF	798		142

[a]Transverse optical lattice mode (see Chapter 9 for other lattice modes).
[b]Matrix isolation result.
( ) Extrapolated values from known radius distances and force constants.

in the salts of the heavier metals were extrapolated from the known radius and force constant data. Klemperer found that the classical ionic model could explain the gaseous molecule spectra. Table 6-33 shows these results. For comparison the transverse lattice mode of the solid alkali metal halides is also shown. Further lattice vibration results for these types of salts are given in Chapter 9. The differences in bonding that must be involved in the two states may easily be observed from the comparisons made for the infrared spectra of the solid and the gas. The shift to higher frequency in the gas spectra reflects the covalent nature of the uncondensed state of the gaseous alkali metal halides.

Recent matrix isolation techniques have provided very interesting results for LiF. Snelson[160] has obtained the infrared spectra of vapor species of LiF. The results gave a symmetrical ^6LiF vibration at 917 cm^{-1} and a ^7LiF vibration at 867 cm^{-1}. The monomer spectra of these molecules were assigned, based on a $C_{\infty v}$ symmetry. Evidence was also presented for a dimer

Table 6-34. Vibrational Frequency Assignments for LiF, Li_2F_2, and Li_3F_3 Species[160]

Molecule	Point group	Symmetry species	Observed frequency, cm^{-1}
^6LiF	$C_{\infty v}$	Σ^+	917
^7LiF	$C_{\infty v}$	Σ^+	867
$^6Li_2F_2$	D_{2h}	b_{3u}	678
		b_{2u}	585
		b_{1u}	303
$^7Li_2F_2$	D_{2h}	b_{3u}	641
		b_{2u}	553
		b_{1u}	287
$^6Li^7LiF_2$	C_{2v}	b_1	662
		a_1	565
		b_2	294
$^3Li_3F_3$			784
			530
			232
			282
$^7Li_3F_3$			736
			309
			230
			267
$^7Li_n{}^6Li_{3-n}F_3$			788
			770
			526
			277
			271

Table 6-35. Vibrational Results for Halogens (A_2) and Interhalogens (AB)

Molecule	Frequency, cm^{-1} IR	R	Reference
F_2 (gas)		892	161
$^{35}Cl_2$ (soln)		548	161
$^{37}Cl_2$ (soln)		533	161
^{35}ClF (gas)	773	758	162
^{37}ClF (gas)	767		161
Br_2 (soln)		312	161
$Br^{35}Cl$ (soln)		434	161
$Br^{37}Cl$ (soln)		425	161
I_2 (soln)		207	161
$I^{35}Cl$ (soln)	375	375	161
$I^{37}Cl$ (soln)	368	367	161
IBr (soln)		251	161

Li_2F_2 and a trimer Li_3F_3. These results are shown in Table 6-34. Table 6-35 lists spectral results for several halogens (A_2) and interhalogens (AB).

6.9. MISCELLANEOUS METAL HALIDE VIBRATIONS

Oxyhalides

Only a few of the oxyhalides have been examined spectroscopically. Table 6-36 tabulates partial data (ν_{MX} stretching vibration for some of these compounds). The trend with an increase in mass of M or X is observed for the ν_{MX} vibration in these molecules. The investigations made concern themselves mostly with the oxyfluorides and oxychlorides and very little with the oxybromides. Caution must be taken in considering these results, for mixing may occur with other modes of vibration, particularly those involving the oxygen atoms.

The chromyl halides have been assigned on the basis of a C_{2v} structure. These results confirm electron diffraction studies. The assignments are tabulated in Table 6-37.

The assignments for the oxyhalides of the type MAB_3 are consistent with a C_{3v} symmetry for the isolated AB_3^- ion. Table 6-38 tabulates these data. Several other oxyhalide assignments are tabulated in Table 6-39.

Xenon Halides

The discovery of the ability of noble gases to form chemical compounds was an important chapter in the recent history of inorganic chemistry. The role that spectroscopy played in establishing the structures of these

Table 6-36. Vibrational Assignments for ν_{MX} Vibration in Several Oxyhalides

Atomic number	Compound	ν_{MX}, cm^{-1}	References
16	SO_2F_2	885, 848	163
22	$K_3[TiO_2F_2]$	521, 510, 472, 429	164
41	$K_2[NbOF_5]$	543	164
41	$K_2[NbO_2F_5]$	547, 497	164
42	$K_2[MoO_3F_4]$	501	164
42	$K_2[MoO_5F_2]$	493	164
73	$K_2[TaO_2F_5]$	513, 469, 459	164
74	$K_2[WO_3F_4]$	480	164
22	$K_2[TiOF_4]$	500–600	165
23	$K_2[VOF_5]$	630, 596, 482	165
23	$K_3[V_2O_4F_5]$	557, 499	165
41	$K_2[Nb^{16}OF_5]\cdot H_2O$	546, 486	165
41	$K_2[Nb^{18}OF_5]\cdot H_2O$	545, 484	165
73	$K[TaOF_4]$	450–550	165
92	$U_2O_5F_2\cdot 2H_2O$	~400	165
50	$K_4[Sn_2F_{10}O]$	530	166
50	$K_2[SnF_4O]$	532, 490	166
33	$K_2[AsO_3F]$	735	167
34	SeO_2F_2	756, 702	168
33	$AsOF_3$	708, 568	169
51	$SbOF_3$	622, 442	169
51	SbO_2F	611	169
15	$POCl_3$	581, 486	170
21	$TiOCl_2$	430, 407	171
23	VO_2Cl	500	172
23	$VOCl_3$	504, 408	170
24	$Cs_2[CrOCl_5]$	319	173
41	$Cs_2[NbOCl_5]$	319	173
42	$Cs_2[MoOCl_5]$	324	173
74	$Cs_2[WOCl_5]$	306	173
42	MoO_2Cl_2	401	174, 174a
74	WO_2Cl_2	383–407	174, 174a
50	$SnOCl_2$	558, 402	175
76	$K_2[OsO_2Cl_4]$	325	176
93	$Cs_2[NpOCl_5]$	275	177
93	$Cs_3[NpO_2Cl_4]$	264	177
93	$Cs_2[NpO_2Cl_4]$	271	177
91	α-Pa_2OCl_8	326, 370	178
91	β-Pa_2OCl_8	324, 370	178
91	$Pa_2O_3Cl_4$	378, 342	178
91	PaO_2Cl	396	178
41	$Rb_2[NbOCl_5]$	339, 327	179
41	$Cs_2[NbOCl_5]$	330, 320	179
42	$Rb_2[MoOCl_5]$	339, 327	179
42	$Cs_2[MoOCl_5]$	329, 320	179

Table 6-36 (Continued)

Atomic number	Compound	ν_{MX}, cm^{-1}	References
74	$Rb_2[WOCl_5]$	339, 317	179
74	$Cs_2[WOCl_5]$	333, 309	179
50	$SnOBr_2$	315, 273	171
21	$TiOBr_2$	412	171
23	$VOBr_2$	360	171
42	MoO_2Br_2	255	174*a*
74	WO_2Br_2	350	174
23	$VOBr_3$	271	109
41	$Cs_2[NbOBr_5]$	241	179
42	$Rb_2[MoOBr_5]$	253, 244	179
42	$Cs_2[MoOBr_5]$	246	179
74	$Rb_2[WOBr_5]$	224	179
74	$Cs_2[WOBr_2]$	220	179
74	$WOCl_4$	403, [400], 379	174a
74	$WOBr_4$	271	174a

[ ] Calculated frequency.

molecules has perhaps been overlooked. Table 6-40 provides spectroscopic data obtained for XeF_2, XeF_4, XeF_6, $XeOF_4$, and XeO_2F_2. Spectral data for XeO_4 and XeO_3 are also included here, since they are related to $XeOF_4$ and XeO_2F_2 and are necessary for the discussion of these compounds.

The structure of XeF_2 in the vapor and solid state is linear, of the $D_{\infty h}$ point group, and has three nonbonding electron pairs at points of an equilateral triangle centered at the xenon atom and at 90° to the F–Xe–F axis. Crystal structure data confirm this. XeF_4 has been found to be a square planar molecule (D_{4h}) in the vapor and solid state. Two nonbonding

Table 6-37. Vibrational Assignments for the Chromyl Halides Based on a C_{2v} Symmetry[180–182]

Species		Vibration	Frequency, cm^{-1}		
			CrO_2F_2	CrO_2Cl_2	
a_1	ν_1	CrO_2 sym. stretch	1006	984[180]	981[181]
	ν_2	CrX_2 sym. stretch	727	465	465
	ν_3	CrO_2 bend	304	357	356
	ν_4	CrX_2 bend	[182]	144	140
a_2	ν_5	torsion	[422]	216	224
b_1	ν_6	CrO_2 asym. stretch	1016	994	995
	ν_7	ϱ_{CrO_2}, CrX_2 wag	274	230	211
b_2	ν_8	CrX_2 asym. stretch	789	497	496
	ν_9	CrO_2 wag, ϱ_{CrX_2}	[257]	263	257

[ ] Estimated bands from combination and overtones.

Table 6-38. Vibrational Assignments for MAB_3 Molecules Based on a C_{3v} Symmetry[170,183–187]

Molecule	Activity	Frequency, cm^{-1}						Type of spectrum
		ν_1	ν_2	ν_3	ν_4	ν_5	ν_6	
VOF_3		720	1060		805			vapor
$VOCl_3$		408	1035	165	504	249	129	liquid
$VOBr_3$		271	1025	120	400	83	212	liquid
$CrFO_3^-$	R	911	637	338	955	370	261	in HF solution
$CrClO_3^-$	R	907	438	295	945	365	209	in acetone solution
$ReClO_3$	R	1001		435	963	344	196	liquid
	IR	1002	293	434	960	344	~196	liquid
$ReBrO_3$	IR	997	195	350	963	332	168	in CCl_4 solution
Using Miller's designation		Sym. MB_3 stretch	MA stretch	MB_3 deformation		Asym. MB_3 stretch	MB_3 deformation degenerate	BMA rock degenerate
			a'				e	
			IR, R				IR, R	

Table 6-39. Vibrational Assignments in Miscellaneous Oxyhalides

Compound	Activity	Frequency, cm^{-1}	Reference
TlOCl	R	$\nu_{sym(TlO_2)}$, 510; $\nu_{asym(TlO_2)}$, 490; ν_{TlX}, 290; δ_{TlO_2}, 243; δ_{ClTlO}, 126	188
TlOBr	R	$\nu_{sym(TlO_2)}$, 480; $\nu_{asym(TlO_2)}$, 430; ν_{TlX}, 224; δ_{TlO_2}, 212.	188
ClO_2F (C_{3v} symmetry)	R	ν_1, 1097; ν_6, 602[a]; ν_3, 533; ν_4, 398; ν_5, 1253; ν_6, 351.	189

[a] ν_{ClF} stretching vibration.

electron pairs are found, one above and one below the plane of the molecule. Evidence confirming this has come from x-ray, neutron, and electron diffraction studies. The discussion of the structure of XeF_6 has been the center of interest in the xenon fluoride area for some time now. The consensus of

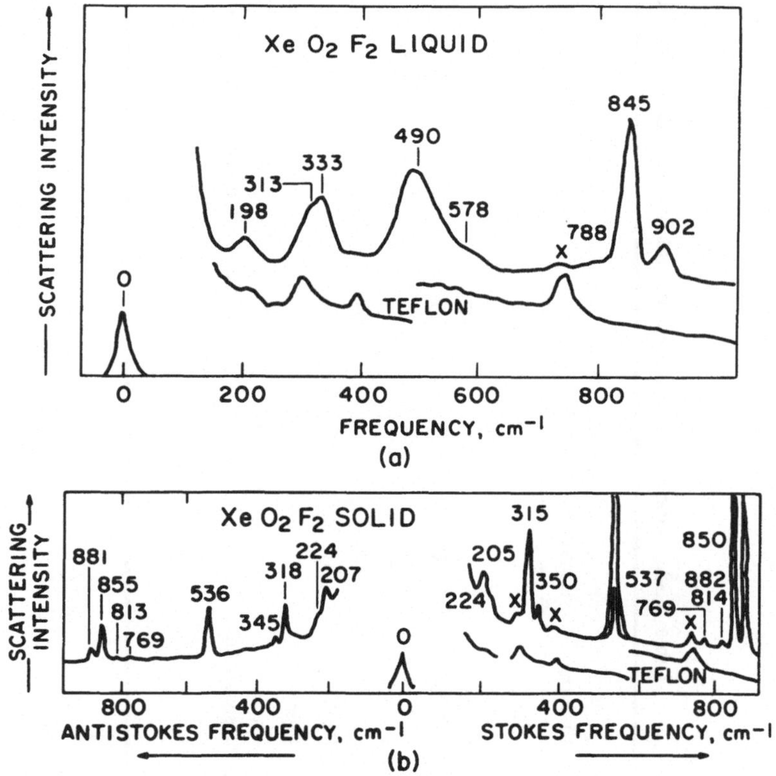

Fig. 6-11. Raman spectra of liquid and solid XeO_2F_2. (Courtesy of American Institute of Physics, New York. From Claassen *et al.*[199])

Table 6-40. Vibrational Assignments for Several Xenon Compounds

Compound	Frequency, cm^{-1} R	IR	Assignment*	References
$XeF_2(D_{\infty h})$	(solid)	(gas)		
	108		libration	
	497		$\nu_1(a_{1g})(\nu_{XeF}$ sym)	190–195
		213	$\nu_2(e_{1u})(\delta_{FXeF}$ bend)	
		555	$\nu_3(a_{1u})(\nu_{XeF}$ asym)	
$XeF_4(D_{4h})^a$				
	543		$\nu_1(a_{1g})(\nu_{XeF}$ sym)	190, 195, 196
	235		$\nu_4(b_{2g})(\delta_{FXeF}$ bend, sym)	
	502		$\nu_2(b_{1g})(\nu_{XeF}$ asym)	
		291	$\nu_3(a_{2u})(\delta_{FXeF}$ bend out-of-plane, asym)	
		586	$\nu_6(e_u)(\nu_{XeF}$, asym)	
		168	$\nu_7(e_u)(\delta_{FXeF}$ bend, asym)	
		(221)	$\nu_5(b_{2u})(\delta_{FXeF}$ bend, out-of-plane, sym)	
XeF_6 (Distorted O_h)	(gas)	(gas)		
	510			190, 195, 197
	587			
		520		
		612		
$XeOF_4$ (C_{4v})	(liquid)	(gas)		
	919(p)	928	$\nu_1(a_1)(\nu_{XeO})$	190, 195, 198
	566(p)	578	$\nu_2(a_1)(\nu_{XeF}$ sym)	
	286(p)	288	$\nu_3(a_1)(\delta_{FXeF}$ sym, out-of-plane)	
	231(d)		$\nu_4(b_1)(\delta_{FXeF}$ asym, out-of-plane)	
	530(d)		$\nu_5(b_2)(\nu_{XeF}$ asym)	
	Inactive		$\nu_6(b_2)(\delta_{FXeF}$ sym, in-plane)	
		609	$\nu_7(e)(\nu_{XeF}$ asym)	
	354(d)	362	$\nu_8(e)(\delta_{FXeO})$	
	161(d)		$\nu_9(e)(\delta_{FXeF}$ asym, in-plane)	
XeO_2F_2 (C_{2v})	(liquid)	(low-temp. matrix)		
	845	848	$\nu_1(a_1)(\nu_{XeO}$ sym)	199
	490		$\nu_2(a_1)(\nu_{XeF}$ sym)	
	333		$\nu_3(a_1)(\delta_{OXeO}$ bend, sym)	
	198		$\nu_4(a_1)(\delta_{FXeF}$ bend, sym)	
	223		$\nu_5(a_2)(\delta_{FXeF}$ bend, asym)	
	902	905	$\nu_6(b_1)(\nu_{XeO}$ asym)	
		324	$\nu_7(b_1)(\delta_{OXeO}$ bend, asym)	
	578	585	$\nu_8(b_2)(\nu_{XeF}$ asym)	
	313	317	$\nu_9(b_2)(\delta_{OXeO}$ bend, asym)	

Table 6-40 (Continued)

Compound	Frequency, cm^{-1} R	IR	Assignment*	References
$XeO_4(T_d)$	(gas)			
	870, 877, 885		$\nu_3(f_2)(\nu_{XeO})$	200
	298, 306, 314		$\nu_4(f_2)(\delta_{OXeO})$	
XeO_3 (C_{3v})	(solution)	(solid)		
	780	770	$\nu_1(a_1)(\nu_{XeO}$ sym)	201
	344	311	$\nu_2(a_1)(\delta_{OXeO}$ sym bend)	
	833	820	$\nu_3(e)(\nu_{XeO}$ asym)	
	317	298	$\nu_4(e)(\delta_{OXeO}$ asym bend)	

[a]See Table 6-19. Legend: d = depolarized, p = polarized.
( ) From overtone; unallowed vibration.

the vibrational data points to the fact that the structure is a distorted octahedron. Gasner and Claassen[198] concluded from their results that the ground state vapor molecules may be in a symmetry lower than O_h or that the molecules have unusual electronic properties which markedly influence the vibrational–rotational region of the spectrum.

The vibrational data for $XeOF_4$ points to a C_{4v} structure. The molecule has a square pyramidal structure with oxygen at the apex of the pyramid; the xenon atom is almost coplanar with the fluorine atoms. The stretching force constant for the Xe–O bond is quite large (7.10 mdyn/Å) and is indicative of an appreciable double-bond character.

The vibrational results for XeO_2F_2 were interpreted in terms of C_{2v} symmetry, where two F atoms are axial to the Xe atom and two O atoms with a lone electron pair are equatorial. Figure 6-11 shows the Raman spectra of liquid and solid XeO_2F_2.

A gaseous XeO_4 compound has been prepared, but the molecule is unstable and difficult to handle. This has caused some difficulty in obtaining vibrational data for it. Only two fundamentals could be assigned in the infrared spectrum of the vapor. The frequency maxima of the *P*, *Q*, and *R* branches for the ν_3 and ν_4 vibrations are given in Table 6-40. The close correlation of the data with those of OsO_4 and IO_4^- lends support to the belief that the molecule is of a T_d symmetry.

A very explosive solid compound XeO_3 has also been prepared. The Raman spectrum in solution and the infrared spectrum of the solid XeO_3 have been determined. Four intense vibrations were found that could be assigned on a C_{3v} symmetry. The frequency correspondence to previous spectral results for TeO_3^{2-} and IO_3^- lend support to this structure.

Table 6-41. Tabulated Assignments of Fundamentals in Xenon Compounds

Compound	Frequency, cm^{-1}				r_{XeF}, Å
	ν_{XeO}	δ_{OXeO}	ν_{XeF}	δ_{FXeF}	
XeF_2			497, 555	213	2.00
XeF_4			502, 543, 586	291, 235, 221, 168	1.953
$XeOF_4$	928	362	530, 566, 609	288, 231	1.955
XeO_2F_2	845, 902	317, 333, 324	490, 578	223, 198	
XeO_4	877	306			
XeO_3	780, 833	317, 344			

Table 6-41 shows a compilation of the fundamental vibrations corresponding to xenon–fluorine and xenon–oxygen vibrations. The position ν_{XeF} is seen to drop rapidly as the XeF distance increases.

The vibration spectral results of krypton fluoride are recorded in Table 6-27.

6.10. COMPLEXES INVOLVING ORGANIC LIGANDS

There are several ways in which metal halide complexes involving organic ligands could be discussed. Advantages and disadvantages may be cited for any method chosen. In this section the discussion will be made in terms of increasing ligand molecules contained in the complexes (effectively, coordination number). Only low-frequency vibrations concerning metal–halogen vibrations will be cited, as those involving metal–ligand vibrations will be discussed in subsequent chapters. It can be noted that there is a lack of Raman data available for these complexes, probably due to their highly colored nature. This situation may be remedied, because of laser Raman excitation developments, but the fact should be kept in mind that in a number of instances these complexes may be unstable in high-powered laser excitation. Since a number of abbreviations will be used for the organic ligands involved, Table 6-42 collects the ligands, their abbreviations, and their chemical structures.

LMX- and LMX_2-Type Complexes

Complexes of the LMX type involve metals with *d* shells all filled or all empty (e.g., Au^+, Hg^{2+}). The complexes may be considered to have a linear symmetry. When the ligand is considered, the symmetry may be considerably lower. Green[202] has made a normal coordinate treatment for CH_3HgI and made assignments on the basis of a C_{3v} symmetry. Table 6-43 records

Table 6-42. Chemical Structures and Abbreviations

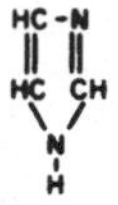

IMIDAZOLE
(IZ)

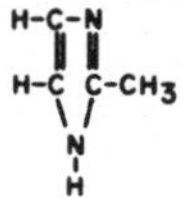

2-METHYLIMIDAZOLE
(MIz)

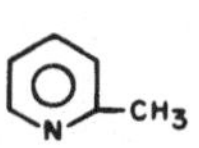

α-PICOLINE
(α-PIC)

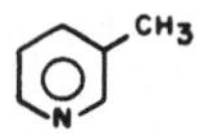

β-PICOLINE
(β-PIC)

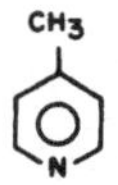

γ-PICOLINE
(γ-PIC)

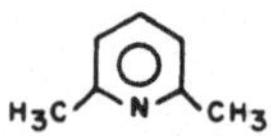

2,6-LUTIDINE
(2,6-LUT)

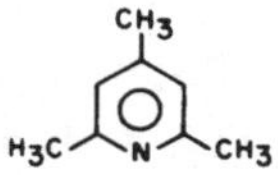

2,4,6-COLLIDINE
(2,4,6-COLL)

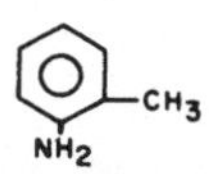

o-TOLUIDINE
(o-TOL)

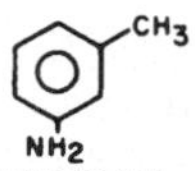

m-TOLUIDINE
(m-TOL)

p-TOLUIDINE
(p-TOL)

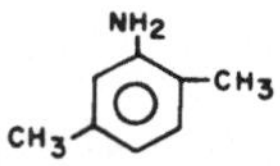

2,5-XYLIDINE
(2,5-XYL)

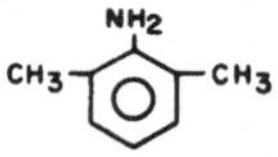

2,6-XYLIDINE
(2,6-XYL)

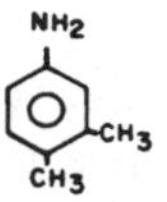

3,4-XYLIDINE
(3,4-XYL)

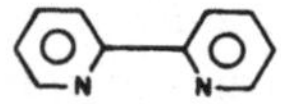

2,2'-BIPYRIDYL
(BIPY)

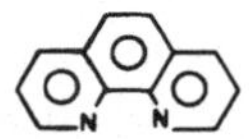

1,10-PHENANTHROLINE
(PHEN)

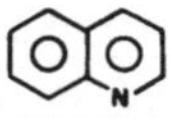

QUINOLINE
(QUIN)

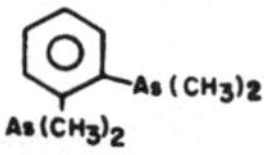

o-PHENYLENEBISDIMETHYLARSINE
(DIARSINE = D)

$(CH_3C_6H_4)_3P$
TRI-p-TOLYLPHOSPHINE
(p-TOL)

$(CH_3OC_6H_4)_3P$
TRI-p-ANISYLPHOSPHINE
(p-ANIS)

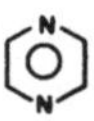

PYRAZINE
(pz)

PYRIMIDINE
(pm)

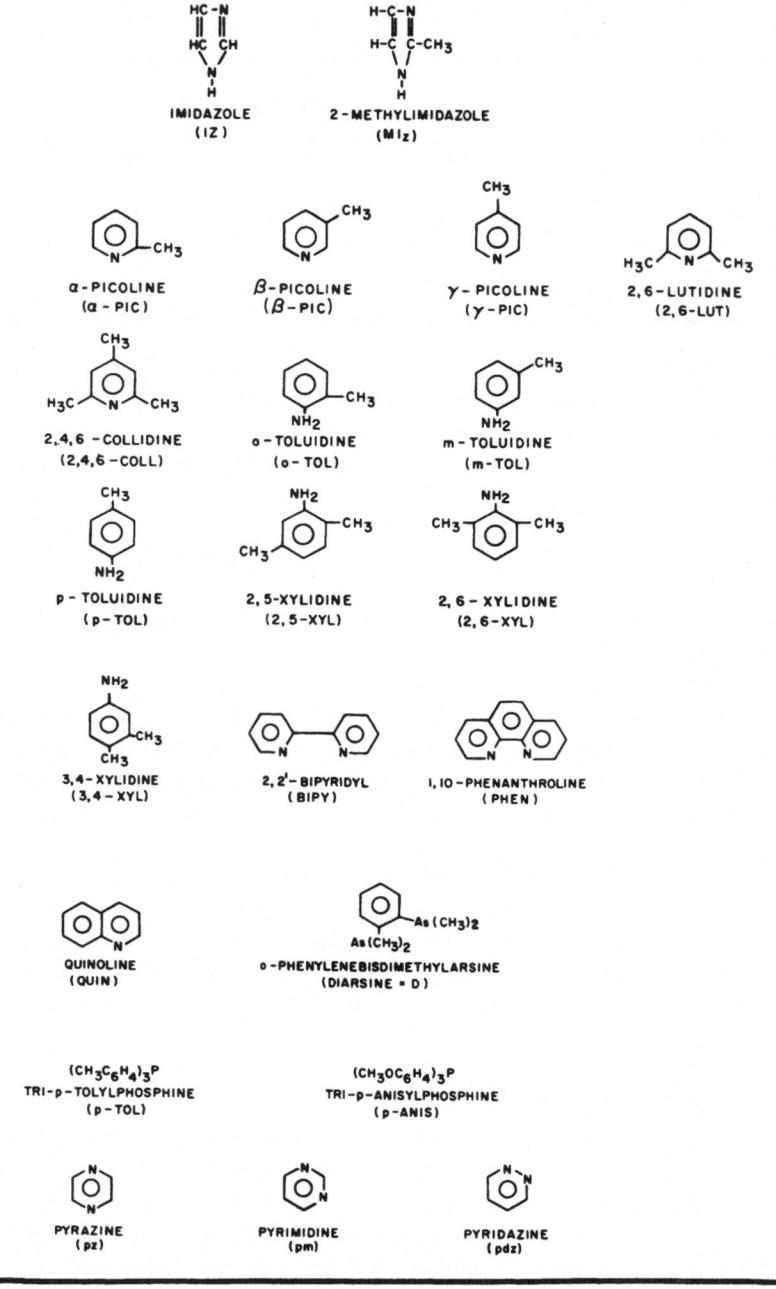

PYRIDAZINE
(pdz)

Table 6-43. Frequency Assignments for Metal Halide Vibrations in LMX- and LMX_2-Type Complexes

	ν_{MX}, cm^{-1}			
Compound	X = Cl	X = Br	X = I	References
Me_3PAuX	311, 305	226		203
Et_3PAuX	312, 305	210		"
φ_3PAuX	329, 323	233, 229		"
Me_2SAuX	324, 319	228		"
Me_3AsAuX	317, 312	210		"
MeHgX	315	214	184, 169	202, 204
EtHgX	314	209		"
*n*PrHgX	322	214		"
*n*BuHgX	316	246		"
C_2H_4HgX	323	216		"
$\varphi C{=}CHgX$	345			"
Cl_3CHgX	335	238		"
φHgX	330	257	170	"
p-$CH_3C_6H_4HgX$	325	246		"
p-$CF_3C_6H_4HgX$	324	230		"
p-ClC_6H_4HgX	330			"
p-$Me_2NC_6H_4HgX$	323	233		"
$C_6H_5CH_2HgX$	307	226		"
		ν_{MX}, cm^{-1}		
(pz)$MnCl_2$		246		204a
(pz)$CoCl_2$		244		"
(pz)$CuCl_2$		306		"
(pz)$CuBr_2$		248		"
(pz)$ZnCl_2$		218		"
(pz)$ZnBr_2$		275, 251		"
(pz)ZnI_2		224		"
(pz)$CdCl_2$		215		"
(pz)$CdBr_2$		186		"
(pz)CdI_2		142		"
(pz)$HgCl_2$		254		"
(pz)$HgBr_2$		208		"
(pm)$MnCl_2$		250		"
(pm)$MnBr_2$		208		"
(pm)$FeCl_2$		262		"
(pm)$CoCl_2$		257		"
(pm)$CoBr_2$		170		"
(pm)$CuCl_2$		309, 303		"
(pm)$CuBr_2$		236		"
(pm)$ZnCl_2$		225		"
(pm)$ZnBr_2$		186		"
(pm)$CdCl_2$		219, 205		"
(pm)$CdBr_2$		205		"
(pm)$HgCl_2$		263		"

Table 6-43 (Continued)

Compound	ν_{MX}, cm^{-1}	References
(pm)$HgBr_2$	205	204a
(pdz)$MnCl_2$	237	″
(pdz)$FeCl_2$	245	″
(pdz)$CoCl_2$	251	″
(pdz)$CuCl_2$	300	″
(pdz)$CuBr_2$	233	″
(pdz)$CdCl_2$	238, 206	″
(pdz)$CdBr_2$	176	″
(pdz)$HgCl_2$	226	″
(pdz)$HgBr_2$	198	″
(terp)$ZnCl_2$	287, 228	236
(terp)$ZnBr_2$	222, 213	″
(terp)ZnI_2	187	″
(terp)$CdCl_2$	269, 255	80

Abbreviations: terp = terpyridyl; φ = phenyl; pz = pyrazine; pm = pyrimidine; pdz = pyridazine.

Table 6-44. Frequency Assignments for Metal Halide Vibrations in $LAuX_3$-Type Complexes

Compound	ν_{asym}(MX),* cm^{-1}	Reference
Py$AuCl_3$	362	205
(2-MePy)$AuCl_3$	362	″
(3-MePy)$AuCl_3$	360	″
(4-MePy)$AuCl_3$	356	″
(2,6-lut)$AuCl_3$	362	″
(3,5-lut)$AuCl_3$	363	″
(2,4-lut)$AuCl_3$	364	″
(4-CNPy)$AuCl_3$	366	″
(Py-d_5)$AuCl_3$	359	″
PyAuBr3	260	″
(2-MePy)$AuBr_3$	262	″
(3-MePy)$AuBr_3$	256	″
(4-MePy)$AuBr_3$	254	″
(2,6-lut)$AuBr_3$	253	″
(3,5-lut)$AuBr_3$	256	″
(2,4-lut)$AuBr_3$	254	″
(4-CNPy)$AuBr_3$	267	″
(Py-d_5)$AuBr_3$	258	″

*In *trans* halogen atoms—C_{2v} symmetry.

ν_{MX} data for these complexes, where M is Au^+ or Hg^{2+}. The gold–chlorine stretching vibration appears to occur in the range 305 to 328 cm^{-1}; the gold–bromine stretching vibration at 210–233 cm^{-1}. The corresponding mercury–chlorine vibrations occur at 307–345 cm^{-1}, mercury–bromine vibrations at 209–257 cm^{-1}, and the mercury–iodine at 169–184 cm^{-1}.

Complexes of the type LMX_2 have recently been investigated in the low-frequency region.[204a] Table 6-43 tabulates data for the ν_{MX} vibration for complexes where L is azine. For the metals Mn^{2+}, Co^{2+}, Fe^{2+}, Cu^{2+}, and Cd^{2+} the complexes are considered to involve polymeric octahedral structures involving ligand and halide bridging. Similar structure are postulated for $ZnCl_2$ with pyrazine. $ZnBr_2$ and ZnI_2 complexes with pyrazine are considered to be tetrahedral with pyrazine bridging. The $ZnCl_2$ and $ZnBr_2$ complexes with pyrimidine are postulated to be of O_h stereochemistry with halide and ligand bridging. For the HgX_2 complexes with pyrazine and pyrimidine, T_d structures are postulated with ligand bridging.

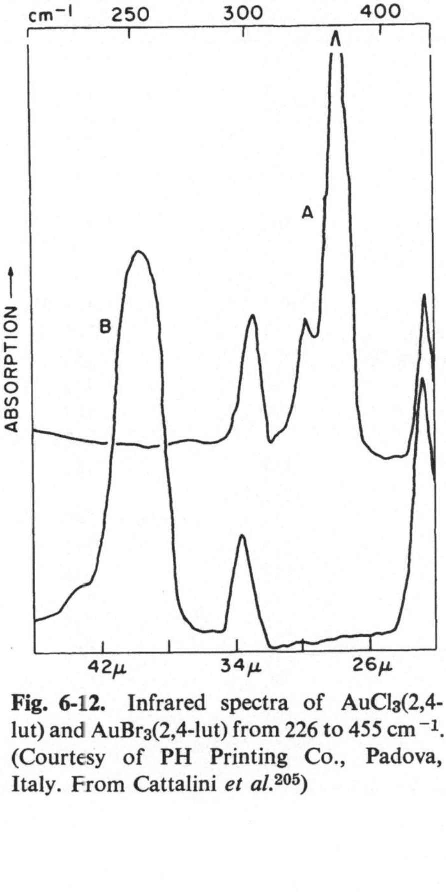

Fig. 6-12. Infrared spectra of $AuCl_3$(2,4-lut) and $AuBr_3$(2,4-lut) from 226 to 455 cm^{-1}. (Courtesy of PH Printing Co., Padova, Italy. From Cattalini *et al.*[205])

LMX_3- and LMX_3^--Type Complexes

Complexes of the type LMX_3 where L is an organic ligand, M is Au^{3+}, Ga^{3+}, Ge^{4+}, Sn^{4+}, or Tl^{3+}, and X is a halogen, have been examined by infrared means in the low-frequency region. The $LAuX_3$-type compounds have been assigned on the basis of a C_{2v} symmetry. In such a symmetry three AuX stretching modes are expected ($2a_1 + b_1$ type). The b_1-type vibration corresponds to an AuX asymmetric stretching vibration of the *trans* halogen atoms. Table 6-44 summarizes the data for various $LAuCl_3$ and $LAuBr_3$ complexes. The asymmetric AuCl vibration appears to remain fairly constant at ~360 cm^{-1}, while the corresponding AuBr vibration is at ~260 cm^{-1}. Low-frequency absorptions occurring at 160–170 cm^{-1} and 120–131 cm^{-1} in the chloride complexes and at 104–115 cm^{-1} and 81–91 cm^{-1} in the

Table 6-45. Frequency Assignments for Metal Halide Vibrations in $LGaX_3$-, $LGeX_3$-, and $LSnX_3$-Type Complexes

Compound	Frequency, cm^{-1}		References
	ν_{asym}(MX)	ν_{sym}(MX)	
H_3PGaCl_3	385	360	206
$(Me)_3PGaCl_3$	373	334	"
φ_3PGaCl_3	385	347	"
$(p\text{-tol})_3PGaCl_3$	381	347	207
$(p\text{-anis})_3PGaCl_3$	380	346	207
{Me_3N, Et_3N, Py} $GaCl_3$	385–392	357–364	209
Et_2OGaCl_3	412	361	"
$(Me_2S,Et_2O)GaCl_3$	398–409	356–359	"
tetrahydrofuran $GaCl_3$	418	358	"
tetrahydrothiophene $GaCl_3$	402	355	"
$MeGeCl_3$	430	403	208
$C_6H_5GeCl_3$	454	412	"
$C_6D_5GeCl_3$		412	"
$C_6H_5SnCl_3$	439	368	208
H_3PGaBr_3	285, 268	223	206
Me_3PGaBr_3	283	238	206
φ_3PGaBr_3	290, 283	232	206, 207
$(p\text{-tol})_3PGaBr_3$	285	234	207
$(p\text{-anis})_3PGaBr_3$	283	235	207
Me_3PGaI_3	238	155	206
φ_3PGaI_3	240	150	206, 207
$(p\text{-tol})_3PGaI_3$	240		207
$(p\text{-anis})_3PGaI_3$	237		207
$(Py)TlI_3$	(136, 155, 168, 176)		209

bromide complexes might correspond to the out-of-plane XAuX bending modes. The other AuX vibrations are more difficult to assign because of ligand vibrations and mixing of modes. Figure 6-12 shows the infrared spectra of (2,4-lut)$AuCl_3$ and (2,4-lut)$AuBr_3$ from 226 to 455 cm^{-1}.

The infrared assignments for the complexes of the type LMX_3 (where M is Ga^{3+}, Ge^{4+}, or Sn^{4+}) were made in terms of a monomeric C_{3v} structure. Two ν_{MX} vibrations are expected of type a_1 and e. In Table 6-45, two modes, the asymmetric and symmetric GaX stretching vibrations, are assigned. For a recent review on the metal–halogen stretching vibrations in various complexes of gallium, indium, and thallium, see Carty.[209]

Anion complexes of the type LMX_3^- have also been studied by infrared techniques. Table 6-46 summarizes the assignments made for the ν_{MX} vibrations in complexes where M is Pd^{2+}, Pt^{2+}, Co^{2+}, Ni^{2+}, or Zn^{2+}. For the Pd^{2+} and Pt^{2+} complexes the point group of the five-atom square planar skeleton is C_{2v}, but it may even be of lower symmetry, perhaps C_s, if the ligand atoms are considered. Three stretching MX modes are predicted by the selection rules; ν_{MX} *trans* to the ligand and ν_{MX} symmetric and asymmetric of the MX_2 group, in which the halogens are *trans* to each other. The *trans* effect is apparent, for the MX_2 group shows MX stretching frequencies at

Table 6-46. Frequency Assignments for Metal Halide Vibrations in LMX_3^--Type Anions

Compound	Frequency, cm^{-1}		Reference
	ν_{MX}	δ_{XMX}	
$[\varphi_3PCoCl_3]^-$	320, 282	112	210
$[PyCoCl_3]^-$	320, 286	118	211
$[\alpha\text{-picCoCl}_3]^-$	309, 287	120	211
$[\varphi_3PNiCl_3]^-$	308, 278	119	210
$[\varphi_3PZnCl_3]^-$	300, 276	115	210
$[PyZnCl_3]^-$	296, 284	118	211
$[\alpha\text{-picZnCl}_3]^-$	290, 284	129, 120	211
$[\varphi_3PCoBr_3]^-$	252, 210	84	210
$[PyCoBr_3]^-$	250, 228	94	211
$[\alpha\text{-picCoBr}_3]^-$	244	97	211
$[\varphi_3PNiBr_3]^-$	242, 212	86	210
$[\alpha\text{-picNiBr}_3]^-$	250	92, 100	211
$[\varphi_3PZnBr_3]^-$	228, 182	86	210
$[PyZnBr_3]^-$	223	94	211
$[\alpha\text{-picZnBr}_3]^-$	216, 189	102, 118	211
$[PyCoI_3)^-$	244, 238	68	211
$[PyNiI_3]^-$	240	76	211
$[PyZnI_3]^-$	204	82	211

Table 6-46 (Continued)

Square planar	$[LMX_3]^-$ $\nu_{asym}(MX_2)$ *trans* Cl–M–Cl	$\nu_{sym}(MX_2)$ *trans* Cl–M–Cl	ν_{MX} *trans* to L	Reference
$[Me_3PPdCl_3]^-$	343	295	265	212
$[Et_3PPdCl_3]^-$	340	293	265	"
$[n\text{-}Pr_3PPdCl_3]^-$	337	293	262	"
$[\varphi_3PPdCl_3]^-$	348	298	271	"
$[Me_3AsPdCl_3]^-$	338		253	253
$[Et_3AsPdCl_3]^-$	338	296	272	"
$[Me_2SPdCl_3]^-$	344	293	307	"
$[Et_2SPdCl_3]^-$	339	288	311	"
$[C_2H_4PdCl_3]^-$	337	288	317	"
$[Me_3PPtCl_3]^-$	332	332	275	212
$[Et_3PPtCl_3]^-$	330	330	271	"
$[n\text{-}Pr_3PPtCl_3]^-$	329	329	270	"
$[\varphi_3PPtCl_3]^-$	333	333	279	"
$[Me_3AsPtCl_3]^-$	329	329	272	"
$[Et_3AsPtCl_3]^-$	328	350[a]	280	"
$[Me_2SPtCl_3]^-$	325		310	"
$[Et_2SPtCl_3]^-$	325		307	"
$[C_2H_4PtCl_3]^-$	330	330	309	213
$[C_3H_6PtCl_3]^-$	330	306		"
$[C_2H_4PtCl_3]^-$	340, 331	306		"
$[olPtCl_3]^-$	330, 335	305		"
$[acPtCl_3]^-$	326	312		"
$[ac_1PtCl_3]^-$	325, 335	314		"

Abbreviations: ol = allyl alcohol; ac = $(Me_2C(OH)C\equiv)_2$; ac₁ = $(Me_2C(OMe)C\equiv)_2$; φ = phenyl.
[a]Less definite assignment.

higher positions and with smaller variation than the MX stretch *trans* to the ligand.

The LMX_3^- complexes involving Co^{2+} and Zn^{2+} are pseudotetrahedral, and the band assignments are made in terms of a C_{3v} or C_{2v} symmetry. Included in the data of Table 6-46 are the bending assignments for these complexes. Not too many such assignments have yet been made.

The far-infrared data for the LMX_3^- anions supports the fact that discrete LMX_3^- ions are involved for these complexes.

L_2MX_2-Type Complexes

The infrared data on L_2MX_2-type complexes is voluminous. Complexes of this type may involve square planar, tetrahedral, and octahedral structures (distorted or nondistorted, polymeric, or monomeric). The metal–halogen data for L_2MX_2 tetrahedral complexes where M is Mn^{2+}, Co^{2+}, Ni^{2+}, or

Table 6-47. Frequency Assignments for the ν_{MX} Vibrations in Tetrahedral Complexes of the Type L_2MX_2

Complex	Frequency, cm^{-1}			References
	ν_{MCl}	ν_{MBr}	ν_{MI}	
$(\varphi_3PO)_2MnX_2$	327, 317, 292	248, 229		214, 215
$(\varphi_3AsO)_2MnX_2$	310	243, 237		216, 217
Py_2CoX_2	339, 299, 276	274	237	"
(α-pic)$_2$CoX$_2$	351, 313	267		"
(β-pic)$_2$CoX$_2$	344, 302	270		"
(γ-pic)$_2$CoX$_2$	339, 306	272		"
(coll)$_2$CoX$_2$	318, 274	278		"
(quin)$_2$CoX$_2$	327, 310	260		"
$(\varphi_3PO)_2CoX_2$	342, 317	249, 233		"
$(\varphi_3P)_2CoX_2$	350, 316	272, 337		"
(*o*-tol)$_2$CoX$_2$	325, 292	245		"
(*m*-tol)$_2$CoX$_2$	320, 298	246		215, 218
(*p*-tol)$_2$CoX$_2$	318, 292	247	217	"
(2,5-xyl)$_2$CoX$_2$	325, 296	238		"
(3,4-xyl)$_2$CoX$_2$	320, 294	241		"
$(Me_3PO)_2CoX_2$	319, 290	246		219
$(Me_3NO)_2CoX_2$	310, 281	232		"
$(Et_3NO)_2CoX_2$	312, 279	237		"
(Iz)$_2$CoX$_2$	322(sym), 312(asym)			220
(MIz)$_2$CoX$_2$	320(sym), 305(asym)	262(sym), 244(asym)	222(sym), 184(asym)	"
(α-pic)$_2$NiX$_2$	328, 294	254		214, 215
(lut)$_2$NiX$_2$	343			214
$(\varphi_3AsO)_2NiX_2$	309, 282			"
$(\varphi_3P)_2NiX_2$	342, 305			"
(quin)$_2$NiX$_2$		258, 251		214, 215
(MIz)$_2$NiX$_2$		241(sym), 227(asym)	214(sym), 190(asym)	220 220
(β-pic)$_2$NiX$_2$			231	215
(Py)$_2$NiX$_2$			229	"
Py_2ZnX_2	329, 291	258		215, 221
(α-pic)p$_2$ZnX$_2$	324, 299	250		"
(β-pic)$_2$ZnX$_2$	327, 284	244		"
(γ-pic)$_2$ZnX$_2$	329, 294	244		"
(2,6-lut)$_2$ZnX$_2$	308, 296, 266	234		"
(2,4,6-coll)$_2$ZnX$_2$	308, 296	256		"
(*o*-tol)$_2$ZnX$_2$	303, 285	250		215, 218
(*m*-tol)$_2$ZnX$_2$	303	232		"
(*p*-tol)$_2$ZnX$_2$	298	227		"
(2,5-xyl)$_2$ZnX$_2$	315, 295	227		"
(2,6-xyl)$_2$ZnX$_2$	315	256		"
(3,4-xyl)$_2$ZnX$_2$	307, 294	227		"
Me_3POZnX_2	310, 292	239		219
Me_3NOZnX_2	304, 279	205		"

Table 6-48. Summary of Infrared Active Vibrations in L_2MX_2-Type Complexes

Complex	Symmetry	Symmetry group	Vibrational species, ν_{MX}	Number of infrared-active ν_{MX} vibrations
L_2MX_2	T_d	C_{2v}	$a_1 + b_1$	2
L_2MX_2	*trans* planar	D_{3h}	$a_g + b_{3u}$	1
L_2MX_2	*cis* planar	C_{3v}	$a_1 + b_1$	2
L_2MX_2	polymeric octahedral	C_i	$2a_g + 2a_u$	2

Zn^{2+}, and X is a halogen are summarized in Table 6-47. Table 6-48 summarizes the number of infrared active MX stretching modes allowed by the selection rules for the various type of L_2MX_2 complexes. For the tetrahedral L_2MX_2 complexes, the symmetry would be C_{2v}, if the ligands are considered, and thus two infrared active ν_{MX} modes are expected. The data substantiates this for the MCl vibration. For the MBr vibration, resolution is not as good, and for the most part only one band is found. Insufficient data are available, as yet, for the MI vibration.

Table 6-49 tabulates ν_{MX} data for several L_2MX_2 and $L_2MX_2^+$ complexes of polymeric octahedral symmetry. Two infrared active ν_{MX} vibrations are expected, but only one is observed in the data available to date.

Table 6-50 records ν_{MX} data for several L_2MX_2 complexes, where M is Cu. Some of these complexes may involve a tetragonal structure with long bonds to halides of other groups as is observed in bis-pyridine copper (II) chloride.[223] This might be considered to be a pseudooctahedral structure with two long Cu–X bands and two short Cu–X bands, due to the Jahn–Teller effect causing the long Cu–Cl bonds to occur.

Table 6-51 records ν_{MX} and δ_{XMX} data for *cis* and *trans* L_2PtX_2, L_2PdX_2, and L_2NiX_2 complexes. The *trans* planar compound should show only one ν_{MX} vibration and the *cis* planar compound two ν_{MX} vibrations. The infrared data corresponds to this expectation, and thus the technique may serve to distinguish between these isomers.

$L_2M_2X_4$-, L_3MX_3-, and L_4MX_2-Type Complexes

Table 6-52 records ν_{MX} infrared data for several L_3MX_3- and L_4MX_2-type complexes for both the *cis* and *trans* isomers in the monomeric octahedral structure. In addition, data is recorded for the bridged complexes $L_2M_2X_4$, where M is Pt^{2+} and Pd^{+2}. A more thorough discussion of these bridged-type vibrations appears in section 6.11.

Table 6-49. Frequency Assignments for Several Octahedral L_2MX_2-Type Complexes

Complex	ν_{MX}, cm^{-1}	Reference
(o-tol)$_2$MnCl$_2$	233	218
(m-tol)$_2$MnCl$_2$	227	"
(p-tol)$_2$MnCl$_2$	227	"
(3,4-xyl)$_2$MnCl$_2$	233	"
Py_2MnCl_2	233	215
(p-tol)$_2$NiCl$_2$	238	"
Py_2NiCl_2	246	"
(o-tol)$_2$NiCl$_2$	236	218
(m-tol)$_2$NiCl$_2$	238	"
(p-tol)$_2$NiCl$_2$	236	"
(2,5-xyl)$_2$NiCl$_2$	224	"
(3,4-xyl)$_2$NiCl$_2$	238	"
D_2FeCl_2	349	222
D_2RuCl_2	316	"
D_2OsCl_2	288	"
D_2TcCl_2	304	"
D_2ReCl_2	279	"
D_2MoCl_2	299	"
Py_2CrCl_2	328, 303	215
Py_2CuCl_2	294, 235	"
Py_2CdCl_2	<200	"
Py_2HgCl_2	292	"
$[D_2NiCl_2]Cl$	238	222
$[D_2CoCl_2]Cl$	384	"
$[D_2RhCl_2]Cl$	349	"
$[D_2IrCl_2]Cl$	320	"
$[D_2TcCl_2]Cl$	335	"
$[D_2ReCl_2]Cl$	313	"
$[D_2NiCl_2]ClO_4$	260	"
$[D_2CoCl_2]ClO_4$	388	"
$[D_2RhCl_2]ClO_4$	358	"
$[D_2IrCl_2]ClO_4$	335	"
$[D_2FeCl_2]ClO_4$	373	"
$[D_2RuCl_2]ClO_4$	340	"
$[D_2OsCl_2]ClO_4$	322	"
$[D_2TcCl_2]ClO_4$	343	"
$[D_2ReCl_2]ClO_4$	325	"
$[D_2CrCl_2]ClO_4$	375	"
(pz)$_2$MnBr$_2 \cdot H_2O$	198	204a
(pz)$_2$FeCl$_2 \cdot 2H_2O$	268	"
(pz)$_2$FeBr$_2$	203	"

D = diarsine (o-phenylenebismethylarsine).

Table 6-50. Frequency Assignments for Several Copper Complexes of the Type L_2CuX_2

Complex	ν_{MX}, cm^{-1}	Reference
$(o\text{-tol})_2CuCl_2$	336, 305, 294	218
$(m\text{-tol})_2CuCl_2$	315, 284	"
$(p\text{-tol})_2CuCl_2$	322, 303	"
$(2,5\text{-xyl})_2CuCl_2$	335, 303, 294	"
$(3,4\text{-xyl})_2CuCl_2$	333, 307, 282	"
$(o\text{-tol})_2CuBr_2$	230	"
$(m\text{-tol})_2CuBr_2$	239	"
$(p\text{-tol})_2CuBr_2$	222	"
$(2,5\text{-xyl})_2CuBr_2$	227	"
$(3,4\text{-xyl})_2CuBr_2$	227	"
$(2,6\text{-xyl})_2CuBr_2$	232	"
Py_2CuCl_2	287	"
$(\alpha\text{-pic})_2CuCl_2$	308	"
$(\beta\text{-pic})_2CuCl_2$	286	"
$(\gamma\text{-pic})_2CuCl_2$	292	"
$(2,6\text{-lut})_2CuCl_2$	305	"
$(2,4,6\text{-coll})_2CuCl_2$	206	"
Py_2CuBr_2	255	"
$(\alpha\text{-pic})_2CuBr_2$	233	"
$(\gamma\text{-pic})_2CuBr_2$	255	"
$(2,6\text{-lut})_2CuBr_2$	233	"
$(2,4,6\text{-coll})_2CuBr_2$	228	"

L_5MX- and L_5MX^{n-}-Type Complexes

Complexes of the type L_5MX and L_5MX^{n-}, where L is carbonyl, X is a halide, and M is a transition metal, have been studied by infrared and x-ray means. Table 6-53 tabulates the ν_{MX} assignments.

LMX_5^{n-}, $L_2MX_4^{n-}$, and $L_3MX_3^{n-}$ Anion-Type Complexes

Table 6-54 tabulates the spectroscopic data for several complexes of the type LMX_5^{n-} and $L_2MX_4^{n-}$ where L may be CO, NO, NH_3, or H_2O. The assignments are made on the basis of a C_{4v} symmetry. Anion complexes of the type $L_2MX_4^{n-}$ (*cis*), where L is CO, are assigned on the basis of a C_{2v} symmetry (*cis*) and are tabulated in Table 6-55. Complexes of the type $L_3MX_3^{n-}$ (*cis*), where L is CO, are assigned on the basis of a C_{3v} symmetry and are recorded in Table 6-55. Table 6-56 summarizes the selection rules for the MX vibrations in these complexes.

2′,2′-Bipyridyl and 1,10-Phenanthroline Complexes

Table 6-57 tabulates data for several 2,2′-bipyridyl and 1,10-phenanthroline complexes of metal halides. Complexes with a bidentate ligand of the

Table 6-51. Frequency Assignments for Various *Cis* and *Trans* Complexes of the L_2MX_2 Type, where M = Pt^{2+}, Pd^{2+}, and Ni^{2+}

Compound	ν_{MX}, cm^{-1}	References
Trans compounds, square planar		
$(Et_2S)_2PtCl_2$	342	224
$(Et_2Se)_2PtCl_2$	337	224
Py_2PtCl_2	343	213, 215, 224
$(Et_3As)_2PtCl_2$	339	224
$(n\text{-}Pr_3As)_2PtCl_2$	337	224
$(n\text{-}Pr_3P)_2PtCl_2$	339	224
$(Et_3P)_2PtCl_2$	340	224
$(Me_3P)_2PtCl_2$	326(339)	203, 212
$(Me_3As)_2PtCl_2$	337	203, 212
$(Et_3As)_2PtCl_2$	337, 328	224a
$(\varphi_3As)_2PtCl_2$	337, 328	224a
Py_2PdCl_2	356	203, 215
$(Et_3P)_2PdCl_2$	355	"
$(n\text{-}Pr_3P)_2PdCl_2$	353	"
$(n\text{-}Bu_3P)_2PdCl_2$	355	"
$(\varphi_3P)_2PdCl_2$	357	"
$(Et_3As)_2PdCl_2$	353	"
$(Me_2S)_2PdCl_2$	359	"
$(Et_2S)_2PdCl_2$	358	"
$(MeCN)_2PdCl_2$	357	225
$(\varphi CN)_2PdCl_2$	366	225
$(Et_3P)_2PtBr_2$	254	224
$(n\text{-}Pr_3P)_2PtBr_2$	259	"
$(Et_2S)_2PtBr_2$	254	"
$(Et_3As)_2PtBr_2$	251	"
$(n\text{-}Pr_3As)_2PtBr_2$	245	"
$(Et_2Se)_2PtBr_2$	241	"
Py_2PtBr_2	251	215, 224
$(Me_3As)_2PdBr_2$	279	212
Cis compounds, square planar		
$(Et_2S)_2PtCl_2$	330, 318	224
$(Et_2Se)_2PtCl_2$	333, 317	"
$(Et_2Te)_2PtCl_2$	302, 282	"
$(n\text{-}Pr_2Te)_2PtCl_2$	306, 291	"
Py_2PtCl_2	343, 328	213, 224
$(Et_3As)_2PtCl_2$	314, 288	224
$(n\text{-}Pr_3As)_2PtCl_2$	310, 286	224
$(Et_3P)_2PtCl_2$	303, 281	203, 224
$(n\text{-}Pr_3P)_2PtCl_2$	307, 277	224
$(n\text{-}Pr_3Sb)_2PtCl_2$	311, 281	224
$(Me_3P)_2PtCl_2$	302, 288, 277	203, 212
$(Me_3As)_2PtCl_2$	312, 293	203, 212
$(MeCN)_2PtCl_2$	360, 345	225
$(\varphi CN)_2PtCl_2$	356, 346	225

Table 6-51 (Continued)

Compund	ν_{MX} cm^{-1}	References
$(Et_3As)_2PtCl_2$	312, 289	224a
$(\varphi_3As)_2PtCl_2$	321, 300	224a
$(Me_3P)_2PdCl_2$	294, 283, 268	203, 212
$(Me_3As)_2PdCl_2$	316, 298	203, 212
$(Et_2S)PtBr_2$	254, 226	224
$(Et_2Te)_2PtBr_2$	217, 208	224
Py_2PtBr_2	219, 211	215, 224
$(n\text{-}Pr_3As)_2PtBr_2$	220	224
$(Et_3P)_2PtBr_2$	212, 194	224
$(n\text{-}Pr_3Sb)_2PtBr_2$	217	224
$(Me_3As)PtBr_2$	211	212

Compound	Frequency, cm^{-1}			References
	$\nu_{asym}(MX)$	$\nu_{sym}(MX)$	δ_{XMX}	
Cis compounds				
Py_2PtCl_2	343	329	163, 108	226, 227
Py_2PtBr_2	252	235		"
Py_2PtI_2	178	167		"
$(Py\text{-}d_5)_2PtCl_2$	341	328	149	"
$(Py\text{-}d_5)_2PtBr_2$	253	224		"
Trans compounds				
Py_2PtCl_2	342		167, 125	"
Py_2PtBr_2	252		120	"
Py_2PtI_2	183		92	"
$(Py\text{-}d_5)_2PtCl_2$	339		165, 124	"
Cis compounds				
Py_2PdCl_2	342	333		"
Trans compounds				
Py_2PdCl_2	353		166, 122	"
Py_2PdBr_2	256		122, 99	"
Py_2PdI_2	176		94, 88	"

Compound	Activity	Frequency, cm^{-1}				Reference
		$\nu_{sym}(MX)$	$\nu_{asym}(MX)$	δ_{MX}	$\delta_{XML}{}^{a}$	
Cis complexes						
$[(CH_3)_3P]_2PdCl_2$	IR	297	272	156	169	227a
	R	296	270	153	193	"
$[(CD_3)_3P]_2PdCl_2$	R	300	273	153	181	"
$[(CH_3)_3As]_2PdCl_2$	IR	316	299	138		"
	R	317	299	136		"
$[CH_3)_3Sb]_2PdCl_2$	IR	304	288	130		"
	R	305	288	133		"

Table 6-51 (Continued)

Compound	Activity	Frequency, cm^{-1} $\nu_{sym}(MX)$	$\nu_{asym}(MX)$	δ_{MX}	δ_{XML}[a]	Reference
$[(CH_3)_3P]_2PtCl_2$	IR	300	303	160		227a
	R	278	280	163	195	"
$[(CH_3)_3As]_2PtCl_2$	IR	314	294	145		"
	R	313	293	145		"
$[(CH_3)_2Sb]_2PtCl_2$	IR	308	298	137		"
	R	307	287	132		"
$[(CH_3)_3P]_2PdBr_2$	IR	208	182	188	135	"
	R	204	178	107	130	"
$[(CH_3)_3P]_2PtBr_2$	IR	204, 194	187	116	138	"
	R	192	189	112	140	"
$[(CD_3)_3P]_2PtBr_2$	IR	212	184	113	146	"
	R	212	184	113	145	"
$[(CH_3)_3As]_2PtBr_2$	IR	211	183	104		"
	R	211	183	105		"
$[(CD_3)_3As]_2PtBr_2$	IR	205	180	104		"
	R	205	181	105		"
$[(CH_3)_3Sb]_2PtBr_2$	IR	207	181	115		"
	R	209	187	117		"

Compound	ν_{MX}, cm^{-1}	Reference
Trans compounds, square planar		
$(Me_3P)_2NiCl_2$	403	203
$(Me_2PPh)_2NiCl_2$	404	"
$(Et_2PPh)_2NiCl_2$	408	"
$(Me_3P)_2NiBr_2$	340	"
$(Me_3P)_2NiI_2$	280	"

[a]Assignments for this bending mode are tentative.

type L–LMX_2 may be either tetrahedral monomeric or dimeric (involving ligand-bridged forms). The tin complexes with bidentate ligands such as 2,2′-bipyridyl and 1,10-phenanthroline are six coordinate monomers in the solid state and in nonpolar solvents.

L_3SnX-, L_2SnX_2-, and $LSnCl_3$-Type Complexes

Table 6-58 records ν_{MX} data for various Sn complexes of the type L_3SnX, L_2SnX_2, and $LSnCl_3$. It may be observed that a decrease in ν_{SnX} stretching frequencies occurs with increasing alkylation. For a summary of vibrational frequencies of anionic tin–halogen complexes see Wharf and Shriver.[241]

Table 6-52. Frequency Assignments for Several Monomeric Octahedral Compounds of the Type $L_2M_2X_4$, L_3MX_3, and L_4MX_2

Compound	ν_{MX}, cm^{-1}	Reference
Py_3CrCl_3	364, 341, 307	215
trans-Py_3RhCl_3	355, 332, 295	"
cis-Py_3RhCl_3	341, 325	"
trans-Py_3IrCl_3	329, 318, 307	"
cis-Py_3IrCl_3	325, 317, 303	"
cis-Py_3InCl_3	276, 242	227
trans-$[Py_2IrCl_4]PyH$	331, 305	215
cis-$[Py_2IrCl_4]PyH$	315, 304, 299	"
trans-Py_4NiCl_2	246	"
trans-Py_4CoCl_2	230	"
trans-$[Py_4RhCl_2]Cl$	364	"
trans-$[Py_4IrCl_2]Cl$	335	"
cis-$[Py_4IrCl_2]Cl$	333, 327	"
Cis compounds		
Py_2PtCl_4	353, 345, 332, 232	228
$(Me_2S)_2PtCl_4$	350–298	"
$(Et_3P)_2PtCl_4$	345, 336, 298, 291, 215	"
$(Et_3As)_2PtCl_4$	346–285	"
Py_2PtBr_4	255, 236, 218, 198	"
$(Et_3P)_2PtBr_4$	246, 209, 192, 144	"
Trans compounds		
$(Me_2S)_2PtCl_4$	333	228
$(Et_3P)_2PtCl_4$	340, 334	"
$(Et_3As)_2PtCl_4$	343, 337	"
$(Me_2S)_2PtBr_4$	245, 236, 220	"
$(Et_3P)_2PtBr_4$	247	"
$(Et_3As)_2PtBr_4$	242	"
$(Me_2S)_2PtI_4$	187, 179	"
$(Et_3P)_2PtI_4$	181	"
$(Et_3As)_2PtI_4$	188	"
Bridged complexes		
$(Me_3P)_2Pd_2Cl_4$	347, 294	212
$(Me_3As)_2Pd_2Cl_4$	339, 303	"
$(Me_3P)_2Pt_2Cl_4$	347, 341	"
$(Me_3As)_2Pt_2Cl_4$	351, 343	"
$(Me_3P)_2Pd_2Br_4$	263	"
$(Me_3As)_2Pd_2Br_4$	268, 202	"
$(Me_3P)_2PtBr_4$	254	"
$(Me_3As)_2Pt_2Br_4$	252	"
$(Me_3P)_2Pd_2I_4$	224	"
$(Me_3As)_2Pd_2I_4$	211	"

Table 6-53. Frequency Assignments for Several L_5MX- and L_5MX^{n-}-Type Complexes

Compound	IR ν_{MX}, cm^{-1}	R ν_{MX}, cm^{-1}	References
$Mn(CO)_5Cl$	295	291	229, 230
$Re(CO)_5Cl$	294	292	"
$[Cr(CO)_5Cl]^-$	257		"
$[Mo(CO)_5Cl]^-$	248		"
$[W(CO)_5Cl]^-$	258	255	"
$[Os(CO)_3Cl_2]_2$	319, 283	328, 294	231
$[Ru(CO)_3Cl_2]_2$	325, 290	320, 283	"
$[Ru(CO)_2Cl_2]_2$	322	300, 258	"
$Mn(CO)_5Br$	218	219	230
$Re(CO)_5Br$	203	204	"
$[Cr(CO)_5Br]^-$	179	175	"
$[Mo(CO)_5Br]^-$	165	165	"
$[W(CO)_5Br]^-$	163	164	"
$[Ru(CO)_3Br_2]_2$	227, 198	233, 207, 204	231
$Mn(CO)_5I$	187	188	230
$Re(CO)_5I$	163	164	"
$[Cr(CO)_5I]^-$	161	161	"
$[Mo(CO)_5I]^-$	146	145	"
$[W(CO)_5I]^-$	139	138	"

Metal Ammine Complexes

Table 6-59 records ν_{MX} data for several metal ammine complexes of the square planar, octahedral, and T_d types.

Miscellaneous

For a summary of far-infrared data on halogen–Ir(III) complexes see Jenkins and Shaw.[247] Infrared studies on platinum halide complexes with glycine,[248] diethylenediamine,[249] and ethylene (Zeise's salt)[250,251] have been reported. For infrared studies of terpyridyl complexes with lanthanide halides, see Basile *et al.*[251a] and Durham *et al.*[252a]

6.11. BRIDGED METAL HALIDE VIBRATIONS

This chapter has discussed the assignments made for metal–halogen vibrations. In most cases these assignments were for terminal or nonbridged metal–halogen vibrations [ν_t(MX)]. For the most part, the assignments can be made unequivocally. For compounds which contain bridging halogens the assignments may be more difficult. It is expected that a bridging vibration would be located at lower frequencies than those found for terminal vibra-

Table 6-54. Metal–Halogen Frequencies for Several Complexes of the Type LMX_5^{n-}, $L_2MX_4^{n-}$, and $L_3MX_3^{n-}$

		Frequency, cm^{-1}			
Compound	Activity	$\nu_3(a_1)$	$\nu_4(a_1)$	$\nu_{10}(e)$	Reference
$Cs_3[Os(CO)Cl_5]$	R	332	316		231
	IR	324	315	306	"
$Cs_2[Os(NO)Cl_5]$	R	326	292		"
	IR	328	292	311	"
$Cs_3[Ru(CO)Cl_5]$	R	319			"
	IR	320			"
$Cs_2[Ru(NO)Cl_5]$	R	320	278		"
	IR	328	285	328	"
$Cs_2[Ir(CO)Cl_5]$	R	326	307		"
	IR	319	305	288	"
$K[Ir(NO)Cl_5]$	R	320	310	340	"
	IR	315	300	335	"
$Cs[Rh(CO)Cl_5]$	R	315	289		"
	IR	310	285	328	"
$Cs_2[Ru(CO)(H_2O)Cl_4]$	IR	311	649	578	"
$[Ru(NO)(NH_3)_5Cl_3]$	R	512	489		"
	IR		482	482	"
$[Ru(NO)(OH)(NH_3)_4]Cl_2$	R	498	597	488	"
	IR	498	595	478	"
$[Ru(NO)Cl(NH_3)_4]Cl_2$	R	506	325	481	"
	IR	500	322	482	"
$Cs_3[Os(CO)Br_5]$	IR	205	217		"
$Cs_2[Os(NO)Br_5]$	R	210	181	225	"
	IR	208		220	"
$Cs_2[Ru(NO)Br_5]$	R	213	175		"
	IR	218		250	"
$Cs_2[Ir(CO)Br_5]$	R	207	190		"
	IR	207		226	"
$K[Ir(NO)_5]$	R	195	181	232	"
	IR			235	"
$Cs_2[Rh(CO)Br_5]$	R	183	175		"
	IR			254	"
$K_2[Ru(NO)I_5]$	R	112	107		"

tions. The sharing of halogens between two metals in a bridged structure causes the bond to be weaker than a terminal metal–halogen bond. Other factors are involved, such as stereochemistry and coordination number of the central metal atom, which contribute toward determining where a particular vibration will be located. The expectation that the bridging metal–halogen frequency will be found at lower energy is realized in the limited number of bridged metal–halogen assignments which have been made.

Table 6-55. Metal–Halogen Frequencies for Several Complexes of the Types $L_2MX_4^{n-}$ and $L_3MX_3^{n-}$

Cis complexes	Activity	Frequency, cm^{-1} $\nu_4(a_1)$	$\nu_5(a_1)$	$\nu_{21}(b_2)$	Reference
$Cs_2[Os(CO)_2Cl_4]$	R	316	281	308	231
	IR	312	276	302	"
$Cs_2[Ru(CO)_2Cl_4]$	R	307	303	275	"
	IR	312	301	268	"
$Cs_2[Os(CO)_2Br_4]$	R	183	164	208	"
	IR			213	"
$Cs_2[Ru(CO)_2Br_4]$	R	180	175	218	"
	IR			211	"
$Cs_2[Os(CO)_2I_4]$	R	135	108	161	"
$Cs_2[Ru(CO)_2I_4]$	R	131	122	156	"
		$\nu_4(a_1)$		$\nu_{13}(e)$	
$Cs[Os(CO)_3Cl_3]$	R	321		287	231
	IR	314		281	"
$Cs[Ru(CO)_3Cl_3]$	R	320		284	"
	IR	315		282	"
$Cs[Os(CO)_3Br_3]$	R	224		201	"
	IR	217		206	"
$Cs[Ru(CO)_3Br_3]$	R	236		208	"
	IR	228		215	"
$Cs[Os(CO)_3I_3]$	R	178		154	"
$Cs[Ru(CO)_3I_3]$	R	196		174	"

Table 6-56. Summary of Selection Rules for LMX_5^{n-}, $L_2MX_4^{n-}$-, and $L_3MX_3^{n-}$-Type Complexes

	Point group	Total number of modes	Symmetry species	ν_{MX} active modes
LMX_5^{n-}	C_{4v}	13	$5a_1 + 2b_1 + b_2 + 5e$	$2a_1 + b_1 + e$* ($\nu_3, \nu_4; \nu_{10}$)
cis-$[L_2MX_4]^{n-}$	C_{2v}	21	$8a_1 + 3a_2 + 6b_1 + 4b_2$	$2a_1 + b_1 + b_2$* ($\nu_4, \nu_5; \nu_{13}; \nu_{21}$)
trans-$L_2MX_4]^{n-}$	D_{4h}	15	$3a_{1g} + 3a_{2u} + b_{1g} + b_{2g} + b_{2u} + 2e_g + 4e_u$	$a_{1g} + b_g + e_u$
$L_3MX_3^{n-}$	C_{3v}	16	$6a_1 + 2a_2 + 8e$	$a_1 + e$ ($\nu_4; \nu_{13}$)

*Only three vibrations observed; for C_{4v} point group $2a_1$ and e vibrations observed; for C_{2v} point group $2a_1$ and b_2 vibrations observed.

Table 6-57. Frequency Assignments for Several Metal Halide Complexes with Bipyridyl and *o*-Phenanthroline

	ν_{MX}, cm^{-1}				
Complex	X = F	X = Cl	X = Br	X = I	References
$[Sc(bipy)_2X_2]X$		342, 320			232
$Ti(bipy)X_4$	634, 562	384, 366			233, 234
$V(bipy)X_4$		366, 356			233
$V(bipy)OX_2$		371			233
β-$Co(bipy)X_2$		331, 304			235
$Zn(bipy)X_2$		323	261	217	236
$Cd(bipy)X_2$		228			237
$Hg(bipy)X_2$		273			237
$Sn(bipy)X_4$		322, 302, 279	256, 240, 216, 192	196, 185, 173	238, 239
$MeSn(bipy)X_3$		289, 267	201, 188, 170	176, 159, 147	"
$EtSn(bipy)X_3$		292, 283, 273	198, 189		"
$BuSn(bipy)X_3$		294, 281, 264	198, 175		"
$Me_2Sn(bipy)X_2$		244	151, 140	145, 139	"
$Et_2Sn(bipy)X_2$		234, 215	163, 141	153, 139	"
$Bu_2Sn(bipy)X_2$		243	169, 145	156, 139	"
$Me_2Sn(bipy)X_2$		246	158	144	"
$Octyl_2Sn(bipy)X_2$		235			"
$[Rh(bipy)_2X_2]X \cdot 2H_2O$		356, 349, 332	207, 192		240
$[Ir(bipy)_2X_2]X \cdot 2H_2O$		337, 331, 309	212, 192		240
$Pd(bipy)X_2$		353, 343	225	181, 166	225, 227
$Pt(bipy)X_2$		351, 338			225
$[Sc(phen)_2X_2]X$		336, 318			232
$Ti(phen)X_4$	645, 570	379, 356			234
$V(phen)X_4$		380, 361			233
$V(phen)OX_2$		366			233
$Zr(phen)_2X_4$	504, 483				234
$Zn(phen)X_2$		324, 314	264, 232		237
$Cd(phen)X_2$		227, 213			237
$Hg(phen)X_2$		279	211, 205		237
$Sn(phen)X_4$		319, 302, 271	240, 222, 196	194, 187, 175	238, 239 "
$MeSn(phen)X_3$		296, 272	200, 194, 184	184, 158, 150	"
$EtSn(phen)X_3$		299, 270	198, 189		"
$BuSn(phen)X_3$		285	199, 192, 177		"
$Me_2Sn(phen)X_2$		247, 239	157, 149	145, 139	"
$Et_2Sn(phen)X_2$		242, 220	163, 141	145, 131	"
$Bu_2Sn(phen)X_2$		242	164, 149	144	"
$Octyl_2Sn(phen)X_2$		235			239
$[Rh(phen)_2X_2]X \cdot 3H_2O$		325, 305	187		64
$[Ir(phen)_2X_2]X \cdot 3H_2O$		336, 313	196		64

Table 6-58. Frequency Assignments for Seveal Tin Complexes of the Type
L_3SnX, L_2SnX_2, and $LSnCl_3$

| Compound | ν_{MX}, cm^{-1} | | | References |
	X = Cl	X = Br	X = I	
Me_3SnX	331	234	189	238
Me_2SnX_2	361, 356	250, 240	204, 186	238, 64
$MeSnX_3$	382, 368	264, 235	207, 174	"
Et_3SnX	337	222	182	"
Et_2SnX_2	359, 352	260, 240	198, 176	"
$EtSnX_3$	377, 366	260, 253		"
n-Bu_3SnX	328	228	184	"
n-Bu_2SnX_2	356, 340	248, 240	196, 180	"
n-$BuSnX_3$	378, 367	261, ~247		"
n-Pr_2SnX_2	349, 338		198, 183	64
n-$OctylSnX_3$	354, 345			64
$\varphi SnCl_3$	438asym, 377asym, 368sym			208

Much of the delay in studying these systems was caused by the instrument cutoff at 200 cm^{-1}, and many of these frequencies are located around 200 cm^{-1} or below. Similar experimental and interpretative difficulties have delayed progress in assigning bridging metal–halogen bending vibrations [$\delta_b(MX)$], which are to be found at even lower frequencies. Thus far, only bridged metal–halogen stretching vibrations [$\nu_b(MX)$] have been assigned.

Assignments for $\nu_b(MX)$ vibrations have been made for compounds of Pt(II)[252-254], Pd(II)[252-254], Al(III)[255-257], and other transition metals [M(II)].[258-262] Table 6-60 tabulates most of the data available at the present time. In most of these complexes the bridging metal-halogen bonds are found in dimer units of the type

$$\begin{array}{ccccc} X' & & X & & X' \\ & \diagdown M \diagdown & & \diagdown M \diagdown & \\ X' & & X & & X' \end{array}$$

The dimer also contains terminal metal–halogen bonds. A comparison of the frequency positions of the assigned terminal and bridging vibrations in such dimers show small differences. Thus the $\nu_b(MX)$ (dimer)/$\nu_t(MX)$ (dimer) ratio ranges from 0.60 to 0.85.

In polymeric compounds such as the type $CoX_2 \cdot L_2$ the assignments for the $\nu_b(CoX)$(polymer) have recently been made.[263] These are located at lower frequencies than those assigned for $\nu_b(CoX)$(dimer). Figure 6-13 shows the infrared spectra of $(py)_2CoCl_2$, monomer and polymer. Thus, in a polymer involving long chains of bridging metal–halogen units, the energy of the

Table 6-59. Frequency Assignments for Several Metal Ammine Complexes of the Square Planar and Octahedral Type

Compound	ν_{MX}, cm^{-1}	References
Trans compounds, square planar		
$Pd(NH_3)_2Cl_2$	332	242, 243, 244
$Pd(ND_3)_2Cl_2$	329	244
$Pt(NH_3)_2Cl_2$	330, 322	242, 245
$Pt(ND_3)_2Cl_2$	337, 326	245
$Pd(NH_3)_2Br_2$	228	246
$Pt(NH_3)_2Br_2$	230	246
$Pd(NH_3)_2I_2$	191	244
$Pt(NH_3)_2Cl_4$	352, 346	228
$Pt(NH_3)_2Br_4$	267	"
$Pt(NH_3)_2I_4$	183	"
Cis compounds, square planar		
$Pd(NH_3)_2Cl_2$	327, 306	243, 244
$Pt(NH_3)_2Cl_2$	324, 317	245
$Pt(ND_3)_2Cl_2$	327, 319	"
$Pd(NH_3)_2Br_2$	258	244
$Pd(NH_3)_2I_2$	195	246
$Pt(NH_3)_2Cl_4$	353, 344, 330, 206	228
$Pt(NH_3)_2Br_4$	255, 238, 219, 183	"
$Pt(NH_3)_2I_4$	215, 189, 180, 159	"
$M(NH_3)_2X_2$*(Octahedral)*		
$Fe(NH_3)_2Cl_2$	~235	242
α-$Co(NH_3)_2Cl_2$	233	"
β-$Co(NH_3)_2Cl_2$	234	"
$Ni(NH_3)_2Cl_2$	240	"
α-$Cu(NH_3)_2Cl_2$	267	"
β-$Cu(NH_3)_2Cl_2$	251	"
$Cd(NH_3)_2Cl_2$	215	"
$Hg(NH_3)_2Cl_2$	~217	"
α-$Cu(NH_3)_2Br_2$	218	"
β-$Cu(NH_3)_2Br_2$	~222, 205	"
Tetrahedral		
$Zn(NH_3)_2Cl_2$	285	242
$Zn(ND_3)_2Cl_2$	285	"
$Zn(NH_3)_2Br_2$	213	"

bridged vibration is smaller than that found for the bridging vibrations in a dimer, and the vibration appears at a lower frequency. Some contribution also comes from the higher coordination involved in the polymer (CN = 6, while the dimer has CN = 4). As a consequence, the ν_b(MX)(polymer)/ν_t(MX)(monomer) ratio is ~0.53. Table 6-61 tabulates these ratios for various complexes. The dimer compound Co_2Cl_4, which has CN = 4, is inter-

Table 6-60. Terminal and Bridging MX Assignments

Compound	Assignments, cm^{-1}	
	$\nu_t(MX)$	$\nu_b(MX)$
$(AlCl_3)_2$(gas)	$506(\nu_1)$	$340(\nu_2)$
	$625(\nu_8)$	$438(\nu_6)$
	$606(\nu_{11})$	$420(\nu_{13})$
	$484(\nu_{16})$	$(301)(\nu_{17})$
$(\varphi_4As)_2Pt_2Cl_6$	$\sim346(\nu_{13})$	$315(\nu_{14})$
	$\sim350(\nu_{16})$	$302(\nu_{17})$
	$237(\nu_{13})$	$196(\nu_{14})$
$(Et_4N)_2Pt_2Br_6$	$243(\nu_{16})$	$211(\nu_{17})$
$(Et_4N)_2PtI_6$	$179(\nu_{13})$	$157(\nu_{14})$
	$194(\nu_{16})$	$144(\nu_{17})$
$(\varphi_4As)_2Pd_2Cl_6$	$\sim353(\nu_{13})$	305
	$\sim353(\nu_{16})$	264
$(n\text{-cetylMe}_3N)_2Pd_2Cl_6$	336	305
	339	263
$(n\text{-Bu}_4N)_2Pd_2Cl_6$	340	302
	349	264
$(Et_4N)_2Pd_2Br_6$	266	192
	262	177
$(Et_4N)_2Pd_2I_6$	222	144
	220	132
$Pt_2Cl_4L_2$[a]	347–368	316–331
		257–301
$Pt_2Br_4L_2$	249–260	209–213
		181–204
$Pt_2I_4L_2$	177–205	164–167
		149–150
$Pd_2Cl_4L_2$	339–336	294–308
		255–301
$Pd_2Br_4L_2$	268–292	189–194
		166–185
$PtCl_4(CO)_2$	368	301
$LiCuCl_3\cdot2H_2O$	295, 272	222, 180
NH_4CuCl_3	311, 280	230
$KCuCl_3$	301, 278	236, 193
$KCuBr_3$	237, 224	168, 122
$CoCl_2\cdot(py)_2$	347, 306	186, 174
$CoCl_2(4\text{-Clpy})_2$	318	185, 165
$CoCl_2(4\text{-Brpy})_2$	318	178, 155
$CoBr_2(4\text{-Brpy})_2$	265	143, 128

[a] L = PCl_3, $P(OEt)_3$, PR_3, TeR_2, pyridine, olefin, AsR_3, SR_2, SeR_2.

esting, as it contains both bridging and terminal CoCl bonds. The ratio ν_b(CoCl) (dimer)/ν_t(CoCl) (dimer) is about 0.74. However, if the ratio is calculated using the ν_t(CoCl)(monomer) value (CN = 2), the result is less

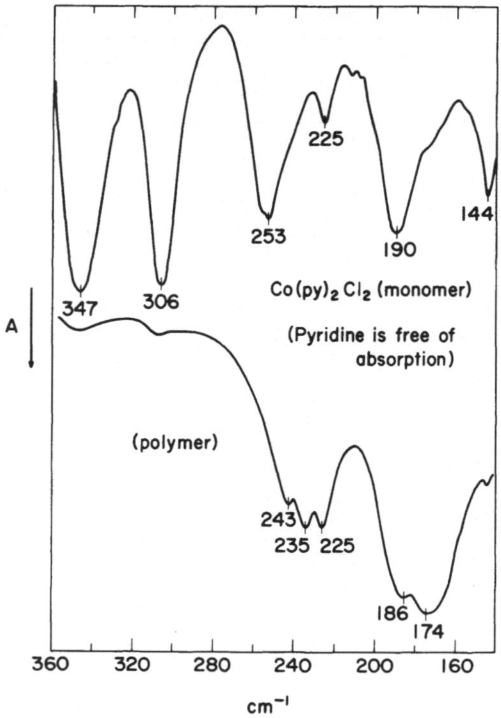

Fig. 6-13. Comparison of low-frequency spectra of $Co(py)_2Cl_2$ monomer and polymer. (Courtesy of American Chemical Society, Washington, D. C. From Postmus *et al.*[263])

than if the ν_t(CoCl)(dimer) value were used. These results are in agreement with those found for the $CoCl_2L_2$ complexes.

Certainly more research is necessary with stretching–bridging vibrations, and in particular, with bending bridging vibrations. It is anticipated that a concerted effort in this direction will be made.

6.12. SUMMARY

The value of obtaining the low-frequency spectra of inorganic and coordination compounds has perhaps been minimized by chemists. Although the tool is not a conclusive one like x-ray analysis, the results obtained may be very informative. Certainly the effort and time expenditure are minimal, and most laboratories are now equipped to go to at least 200 cm^{-1}. In the case of intractable solids which do not lend themselves to study by other

Table 6-61. Comparison of Bridged MX Stretching Vibration with the Terminal MX Stretching Vibration

Metal			$\nu_b(MX)/\nu_t(MX)^e$	$\nu_b(MX)/\nu_t(MX)^f$
Dimer				
		X = halogen		
Pt(II)	$R_2Pt_2X_6$[d]	Cl	0.88[a]	
		Br	0.85[a]	
		I	0.81[a]	
Pt(II)	$Pt_2X_4L_2$[d]	Cl	0.83[b]	
		Br	0.81[b]	
		I	0.84[b]	
Pd(II)	$R_2Pd_2X_6$[d]	Cl	0.83[a]	
		Br	0.70[a]	
		I	0.62[a]	
Pd(II)	$Pd_2X_4L_2$	Cl	0.80[b]	
		Br	0.70[b]	
Cu(II)	$MCuX_3$	Cl	0.74[a]	
		Br	0.63[a]	
Co(II)	Co_2X_4	Cl	0.74[b]	0.60
Zn(II)	Zn_2X_4	Cl	0.71[b]	
Al(III)	$(AlX_3)_2$	Cl	0.68[c]	
Polymer				
Co(II)	CoL_2X_2	Cl		0.55
		Br		0.50

[a] Two $\nu_t(MX)$ and two $\nu_b(MX)$ vibrations are assigned. Average of each of the two vibrations is used to obtain this column.
[b] One $\nu_t(MX)$ and two $\nu_b(MX)$ vibrations are assigned, and the average of the two $\nu_b(MX)$ is used.
[c] Five $\nu_t(MX)$ and five $\nu_b(MX)$ are assigned. Five $\nu_b(MX)/\nu_t(MX)$ values are obtained and these are averaged.
[d] For nature of R and L see Table 6-60.
[e] This ratio is taken using the $\nu_b(MX)$ and $\nu_t(MX)$ vibrations in the dimer.
[f] This ratio is taken using the $\nu_b(MX)$(polymer) and $\nu_t(MX)$ (monomer) vibrations.

means the far-infrared spectrum may provide the only means of obtaining some structural information concerning the molecule.

Clark has cited various relationships that result from a study of the far-infrared data of inorganic and coordination complexes. Table 6-1 has tabulated the factors involved. More information has become available since Clark wrote his article in 1965, and it may be worthwhile to discuss these relationships in terms of the additional data accumulated.

Stereochemistry and Coordination Number

With the increase in far-infrared activity, it has become more readily apparent that a change in the stereochemistry (e.g., $T_d \rightarrow O_h$) and coordination number (4→6), there is a decrease in the metal halide stretching vibration [both ν_1 (symmetric) and ν_3 (asymmetric)]. Table 6-62 tabulates various data indicating this trend. This trend persists in the bromides as well as the

Table 6-62. Effect of Stereochemistry and Coordination Number on ν_{MX} Vibrations

Compound	Frequency, cm^{-1}		Stereochemistry
	ν_1	$\nu_3(T_d)$ vs. $\nu_4(O_h)$	
$TiCl_4$	389	495	T_d
$TiCl_6^{2-}$	321–331	188–193	O_h
$ZrCl_4$	383		T_d
$ZrCl_6^{2-}$	323–333	146–156	O_h
$HfCl_4$		393	T_d
$HfCl_6^{2-}$	328–333	138–147	O_h
$SnCl_4$	368	403	T_d
$SnCl_6^{2-}$	311–318	161–177	O_h
$SnBr_4$	222	281	T_d
$SnBr_6^{2-}$	183–185	94–122	O_h
SnI_4	149	216	T_d
SnI_6^{2-}		48	O_h
GeF_4	738	800	T_d
GeF_6^{2-}	627	350	O_h
	ν_3		
$NiCl_2$	521		$D_{\infty h}$
$NiCl_4^{2-}$	289		T_d
$NiBr_2$	415		$D_{\infty h}$
$NiBr_4^{2-}$	224, 231		T_d
	ν_1		
PF_3	892		C_{3v}
PF_5	817		D_{3h}
PCl_3	507		C_{3v}
PCl_5	393		D_{3h}
$SbCl_3$	377		C_{3v}
$SbCl_5$	353		D_{3h}
$HgCl_2$	360		$D_{\infty h}$
$HgCl_3^-$	282–294		D_{3h}
$HgBr_2$	225		$D_{\infty h}$
$HgBr_3^-$	179		D_{3h}
HgI_2	156		$D_{\infty h}$
HgI_3	125		D_{3h}
	ν_{MX}		
$(\varphi_3PO)_2MnCl_2$	327, 317, 292		T_d
$(o\text{-tol})_2MnCl_2$	233		O_h
$(M_3P)_2NiCl_2$ (*trans*)	403		D_{4h}
$(\alpha\text{-pic})_2NiCl_2$	328, 294		T_d
$(o\text{-tol})_2NiCl_2$	236		O_h

Table 6-63. Number of Infrared Active Vibrations Expected for *Cis* and *Trans* Isomers of Complexes of the Type L_2MX_2, L_3MX_3, and L_4MX_2

Complex	Isomer	Symmetry	ν_{MX}
L_2MX_2	*trans*	D_{3h}	1
L_2MX_2	*cis*	C_{3v}	2
L_3MX_3	*trans*	C_{2v}	3
L_3MX_3	*cis*	C_{3v}	2
L_4MX_2	*trans*	C_{4h}	1
L_4MX_2	*cis*	C_{2v}	2

iodides. Further, it persists in molecules possessing a stereochemistry different than T_d and O_h and in molecules containing an organic ligand L. This is a very valuable trend, for by locating and properly assigning a metal halide stretching vibration valuable information on the nature of the structure of the molecule may be obtained.

The far-infrared region may also be helpful in elucidating and distinguishing between *cis* and *trans* isomers of the square planar type (e.g., L_2MX_2 where M is Pt or Pd). The *trans* isomer should have only one infrared active ν_{MX} vibration, while the *cis* isomer would have two ν_{MX} vibrations. Similar identification of *cis* and *trans* isomers may be made for complexes of the type L_3MX_3 and L_4MX_2. Table 6-63 summarizes the infrared selection rules for various *cis* and *trans* isomers.

Table 6-64. Effect of Oxidation Number on ν_{MX} Vibration

Compound	Stereo-chemistry	CN	Oxidation number	ν_{MX}	Frequency, cm^{-1}			
					M=Ru	M=Re	M=Os	M=Ir
MF_6^{2-}	O_h	6	4	ν_3	588	541	548	568
MF_6^-	O_h	6	5	ν_3	640	627	616	667
MF_6	O_h	6	6	ν_3	735	715	720	719
$FeCl_4^{2-}$	T_d	4	2	ν_3	286			
$FeCl_4^-$	T_d	4	3	ν_3	377			
$FeBr_4^{2-}$	T_d	4	2	ν_3	219			
$FeBr_4^-$	T_d	4	3	ν_3	289			
$GeCl_4$	T_d	4	4	ν_3	451			
$GaCl_4^-$	T_d	4	3	ν_3	373			
$ZnCl_4^{2-}$	T_d	4	2	ν_3	298			
$GeBr_4$	T_d	4	4	ν_3	328			
$GaBr_4^-$	T_d	4	3	ν_3	278			
$ZnBr_4^{2-}$	T_d	4	2	ν_3	210			
GeI_4	T_d	4	4	ν_3	264			
GaI_4^-	T_d	4	3	ν_3	222			
ZnI_4^{2-}	T_d	4	2	ν_3	170			

Table 6-65. Effect of Ligand Field Stabilization Energy on the ν_{MX} Vibration of Transition Metal Complexes

Complex	ν_3, cm^{-1}
T_d symmetry	
$MnCl_4^{2-}$	284
$FeCl_4^{2-}$	286
$CoCl_4^{2-}$	297
$NiCl_4^{2-}$	289
$ZnCl_4^{2-}$	277
$MnBr_4^{2-}$	221
$FeBr_4^{2-}$	219
$CoBr_4^{2-}$	231
$NiBr_4^{2-}$	224
$ZnBr_4^{2-}$	210
MnI_4^{2-}	185
FeI_4^{2-}	186
CoI_4^{2-}	192
NiI_4^{2-}	189
ZnI_4^{2-}	170
O_h symmetry	
$ReCl_6^{2-}$	300–332
$OsCl_6^{2-}$	305–340
$IrCl_6^{2-}$	316–333
$PtCl_6^{2-}$	330–345
$ReBr_6^{2-}$	217
$PtBr_6^{2-}$	240–244

Note: CuX_4^{2-} distorted T_d structure noncomparable.

Table 6-66. Effect of Counter-Ion on ν_{MX} Vibration

Compound	Frequency, cm^{-1}
	ν_{CuX} (D_{2d} symmetry)
Cs_2CuCl_4	292, 257
$(Me_4N)_2CuCl_4$	281, 237
$(Et_4N)_2CuCl_4$	267, 248
Cs_2CuBr_4	224, 189
$(Et_4N)_2CuBr_4$	216, 174
	ν_3 (O_h symmetry)[32]
K_2MoCl_6	340
Rb_2MoCl_6	334
Cs_2MoCl_6	325
	ν_3 (T_d symmetry)
$(NH_4)_3ZnCl_5$	298
Cs_2ZnCl_4	292
$(\varphi_4As)_2ZnCl_4$	272

Oxidation Number

The effect of oxidation number on the ν_{MX} stretching vibration is illustrated in Table 6-64. It is rather definite that the ν_{MX} increases in frequency with an increase in the oxidation number. This is an expected consequence. The oxidation effect overrides the mass effect, and with an increase in oxidation number the ν_{MX} frequency shows a rather pronounced increase. This is illustrated in Table 6-64 for the GeX_4, GaX_4^-, and ZnX_4^{2-} series. Further, the effect appears to hold for ν_{MX} regardless of the nature of X. The fact that insufficient data are available at present makes comparisons for other than T_d and O_h structures impossible.

Table 6-67. Effect of Mass on ν_{MX} Vibration

Compound	Frequency, cm^{-1}			Symmetry
	ν_1		ν_3	
TiF_6^{2-}			560	O_h
$TiCl_6^{2-}$	321–331		302–330	
$TiBr_6^{2-}$	192		268	
		ν_1		
PF_5		817		D_{3h}
PCl_5		393		
		ν_3		
ZrF_4		668		T_d
$ZrCl_4$		421–423		
$CoCl_4^{2-}$		297		
$CoBr_4^{2-}$		231		
CoI_4^{2-}		192, 197		
		ν_1		
$PtCl_4^-$		333		D_{4h}
$PtBr_4^-$		205		
PtI_4^-		142		
		ν_1		
PF_3		892		D_{3h}
PCl_3		507		
PBr_3		392		
PI_3		303		
		ν_3		
$NiCl_2$		521		$D_{\infty h}$
$NiBr_2$		415		
$CdCl_2$		409		
$CdBr_2$		315		
CdI_2		265		

Effect of Ligand Field Stabilization Energy on ν_{MX} in Transition Metal Complexes

Table 6-65 presents data for the MCl_4^{2-}, MBr_4^{2-}, and MI_4^{2-} complexes. It may be observed that a maximum in the ν_{MX} vibration occurs for Co, which has a maximum in ligand field stabilization energy. The relationship also holds for the MX_6^{2-} complexes with O_h symmetry. In the series $ReCl_6^{2-}$, $OsCl_6^{2-}$, $IrCl_6^{2-}$, $PtCl_6^{2-}$ the maximum in frequency position for the ν_3 metal-chloride stretching is shown by platinum. The relationship may also hold for the bromide complexes.

Effect of Counter-Ion on ν_{MX} Vibration

The examples that can be cited to illustrate this effect are not as numerous. However, it does appear that the larger the counter-ion size, the lower the frequency of the ν_{MX} vibration. Table 6-66 tabulates some data to illustrate this effect.

Effect of Mass on ν_{MX} Vibration

The effect of mass on the position of the ν_{MX} vibration may be more difficult to illustrate in the presence of the other factors involved. However, many examples may be cited in which the mass effect shifts the ν_{MX} vibration to a lower frequency. Table 6-67 shows a summary of far-infrared data and its relationship to mass for different symmetry groups.

Differentiation Between Bridged and Nonbridged ν_{MX} Vibrations

Section 6.11 discussed the difference in far-infrared spectra between complexes containing bridged and/or nonbridged halides. Knowing whether the complex contains bridging or nonbridging halides helps in determining the stereochemistry of the molecule. At present, in most of the complexes studied involving bridging halides, the frequency position of the ν_bMX is lower than that found for the nonbridging halides, ν_tMX.

BIBLIOGRAPHY

1. R. J. H. Clark, *Spectrochim. Acta* **21**, 955 (1965).
2. R. C. Lord, M. A. Lynch, W. S. C. Schumb, and R. J. Slowinski, *J. Am. Chem. Soc.* **72**, 522 (1950).
3. H. H. Claassen and H. Selig, *J. Chem. Phys.* **43**, 103 (1965).

3a. H. H. Claassen and H. Selig, *J. Chem. Phys.* **49**, 1803 (1969).

4. O. Glemser, H. W. Roesky, K. H. Hellberg, and H. U. Werther, *Chem. Ber.* **99**, 2652 (1966).
5. R. D. Peacock and D. W. A. Sharp, *J. Chem. Soc.*, 2762 (1959).
6. K. W. Bagnall, D. Brown, and J. G. H. du Preez, *J. Chem. Soc.*, 2603 (1964).
7. K. W. Bagnall and D. Brown, *J. Chem. Soc.*, 3021 (1964).

8. D. Brown and J. F. Easey, *J. Chem. Soc.* (A), 254 (1966).
9. R. J. H. Clark, R. H U. Negrotti, and R. J. Nyholm, *Chem. Commun.*, 486 (1966).
9a. E. L. Gasner and B. Frlec, *J. Chem. Phys.* **49**, 5135 (1968).
10. B. Weinstock and G. L. Goodman, *Advan. Chem. Phys.* **11**, 695 (1966).
11. O. Glemser, H. Roesky, and K. H. Hellberg, *Angew. Chem.* **75**, 346 (1963).
12. H. H. Claassen, H. Selig, and J. G. Malm, *J. Chem. Phys.* **36**, 2888 (1962).
13. D. F. Evans and P. A. W. Dean, *J. Chem. Soc.* (A), 698 (1967).
14. J. E. Griffiths and D. E. Irish, *Inorg. Chem.* **3**, 1134 (1964).
15. M. J. Reisfeld, *J. Mol. Spectry.* **29**, 120 (1969).
16. L. A. Woodward and M. J. Ware, *Spectrochim. Acta* **19**, 775 (1963); **20**, 711 (1964).
17. L. A. Woodward and L. E. Anderson, *J. Inorg. Nucl. Chem.* **3**, 326 (1956).
18. J. Weidlein and K. Dehnicke, *Z. Anorg. Allgem. Chem.* **337**, 113 (1965).
19. O. L. Keller, *Inorg. Chem.* **2**, 783 (1963).
20. O. L. Keller, and A. Chetham-Strode, *Inorg. Chem.* **5**, 367 (1966).
21. M. J. Reisfeld and G. A. Crosby, *Inorg. Chem.* **4**, 65 (1965).
22. B. J. Brisdon, G. A. Ozin, and K. A. Walton, *J. Chem. Soc.* (A), 342 (1969).
23. R. J. H. Clark, L. Maresca, and R. J. Puddephatt, *Inorg. Chem.* **7**, 1603 (1968).
24. D. M. Adams and D. C. Newton, *J. Chem. Soc.* (A), 2262 (1968).
25. I. R. Beattie, G. P. McQuillan, L. Rule, and M. Webster, *J. Chem. Soc.*, 1514 (1963).
26. D. M. Adams, J. Chatt, J. M. Davidson, and J. Garratt, *J. Chem. Soc.*, 2189 (1963).
27. N. N. Greenwood and B. P. Straughan, *J. Chem. Soc.* (A), 962 (1966).
28. J. Hiraishi, I. Nakagawa and T. Shimanouchi, *Spectrochim. Acta* **20**, 819 (1964).
29. M. Debeau and J. P. Mathieu, *Compt. Rend.* **260**, 5229 (1965).
30. J. A. Creighton and J. H. S. Green, *J. Chem. Soc.* (A), 808 (1968).
31. J. A. Creighton and L. A. Woodward, *Trans. Faraday Soc.* **58**, 1077 (1962).
32. D. M. Adams, H. A Gebbie, and R. D. Peacock, *Nature* **199**, 278 (1963).
33. D. M. Adams and H. A. Gebbie, *Spectrochim. Acta* **19**, 925 (1963).
34. L. A. Woodward and J. A. Creighton, *Spectrochim. Acta* **17**, 594 (1961).
35. D. M. Adams, *Proc Chem. Soc.*, 335 (1961).
36. M. Le Postelloc, J. D. Mathieu, and H. Poulet, *J. Chim. Phys.* **60**, 1319 (1963).
37. D. W. James and M. J. Nolan, *Inorg. Nucl. Chem. Letters* **4**, 97 (1968).
38. D. M. Adams and D. M. Morris, *J. Chem. Soc.* (A), 694 (1968).
39. K. W. Bagnall and D. Brown, *J. Chem. Soc.*, 3021 (1964).
40. M. Webster, *Chem. Rev.* **66**, 87 (1966).
41. K. W. Bagnall, D. Brown, and J. G. H. du Preez, *J. Chem. Soc.*, 2603 (1964)
42. G. L. Carlson, *Spectrochim. Acta* **19**, 1291 (1963).
43. L. A. Woodward and M. J. Taylor, *J. Chem. Soc.*, 4473 (1960).
44. H. H. Claassen and H. Selig, *J. Chem. Phys.* **44**, 4039 (1966).
45. J. E. Griffiths, R. R. Carter, and R. R. Holmes, *J. Chem. Phys.* **41**, 863 (1965); **42**, 2632 (1966).
46. J. K. Wilmhurst, *J. Mol. Spectry.* **5**, 343 (1960).
47. K. A. Jensen, *Z. Anorg. Allgem. Chem.* **250**, 257 (1943).
48. K. Dehnicke and J. Weidlein, *Z. Anorg. Allgem. Chem.* **323**, 267 (1963).
49. S. Blanchard, *J. Chim. Phys.* **62**, 919 (1965).
50. I. R. Beattie, T. Gilson, K. Livingston, V. Fawcett, and G. A. Ozin, *J. Chem. Soc.* (A), 712 (1967).
51. G. M. Begun, W. H Fletcher, and D. F. Smith, *J. Chem. Phys.* **42**, 2236 (1965).
52. C. V. Stephenson and E. A. Jones, *J. Chem. Phys.* **20**, 1830 (1952).

53. T. G. Burke and E. A. Jones, *J. Chem. Phys.* **19,** 1611 (1951).
54. L. Stein, *J. Am. Chem. Soc.* **81,** 1273 (1959).
55. H. A. Szymanski, R. Yelin, and L. Marabella, *J. Chem. Phys.* **47,** 1877 (1967).
56. N. N. Greenwood, A. C. Sarma, and B. P. Straughan, *J. Chem. Soc.* (A), 1446 (1966).
57. A. T. Caunt, L. N. Short, and L. A. Woodward, *Trans. Faraday Soc.* **48,** 873 (1952).
58. P. J. H. Woltz and A. H. Nielsen, *J. Chem. Phys.* **20,** 307 (1952).
59. A. Büchler, J. B. Berkowitz-Mattuck, and D. H. Dugre, *J. Chem. Phys.* **34,** 2202 (1961).
60. M. F. A. Dove, J. A. Creighton, and L. A. Woodward, *Spectrochim. Acta* **18,** 267 (1962).
61. M. L. Delwaulle and F. Francois, *Bull. Soc. Chim. France* **13,** 205 (1946).
61a. M. L. Delwaulle and F. Francois, *J. Phys. Radium* **7,** 15 (1946).
62. Landolt-Börnstein, *Physikalische-Chemische Tabellen, 2 Teil* (1951).
63. L. P. Lindeman and M. K. Wilson, *Spectrochim. Acta* **9,** 47 (1957).
64. F. K. Butcher, W. Gerrard, E. F. Mooney, R. G. Rees, H. A. Willis, A. Anderson, and H. A. Gebbie, *J. Organometallic Chem.* **1,** 4311 (1964).
65. J. T. Neu and W. G. Gwinn, *J. Am. Chem. Soc.* **70,** 3463 (1948).
66. E. L. Grubb and R. L. Belford, *J. Chem. Phys.* **39,** 244 (1963).
67. K. Dehnicke and J. Weidlein, *Chem. Ber.* **98,** 1087 (1965).
68. H. Gerding and H. Houtgraaf, *Rec. Trav. Chim.* **72,** 21 (1953).
69. L. A. Woodward and A. A. Nord, *J. Chem. Soc.*, 3721 (1956).
70. G. M. Bancroft, A. G. Maddock, W. K. Ong, and R. H. Prince, *J. Chem. Soc.* (A), 723 (1966).
71. M. L. Delwaulle, *Compt. Rend.* **240,** 2132 (1955).
72. A. Sabatini and L. Sacconi, *J. Am. Chem. Soc.* **86,** 17 (1964).
73. G. T. Janz and D. W. James, *J. Chem. Phys.* **38,** 902 (1963).
74. F. A. Miller and G. L. Carlson, *Spectrochim. Acta* **16,** 6 (1960).
75. M. L. Delwaulle and F. Francois, *Compt. Rend.* **220,** 173 (1945).
76. G. Herzberg, *Molecular Spectra and Molecular Structure, Vol. 2*, Van Nostrand, New York (1960) p. 167.
77. B. Maszynska, *Compt. Rend.* **256,** 1261 (1963).
78. L. A. Woodward and P. T. Bill, *J. Chem. Soc.*, 1699 (1955).
78a. J. J. Avery, C. D. Burbridge, and D. L. Goodgame, *Spectrochim. Acta* **24A,** 1721 (1968).
78b. C. O. Quicksall and T. G. Spiro, *Inorg. Chem.* **5,** 2232 (1966).
79. M. L. Delwaulle, *Compt. Rend.* **238,** 2522 (1954).
80. G. E. Coates and D. Ridley, *J. Chem. Soc.*, 166 (1964).
81. R. A. Rolfe, D. E. Sheppard, and L. A. Woodward, *Trans. Faraday Soc.* **50,** 1275 (1954).
82. H. Stammreich and R. Forneris, *J. Chem. Phys.* **25,** 1278 (1956).
83. L. A. Woodward and G. H. Singer, *J. Chem. Soc.*, 716 (1958).
84. D. A. Long and J. Y. H. Chan, *Trans. Faraday Soc.* **58,** 2325 (1962).
85. K. Nakamoto, *Infrared Spectra of Inorganic and Coordination Compounds*, J. Wiley and Sons, New York (1963).
86. J. R. Ferraro and J. S. Ziomek, *Introductory Group Theory and Its Application to Molecular Structure*, Plenum Press, New York (1969).
87. D. W. James and M. J. Nolan, in *Progress in Irorganic Chemistry, Vol. 9* (F. A. Cotton, Ed.), Interscience Publishers, New York (1968) p. 235.
88. J. H. Fertel and C. H. Perry, *J. Phys. Chem. Solids* **26,** 1773 (1965).

89. D. M. Adams, *Metal-Ligand and Related Vibrations*, St. Martin's Press, New York (1968).
90. A. Sabatini, L. Sacconi, and V. Schettino, *Inorg. Chem.* **3**, 1775 (1964).
90a. P. Murray-Rust, P. Day, and C. K. Prout, *Chem. Commun.*, 277 (1966).
90b. W. E. Hatfield and T. S. Piper, *Inorg. Chem.* **3**, 841 (1964).
91. P. J. Hendra, *Nature* **212**, 179 (1966).
92. C. H. Perry, D. P. Athans, E. F. Young, J. R. Durig, and B. R. Mitchell, *Spectrochim. Acta* **23A**, 1137 (1967).
93. P. J. Hendra, *J. Chem. Soc.* (A), 1298 (1967).
94. J. Hiraishi and T. Shimanouchi, *Spectrochim. Acta* **22**, 1483 (1966).
95. P. J. Hendra, *Spectrochim. Acta* **23A**, 1635 (1967).
96. H. Stammreich and R. Forneris, *Spectrochim. Acta* **16**, 363 (1960).
97. D. M. Adams and H. A. Gebbie, *Spectrochim. Acta* **19**, 925 (1963);
97a. D. M. Adams and D. M. Morris, *Nature* **2**, 283 (1965).
98. P. J. Hendra and P. J. D. Park, *Spectrochim. Acta* **23A**, 1635 (1967).
99. H. Poulet, P. Delorme, and J. P. Mathieu, *Spectrochim. Acta* **20**, 1855 (1964).
100. J. R. Ferraro, *J. Chem. Phys.* **53**, 117 (1970).
101. J. D. S. Goulden, A. Maccoll, and D. J. Miller, *J. Chem. Soc.*, 1635 (1950).
102. G. E. Coates and C. Parkin, *J. Chem. Soc.*, 421 (1963).
103. D. W. James and M. J. Nolan, in *Progress in Inorganic Chemistry, Vol. 9* (F. A. Cotton, Ed.), Interscience Publishers, New York (1968) p. 237.
104. I. W. Levin and S. Abramowitz, *J. Chem. Phys.* **44**, 2562 (1966).
105. E. L. Pace and L. Pierce, *J. Chem. Phys.* **23**, 1248 (1955).
106. M. K. Wilson and S. R. Polo, *J. Chem. Phys.* **20**, 1716 (1952).
107. H. J. Gutowsky and A. D. Liehr, *J. Chem. Phys.* **20**, 1652 (1952).
108. P. W. Davis and R. A. Oetjen, *J. Mol. Spectry.* **2**, 253 (1958).
109. F. A. Miller and W. K. Baer, *Spectrochim. Acta* **17**, 112 (1961).
110. T. R. Manley and D. A. William, *Spectrochim. Acta* **21**, 1773 (1965).
111. E. F. Gross and I. M. Ginsburg, *Opt. Spectry. USSR* **1**, 710 (1956).
112. J. C. Evans, *J. Mol. Spectry.* **4**, 435 (1960).
113. H. Stammreich, R. Forneris, and Y. Tovares, *J. Chem. Phys.* **25**, 580 (1956).
114. M. L. Delwaulle, *Compt. Rend.* **228**, 1585 (1949).
115. J. D. Donaldson, J. F. Knifton, J. O'Donoghue, and S. D. Ross, *Spectrochim. Acta* **22**, 1173 (1966).
116. W. Sawodny and K. Dehnicke, *Z. Anorg. Allgem. Chem.* **349**, 169 (1967).
117. N. N. Greenwood, B. P. Straughan, and A. E. Wilson, *J. Chem. Soc.* (A), 1479 (1966).
118. N. N. Greenwood, T. S. Srivastava, and B. P. Straughan, *J. Chem. Soc.* (A), 699 (1966).
119. Z. Kecki, *Spectrochim. Acta* **18**, 1165 (1962).
119a. G. J. Janz and D. W. James, *J. Chem. Phys.* **38**, 902 (1957).
120. E. L. Short, D. N. Waters, and D. F. C. Morris, *J. Inorg. Nucl. Chem.* **26**, 902 (1964).
121. L. C. McCary, R. C. Paule, and J. L. Margrave, *J. Phys. Chem.* **67**, 1086 (1963).
122. W. Klemperer, *J. Chem. Phys.* **24**, 353 (1956).
123. H. H. Claassen, B. Weinstock, and J. G. Malm, *J. Chem. Phys.* **28**, 285 (1958).
123a. G. N. Kustova and L. R. Batsanova, *Zh. Prikl. Spektroskopii Akad. Nauk Belorusski* SSR **4**, 75 (1966).
124. G. C. Hayward and P. J. Hendra, *J. Chem. Soc.* (A), 643 (1967).
125. A. Büchler and W. Klemperer, *J. Chem. Phys.* **29**, 121 (1958).

125a. A. Snelson, *J. Phys. Chem.* **72,** 250 (1968).
126. S. P. Randall, F. T. Greene, and J. L. Margrave, *J. Phys. Chem.* **63,** 758 (1959).
127. G. E. Leroi, T. C. Jones, J. T. Hougen, and W. Klemperer, *J. Chem. Phys.* **36,** 2879 (1962).
128. A. Büchler, W. Klemperer, and A. G. Emslie, *J. Chem. Phys.* **36,** 2499 (1962).
129. W. Klemperer, *J. Electrochem. Soc.* **110,** 1023 (1963).
130. W. Klemperer and L. Lindeman, *J. Chem. Phys.* **25,** 397 (1956).
131. K. R. Thompson and K. D. Carlson, *J. Chem. Phys.* **49,** 4379 (1968).
132. W. Klemperer, *J. Chem. Phys.* **25,** 1066 (1956).
133. G. J. Janz and D. W. James, *J. Chem. Phys.* **38,** 902 (1963).
134. J. A. Creighton and E. R. Lippincott, *J. Chem. Soc.*, 5134 (1963).
135. G. E. Coates and C. Parkin, *J. Chem. Soc.*, 421 (1963).
136. W. B. Person, G. R. Anderson, J. N. Fordemwalt, H. Stammreich, and R. Forneris, *J. Chem. Phys.* **35,** 908 (1941).
137. G. C. Hayward and P. J. Hendra, *Spectrochim. Acta* **23A,** 2309 (1969).
138. S. G. W. Ginn and J. L. Wood, *Trans. Faraday Soc.* **62,** 777 (1966).
139. J. C. Evans and G. Y. S. Lo, *J. Chem. Phys.* **44,** 3638 (1966).
140. K. O. Criste and J. P. Guertin, *Inorg. Chem.* **4,** 1785 (1965).
141. H. H. Claassen, G. L. Goodman, J. G. Malm, and F. Schreiner, *J. Chem. Phys.* **42,** 1229 (1965).
142. A. Snelson, *J. Phys. Chem.* **70,** 3208 (1966).
143. D. M. Adams, M. Goldstein, and E. F. Mooney, *Trans. Faraday Soc.* **59,** 2228 (1963).
144. Y. Mikawa, R. J. Jakobsen, and J. W. Brasch, *J. Chem. Phys.* **45,** 4528 (1968).
145. A. J. Melveger, R. K. Khanna, B. R. Guscott, and E. R. Lippincott, *Inorg. Chem.* **7,** 1630 (1968).
146. H. Braune and G. Englebricht, *Z. Physik. Chem.* **B19,** 303 (1932).
147. M. Wherli, *Helv. Phys. Acta* **11,** 339 (1938).
148. J. H. R. Clarke, C. Solomons, and K. Balasubrahmangam, *Rev. Sci. Instr.* **38,** 655 (1967).
149. G. Randi and F. Gesmundo, *Lincei-Rend. Sc. Fis. Mat. Nat.* **41,** 200 (1966).
150. H. Poulet and J. P. Mathieu, *J. Chim. Phys.* **60,** 442 (1963).
151. M. Goldstein, *Spectrochim. Acta* **22,** 1389 (1966).
151a. D. M. Adams, M. Goldstein, and E. F. Mooney, *Trans. Faraday Soc.* **59,** 2228 (1968).
151b. J. R. Durig, K. K. Lau, G. Nagarajon, M. Walker, and J. Bragin, *J. Chem. Phys.* **50,** 2130 (1969).
151c. R. P. J. Cooney, J. R. Hall, and M. A. Hooper, *Australian J. Chem.* **21,** 2145 (1968).
152. G. J. Janz and D. W. James, *J. Chem. Phys.* **38,** 902 (1963).
153. M. L. Delwaulle, *Compt. Rend.* **206,** 1965 (1938).
154. F. Francois, *Compt. Rend.* **207,** 425 (1938).
155. G. Allen and E. Warhurst, *Trans. Faraday Soc.* **54,** 1786 (1958).
156. W. Klemperer, *J. Electrochem. Soc.* **110,** 1023 (1963).
157. S. A. Rice and W. Klemperer, *J. Chem. Phys.* **27,** 573 (1957).
158. D. E. Mann, G. V. Calder, K. S. Seshadri, D. White, and M. J. Linevsky, *J. Chem. Phys.* **46,** 1138 (1967).
159. W. Klemperer, W. G. Norris, A. Büchler, and A. G. Enslie, *J. Chem. Phys.* **33,** 1536 (1960).
159a. A. Büchler and W. Klemperer, *J. Chem. Phys.* **29,** 121 (1958).

160. A. Snelson, *J. Chem. Phys.* **46,** 3652 (1967).
161. H. Stammreich, R. Forneris, and Y. Tavares, *Spectrochim. Acta* **17,** 1173 (1961).
162. E. A. Jones, T. F. Parkinson, and G. T. Burke, *J. Chem. Phys.* **18,** 235 (1950).
163. D. R. Lide, D. E. Mann and J. J. Comeford, *Spectrochim. Acta* **21,** 497 (1965).
164. W. P. Griffith, *J. Chem. Soc.*, 5248 (1964).
165. Y. Y. Kharitonov and Y. A. Busalev, *Izvest. Akad. Nauk SSSR, Otd. Khim. Nauk*, 808 (1964).
166. L. Kolditz and H. Preiss, Z. *Anorg. Allgem. Chem.* **325,** 263 (1963).
167. K. Martin and E. Steger, *Z. Anorg. Allgem. Chem.* **345,** 306 (1966).
168. T. Birchall and R. J. Gillespie, *Spectrochim. Acta* **22,** 681 (1966).
169. K. Dehnicke and J. Weidlein, *Z. Anorg. Allgem. Chem.* **342,** 225 (1966).
170. F. A. Miller and L. K. Cousins, *J. Chem. Phys.* **26,** 329 (1957).
171. K. Dehnicke, *Chem. Ber.* **98,** 290 (1965).
172. K. Dehnicke, *Chem. Ber.* **97,** 3354 (1964).
173. D. Brown, *J. Chem. Soc.*, 4944 (1964).
174. C. G. Barraclough and J. Stals, *Australian J. Chem.* **19,** 741 (1966).
174a. D. M. Adams and R. G. Churchill, *J. Chem. Soc.* (A), 2310 (1968).
175. K. Dehnicke, *Z. Anorg. Chem.* **308,** 72 (1961).
176. W. P. Griffith, *J. Chem. Soc.*, 245 (1964).
177. K. W. Bagnall and J. B. Laidler, *J. Chem. Soc.* (A), 516 (1966).
178. D. Brown and P. J. Jones, *J. Chem. Soc.*, 874 (1966).
179. A. Sabatini and J. Bertini, *Inorg. Chem.* **5,** 204 (1966).
180. H. Stammreich, K. Kawai, and Y. Tavares, *Spectrochim. Acta* **15,** 438 (1959).
181. F. A. Miller, G. L. Carlson, and W. B. White, *Spectrochim. Acta* **15,** 708 (1959).
182. W. E. Hobbs, *J. Chem. Phys.* **28,** 1220 (1958).
183. H. Stammreich, O. Sala, and K. Kawai, *Spectrochim. Acta* **17,** 226 (1961).
184. H. Stammreich, O. Sala, and D. Bassi, *Spectrochim. Acta* **19,** 593 (1963).
185. S. Blanchard, *J. Chim. Phys.* **61,** 747 (1964).
186. F. A. Miller and W. K. Baer, *Spectrochim. Acta* **26,** 329 (1957).
187. F. A. Miller and G. L. Carlson, *Spectrochim. Acta* **16,** 1148 (1960).
188. S. Kauffmann and K. Dehnicke, *Z. Anorg. Allgem. Chem.* **347,** 318 (1966).
189. D. F. Smith, G. M. Begun, and W. F. Fletcher, *Spectrochim. Acta* **20,** 1763 (1964).
190. H. H. Hyman, Ed., *Noble Gas Compounds*, University of Chicago Press, Chicago (1963).
191. P. A. Agron, G. M. Begun, H. A. Levy, A. A. Mason, C. F. Jones, and D. F. Smith, *Science* **139,** 842 (1963).
192. D. F. Smith, *J. Chem. Phys.* **38,** 270 (1963).
193. J. L. Weeks, C. L. Chernick, and M. S. Matheson, *J. Am. Chem. Soc.* **84,** 4612 (1962).
194. D. E. Milligan and D. Sears, *J. Am. Chem. Soc.* **85,** 823 (1963).
195. J. H. Holloway, *Noble Gas Chemistry*, Methuen (1969).
196. H. H. Claassen, C. L. Chernick, and J. G. Malm, *J. Am. Chem. Soc.* **85,** 1927 (1963).
197. E. L. Gasner and H. H. Claassen, *Inorg. Chem.* **6,** 1937 (1967).
198. D. F. Smith, *Science* **140,** 899 (1963).
199. H. H. Claassen, E. L. Gasner, and H. Kim, *J. Chem. Phys.* **49,** 253 (1968).
200. H. Selig, H. H. Claassen, C. L. Chernick, J. G. Malm, and J. L. Huston, *Science* **143,** 1322 (1964).
201. H. H. Claassen and G. Knapp, *J. Am. Chem. Soc.* **86,** 2341 (1964).
202. J. H. S. Green, *Spectrochim. Acta* **24A,** 863 (1968).

203. G. E. Coates and C. Parkins, *J. Chem. Soc.*, 421 (1963).
204. G. E. Coates and D. Ridley, *J. Chem. Soc.*, 166 (1964).
204a. J. R. Ferraro, W. Wozniak, and G. Roch, *Ric. Sci.* **38,** 433 (1968).
204b. J. R. Ferraro, J. Zipper, and W. Wozniak, *Appl. Spectry.* **23,** 160 (1969).
205. L. Cattalini, R. J. H. Clark, A. Orio, and C. K. Poon, *Inorg. Chim. Acta* **2,** 62 (1968).
206. A. Balls, N. N. Greenwood, and B. P. Straughan, *J. Chem. Soc.* (A), 753 (1968).
207. A. J. Carty, *Can. J. Chem.* **45,** 3187 (1967).
208. J. R. Durig, C. W. Sink, and S. F. Bush, *J. Chem. Phys.* **45,** 66 (1966).
209. A. J. Carty, *Coord. Chem. Rev.* **4,** 29 (1969).
210. J. Bradbury, K. P. Forest, R. H. Nuttall, and D. W. A. Sharp, *Spectrochim. Acta* **23A,** 2701 (1967).
211. D. H. Brown, K. P. Forest, R. H. Nuttall, and D. W. A. Sharp, *J. Chem. Soc.* (A), 2146 (1968).
212. R. J. Goodfellow, P. L. Goggins, and D. A. Duddell, *J. Chem. Soc.* (A), 504 (1968).
213. A. D. Allen and T. Theophanides, *Can. J. Chem.* **42,** 1551 (1964).
214. J. R. Allan, D. H. Brown, R. H. Nuttall, and D. W. A. Sharp, *J. Inorg. Nucl. Chem.* **27,** 1305 (1965).
215. R. J. H. Clark and C. S. Williams, *Inorg. Chem.* **4,** 350 (1965).
216. R. J. H. Clark and C. S. Williams, *Chem. Ind. London*, 1317 (1964).
217. N. S. Gill and H. J. Kingdon, *Aust. J. Chem.* **19,** 2197 (1966).
218. I. S. Ahuja, D. H. Brown, R. H. Nuttall, and D. W. A. Sharp, *J. Inorg. Nucl. Chem.* **27,** 1625 (1965).
219. S. H. Hunter, V. M. Langford, G. A. Rodley, and C. J. Wilkins, *J. Chem. Soc.* (A), 305 (1968).
220. W. J. Eilbeck, F. Holmes, C. E. Taylor, and A. E. Underhill, *J. Chem. Soc.* (A), 128 (1968).
221. J. R. Allan, D. H. Brown, R. H. Nuttall, and D. W. A. Sharp, *J. Chem. Soc.* (A), 1031 (1966).
222. J. Lewis, R. J. Nyholm, and G. A. Rodley, *J. Chem. Soc.*, 1483 (1965).
223. J. D. Dunitz, *Acta Cryst.* **10,** 307 (1957).
224. D. M. Adams, J. Chatt, J. Gerratt, and A. D. Westland, *J. Chem. Soc.*, 734 (1964).
224a. M. S. Taylor, A. L. Odell, and H. A. Raethel, *Spectrochim. Acta* **24A,** 1855 (1968).
225. R. A. Walton, *Spectrochim. Acta* **21,** 1795 (1965).
226. J. R. Durig, B. R. Mitchell, D. W. Sink, J. N. Willis, and A. J. Wilson, *Spectrochim. Acta* **23A,** 1121 (1967).
227. R. A. Walton, *J. Chem. Soc.* (A), 61 (1969).
227a. P. J. D. Park and P. J. Hendra, *Spectrochim. Acta* **25A,** 227 (1969).
228. D. M. Adams and P. J. Chandler, *J. Chem. Soc.* (A), 1009 (1967).
229. M. A. Bennett and R. J. H. Clark, *J. Chem. Soc.*, 5560 (1964).
230. R. J. H. Clark and B. C. Crosse, *J. Chem. Soc.* (A), 224 (1969).
231. M. J. Cleare and W. P. Griffith, *J. Chem. Soc.* (A), 372 (1969).
232. N. P. Crawford and G. A. Melson, *J. Chem. Soc.* (A), 427 (1969).
233. R. J. H. Clark, *J. Chem. Soc.*, 1377 (1963).
234. R. J. H. Clark and W. Errington, *J. Chem. Soc.* (A), 258 (1967).
235. R. J. H. Clark and C. S. Williams, *Spectrochim. Acta* **23A,** 1055 (1967).
236. C. Postmus, J. R. Ferraro, and W. Wozniak, *Inorg. Chem.* **6,** 2030 (1967).
237. G. E. Coates and D. Ridley, *J. Chem. Soc.*, 166 (1964).
238. R. J. H. Clark, A. G. Davies, and R. J. Puddephatt, *J. Chem. Soc.* (A), 1828 (1968).
239. R. J. H. Clark and C. S. Williams, *Spectrochim. Acta* **21,** 1861 (1965).

240. R. D. Gillard and B. T. Heaton, *J. Chem. Soc.* (A), 451 (1969).
241. I. Wharf and D. F. Shriver, *Inorg. Chem.* **8**, 914 (1969).
242. R. J. H. Clark and C. S. Williams, *J. Chem. Soc.* (A), 1425 (1966).
243. R. Layton, D. W. Sink, and J. R. Durig, *J. Inorg. Nucl. Chem.* **28**, 1965 (1966).
244. C. H. Perry, D. P. Athans, E. F. Young, J. R. Durig, and B. R. Mitchell, *Spectrochim. Acta* **23A**, 1137 (1966).
245. K. N. Nakamoto, P. J. McCarthy, J. Fujita, R. A. Condrate, and G. T. Behnke, *Inorg. Chem.* **4**, 36 (1965).
246. P. J. Hendra and N. Sadasivan, *Spectrochim. Acta* **21**, 1271 (1965).
247. J. M. Jenkins and B. L. Shaw, *J. Chem. Soc.*, 6789 (1965).
248. J. A. Kieft and K. N. Nakamoto, *J. Inorg. Nucl. Chem.* **29**, 2561 (1967).
249. G. W. Watt and D. S. Klett, *Spectrochim. Acta* **20**, 1053 (1964).
250. J. Hiraishi, *Spectrochim. Acta* **25A**, 749 (1969).
251. J. Pradilla-Sorzano and J. P. Fackler, *J. Mol. Spectry.* **22**, 80 (1967).
251a. L. J. Basile, D. L. Gronert, and J. R. Ferraro, *Spectrochim. Acta* **24A**, 1707 (1964).
252. D. M. Adams, P. J. Chandler, and R. G. Churchill, *J. Chem. Soc.* (A), 1272 (1967).
252a. D. A. Durham, G. H. Frost, and F. A. Hart, *J. Inorg. Nucl. Chem.* **31**, 833 (1969).
253. D. M. Adams and P. J. Chandler, *Chem. Commun.*, 69 (1966).
254. R. J. Goodfellow, P. L. Goggin and L. M. Venanzi, *J. Chem. Soc.* (A), 1897 (1967).
255. W. Klemperer, *J. Chem. Phys.* **24**, 353 (1956).
256. H. Gerding and E. Smit, *Z. Physik. Chem.* **B50**, 171 (1941); **B51**, 217 (1942).
257. T. Onishi and T. Shimanouchi, *Spectrochim. Acta* **20**, 325 (1964).
258. G. E. Leroi, T. C. Jones, J. T. Hougen, and W. Klemperer, *J. Chem. Phys.* **36**, 2879 (1962).
259. K. R. Thompson and K. D. Carlson, *J. Chem. Phys.* **49**, 4379 (1968).
260. D. M. Adams and P. J. Lock, *J. Chem. Soc.* (A), 620 (1967).
261. M. J. Campbell, M. Goldstein, and R. Greschskowiak, *Chem. Commun.*, 778 (1967).
262. D. M. Adams and P. J. Chandler, *J. Chem. Soc.* (A), 588 (1969).
263. C. Postmus, J. R. Ferraro, A. Quattrochi, and K. N. Nakamoto, *Inorg. Chem.* **8**, 1851 (1969).

Review Articles and General Bibliography

1. D. M. Adams, *Metal–Ligand and Related Vibrations*, St. Martin's Press, New York (1968).
2. R. J. H. Clark, *Halogen Chemistry* (V. Gutmann, Ed.), Academic Press, New York (1967), pp. 85–121.
3. D. W. James and M. J. Nolan, in *Progress in Inorganic Chemistry*, *Vol. 9* (F. A. Cotton, Ed.), Interscience Publishers, New York (1968) pp. 195–275.
4. R. H. Nuttall, *Talanta* **15**, 157 (1968).
5. R. E. Hester, *Coord. Chem. Rev.* **2**, 319 (1967).
6. R. J. H. Clark, *Record Chem. Prog.*, **20**(4), 269–282 (1965).
7. J. W. Brasch, Y. Mikawa and R. J. Jakobsen, *Applied Spectroscopy Reviews, Vol. 1* (E. G. Brame, Ed.), M. Dekker, Inc., New York (1968), pp. 187–235.
8. J. R. Ferraro, in *Far-Infrared Properties of Solids* (S. S. Mitra and S. Nudelman, Eds.), Plenum Press (1970).
9. P. J. Hendra and P. M. Stratton, *Chem. Rev.* **69**, 325 (1969).
10. D. M. Adams, in *Molecular Spectroscopy* (P. Hipple, Ed.), Elsevier Publishing Co., London (1968).

Chapter 7

METAL-NITROGEN VIBRATIONS*

7.1. INTRODUCTION

In recent years considerable effort has been directed toward complexes involving metal–nitrogen bonds. Infrared assignments for the metal–nitrogen (ν_{MN}) vibration in various coordination complexes have become more numerous. Assignments for the metal–nitrogen bending vibrations have received less attention. In general, the infrared intensity of the ν_{MN} vibration is weak to medium in nature, and thus has been the major cause of some confusion in the assignments. Raman investigations of these compounds have been minimal in nature for experimental reasons already mentioned in Chapter 6.

The factors determining the position of the ν_{MN} vibration are the same as those contributing to the position of the ν_{MX} vibration and have previously been discussed. In the case of the metal–nitrogen bond, the bond orders may vary, as was found for the metal–oxygen bond. For metal–nitrogen triple bonds, the stretching frequency may extend as high as $\sim$1100 cm^{-1}. On the other hand, metal–nitrogen links with a bond order of one appear in the lower-frequency range of 200–300 cm^{-1}.

The confusion in the metal–nitrogen assignments has centered on the metal–nitrogen vibration in metal (III) ammine complexes. Much of this has been due to the low intensity of this vibration. Deuteration studies and theoretical studies such as the normal coordinate treatment have been very helpful in solving this problem. In particular, the works of Sacconi,[1] Griffith,[2] and Shimanouchi *et al.*[3,4] have been very valuable.

7.2. METAL AMMINE COMPLEXES

The metal ammine complexes involve predominantly sigma bonding. However, a π-type overlap for *trans* $[Pt(NH_3)_2Cl_2]$ has been postulated

---

* With Louis J. Basile.

Table 7-1. Vibrational Frequencies for Several Hexacoordinated Transition Metal(III) Complexes with NH_3 (Type $M(NH_3)_6{}^{3+}$)

	Frequency,[a] cm^{-1}								
O_h symmetry	$\nu_1(a_{1g})$ R	$\nu_2(e_g)$ R	$\nu_3(f_{1u})$ IR	$\nu_4(f_{1u})$ IR	$\nu_5(f_{2g})$ R	$\nu_6(f_{2u})$ IA	Lattice	ρ_{NH_3}	References
$[Cr(NH_3)_6]Cl_3$	495	440	495, 470, 459	285				760	2–4, 7–9
$[Co(NH_3)_6]Cl_3$			499, 476, 449	332			155	830	2–4, 7–9
$[Co(ND_3)_6]Cl_3$			465, 446, 419	296			153	662	2–4, 7–9
$[Co(NH_3)_6]Br_3$			496, 474, 444	332			118	820	2–4, 7–9
$[Co(NH_3)_6]I_3$			464	319			99	803	2–4, 7–9
$[Ru(NH_3)_6]Cl_3$	500	475	463	283, 263	248			788	2, 7–9
$[Ru(ND_3)_6]Cl_3$	466	430	417		228			611	2, 7–9
$[Rh(NH_3)_6]Cl_3$	514	483	472	302, 287	240			845	2, 6
$[Rh(ND_3)_6]Cl_3$	489	455	433	278, 256	220			658	2
$[Os(NH_3)_6]OsBr_6$			452	256				818	2
$[Os(ND_3)_6]OsBr_6$			410					621	2
$[Ir(NH_3)_6]Cl_3$	527	560	475	279, 264	262			857	2
$[Ir(ND_3)_6]Cl_3$	498	471	440	255, 235	245			662	2
$[Pt(NH_3)_6]^{4+}$	603[b]	556[b]							6, 10

[a] In O_h symmetry the vibration modes are the following: ν_1(MN str.); ν_2(NMN str.); ν_3(NMN def.); ν_4(MN str.); ν_5(NMN def.); ν_6(NMN def.).
[b] Measured in solution of H_2O.

Table 7-2. Vibrational Frequencies for Several Hexacoordinated Cobalt Complexes with NH_3 (Type $[Co(NH_3)_5X]^{2+}$)

	Frequency,[a] cm^{-1}											
	$\nu_1(a_1)$	$\nu_2(a_1)$	$\nu_3(a_1)$	$\nu_4(a_1)$	$\nu_5(b_1)$	$\nu_6(b_1)$	$\nu_7(b_2)$	$\nu_8(e)$	$\nu_9(e)$	$\nu_{10}(e)$	$\nu_{11}(e_1)$	
C_{4v} symmetry	IR, R	IR, R	IR, R	IR, R	R	R	R	IR, R	IR, R	IR, R	IR, R	Reference
$[Co(NH_3)_5F]Br_2$	504, 493			310				449	310			11
$[Co(NH_3)_5Cl]Cl_2$	487, 479			283				457	325			1
$[Co(NH_3)_5Br]Br_2$	487							462	324			1
$[Co(NH_3)_5I]I_2$	428							473	319			1

[a] In C_{4v} symmetry the vibration modes are the following: ν_1(CoX str.); ν_2(CoN str.); ν_3(CoN str.); ν_4(NCoN, NCoX def.); ν_5(CoN str.); ν_6(NCoN, NCoX def.); ν_7(NCoN def.); ν_8(CoN str.); ν_9(NCoN, NCoX def.); ν_{10}(NCoN, NCoX def.); ν_{11}(NCoN def.).

by Nakamoto,[5] was based on two platinum–nitrogen frequencies in the infrared, one of which was attributed to the overlap between the metal π orbital and suitably placed hydrogen atoms.

Tables 7-1 and 7-2 show a collection of low-frequency data for several hexacoordinated transition metal(III) complexes with ammonia, of octahedral and C_{4v} symmetries. (See Fig. 6-1 for the normal modes of vibrations for molecules in O_h symmetry.) In the case of the $M(NH_3)_6{}^{3+}$ complexes, the metal–nitrogen symmetrical stretching vibration (ν_1) is in the range 466–527 cm^{-1}; the degenerate stretch (ν_2) is at 440–560 cm^{-1}, while the NMN bending vibration (ν_3) is in the range 419–499 cm^{-1}. The metal–nitrogen vibrations ν_4 and ν_5 range from 256 to 332 cm^{-1} and 220 to 262 cm^{-1} respectively. Some lattice vibrations have been assigned for the $Co(NH_3)_6X_3$ complexes in the region of 99–150 cm^{-1}. The coordinated ammonia rocking frequency is found at 760–857 cm^{-1} and shifts in the deuterated ammonia complexes to 611–662 cm^{-1}. The ratio of the intensities of the Raman vibrations ν_1:ν_2:ν_5 is 10:4:2. It may also be observed that ν_1 and ν_2 increase, as does ρ_{NH_3}, as one goes from ruthenium to rhodium to iridium. Figure 7-1 shows the infrared spectra of several Co(III)–ammine complexes in the CsBr region.

Table 7-3 and 7-4 summarize the data for the hexacoordinated transition metal(II) complexes with ammonia of octahedral symmetry. The metal–

Table 7-3. Vibrational Frequencies for Several Hexacoordinated Transition Metal(II) Complexes with NH_3 (Type $M(NH_3)_6{}^{2+}$)

O_h symmetry	Frequency, cm^{-1}			References
	ν_{MN}	δ_{NMN}	ρ_{NH_3}	
$[Mn(NH_3)_6]Cl_2$	307		617	1
$[Fe(NH_3)_6]Cl_2$	321		641	"
$[Co(NH_3)_6]Cl_2$	318	192	634	1, 3, 4
$[Ni(NH_3)_6]Cl_2$	330	215	678	"
$[Ni(ND_3)_6]Cl_2$	318	206	517	1
$[Zn(NH_3)_6]Cl_2$	300		645	"
$[Cd(NH_3)_6]Cl_2$	298		613	"
$[Mn(NH_3)_6]Br_2$	299		606	"
$[Fe(NH_3)_6]Br_2$	315		625	"
$[Co(NH_3)_6]Br_2$	318		647	"
$[Ni(NH_3)_6]Br_2$	327	217	672	"
$[Zn(NH_3)_6]Br_2$	294		636	"
$[Cd(NH_3)_6]Br_2$	291		594	"
$[Mn(NH_3)_6]I_2$	295		592	"
$[Fe(NH_3)_6]I_2$	306		617	"
$[Co(NH_3)_6]I_2$	312		626	"
$[Ni(NH_3)_6]I_2$	322	216	654	"
$[Zn(NH_3)_6]I_2$	282		621	"
$[Cd(NH_3)_6]I_2$	277		585	"

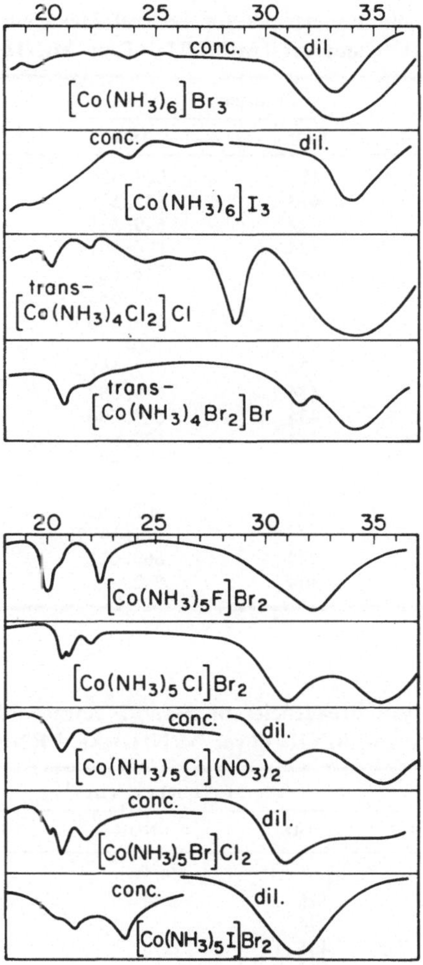

Fig. 7-1. (a) Infrared spectra of cobalt hexammines and tetrammines in the CsBr region. (b) Infrared spectra of cobalt pentammines in the CsBr region.

nitrogen stretching vibrations in the $M(NH_3)_6^{2+}$ complexes are in the range 277–330 cm^{-1}, considerably lower than those found for the trivalent metal complexes. This reflects the change expected from the oxidation number effect. Few data concerning the bending vibrations are available, but from the limited data it appears as if the δ_{NMN} vibration is located at 192–217 cm^{-1}. The ammonia rocking vibration, as expected, is found at a lower range in the

Table 7-4. Vibrational Frequencies for Several Hexacoordinated Transition Metal(II) Complexes with NH_3 (Type $M(NH_3)_2X_2$)

O_h symmetry	Frequency, cm^{-1}		Reference
	ν_{MN}	ρ_{NH_3}	
$Fe(NH_3)_2Cl_2$	415	624, 554	12
α-$Co(NH_3)_2Cl_2$	423	645, 571	″
β-$Co(NH_3)_2Cl_2$	422	639, 575	″
$Ni(NH_3)_2Cl_2$	435	676, 591	″
α-$Cu(NH_3)_2Cl_2$	480	716, 666	″
β-$Cu(NH_3)_2Cl_2$	482	718, 666	″
$Cd(NH_3)_2Cl_2$	375	608, 560	″
$Hg(NH_3)_2Cl_2$		720	″
$Fe(NH_3)_2Br_2$	424	617, 567	″
α-$Co(NH_3)_2Br_2$	436	639, 583	″
β-$Co(NH_3)_2Br_2$	434	640, 590	″
$Ni(NH_3)_2Br_2$	434	670, 602	″
α-$Cu(NH_3)_2Br_2$	488	721, 663	″
β-$Cu(NH_3)_2Br_2$	486	720, 658	″
$Cd(NH_3)_2Br_2$	368	602, 562	″
$Hg(NH_3)_2Br_2$		721, ~700	″
$Ni(NH_3)_2I_2$	434	660, 621	″
$Cd(NH_3)_2I_2$	364	582	″

Table 7-5. Vibrational Frequencies of Several Tetracoordinated Pd and Pt Complexes with NH_3 (Type $M(NH_3)_2X_2$, $M(NH_3)_2X_4$)

D_{4h} symmetry	Frequency, cm^{-1}			References
	ν_{MN}	δ_{NMN}	ρ_{NH_3}	
Trans				
$Pd(NH_3)_2Cl_2$	496	244	756	12, 13, 15
$Pd(NH_3)_2Br_2$	494			12, 13, 15
$Pd(NH_3)_2(SCN)_2$	485	276	822, 760	12
$Pd(NH_3)_2I_2$	486			15
$Pt(NH_3)_2Cl_2$	509	244	825, 804	3,10, 12, 15
$Pt(NH_3)_2Br_2$	531, 506	228		14
$Pt(NH_3)_2(SCN)_2$	508	(268), 233	863, 838	12
$[Pt(NH_3)_2]Cl_4$	554, 517	253		14
$[Pt(NH_3)_2]Br_4$	531, 506	228		14
$[Pt(NH_3)_2]I_4$	516, 507	249		14
Cis				
$Pd(NH_3)_2Cl_2$	495, 476	245, 218	752	12, 13, 16
$Pd(NH_3)_2Br_2$	480, 460	225	746	13
$Pt(NH_3)_2Cl_2$	510	250	795	10
$Pt(NH_3)_2Br_2$	492, 485	219		10
$Pt(NH_3)_2(SCN)_2$	496	(283), 222	802, 773	12
$[Pt(NH_3)_2]Cl_4$	558, 518	240		14
$[Pt(NH_3)_2]I_4$	425	182		14

Table 7-6. Vibrational Frequencies for Several Tetracoordinated Metal(II) Complexes with NH_3 (Type $M(NH_3)_2X_2$)

T_d symmetry	Frequency, cm^{-1}		Reference
	ν_{MN}	ρ_{HN_3}	
$Zn(NH_3)_2Cl_2$	421	685, 661, 638	12
$Zn(ND_3)_2Cl_2$	392	585, 500	"
$Zn(NH_3)_2Br_2$	414	680, 660, 637	"
$Zn(NH_3)_2I_2$	399	676, 648	"
α-$Co(NH_3)_2I_2$	430	659, 638, 623	"

divalent metal complexes, 585–654 cm^{-1}. Again, this vibration shifts to lower frequency in the deuterated ammine complexes.

For the $M(NH_3)_2X_2$ complexes the ν_{MN} vibration ranges from 415 to 486 cm^{-1} and appears to increase in frequency Fe<Co<Ni<Cu. For the $Cd(NH_3)_2X_2$ complexes the vibration is much lower, at ~370 cm^{-1}. The ammonia rocking vibration is located at 554–721 cm^{-1}.

Table 7-5 records the data for the tetracoordinated Pd(II) and Pt(II) complexes with ammonia having D_{4h} symmetry. The effect of coordination number in both the *trans* and *cis* complexes is readily observable when compared with the hexacoordinated M(II) ammine complexes. The ν_{MN} frequencies range from 485 to 554 cm^{-1}, the δ_{NMN} is at 228–276 cm^{-1}, and the ρ_{NH_3} is at 756–863 cm^{-1} in the *trans* compounds. In the *cis* compounds the ν_{MN} frequencies range from 425 to 558 cm^{-1}, the δ_{NMN} from 182 to 283 cm^{-1} and the ρ_{NH_3} from 746 to 802 cm^{-1}.

Table 7-6 summarizes the data for the tetracoordinated metal(II) complexes with ammonia (T_d symmetry), and Table 7-7 lists the data for tetracoordinated metal(II) complexes with ammonia of the type $[M(NH_3)_4X_2]$ and possessing D_{4h} symmetry. Table 7-8 compares the various metal–nitrogen ligand vibrations in the various ammine complexes.

Table 7-7. Vibrational Frequencies of Several Tetracoordinated Metal(II) Complexes with NH_3 (Type $M(NH_3)_4X_2$)

D_{4h} symmetry	Frequency, cm^{-1}			References
	ν_{MN}	δ_{NMN}	ρ_{NH}	
$[Pd(NH_3)_4]Cl_2$	491–494	295, 245 160	797–830	1, 3, 10, 15
$[Pd(NH_3)_4]Br_2$	490	273, 190	825	13
$[Pd(NH_3)_4][PdCl_4]$	488	263	778	12
$[Pt(NH_3)_4]Cl_2$	510	297	842	3, 10, 13
$[Pt(NH_3)_4][PtCl_4]$	500	264, 237	844, 823	12
$[Pt(NH_3)_4][Pt(SCN)_4]$	506, 498	276, 228	868, 814	12
$[Cu(NH_3)_4]SO_4$	420	250	713	3, 10

Table 7-8. Summary of Metal–Nitrogen Ligand Vibrations in Various Ammine Complexes

Symmetry	Vibration	Frequency, cm^{-1} M(III)	Frequency, cm^{-1} M(II)
O_h	ν_{MN}	466–527	277–330
$M(NH_3)_6{}^{n+}$	ν_{NMN}	440–560	
type	ν_{MN} (degenerate)	256–332	
	δ_{NMN}	417–499	192–217
	δ_{NMN}	220–332	
	ρ_{NH_3}	760–857	585–654
	ρ_{ND_3}	611–662	517
$M(NH_3)_2X_2$	ν_{MN}		415–486
type	ρ_{NH_3}		554–721
D_{4h}			
Trans	ν_{MN}		485–531 ($M(NH_3)_2X_2$) 506–531 ($M(NH_3)_2X_4$)
$M(NH_3)_2X_2$	δ_{NMN}		228–276 ($M(NH_3)_2X_2$ and $M(NH_3)_2X_4$)
$M(NH_3)_2X_4$	ρ_{NH_3}		756–863 ($M(NH_3)_2X_2$)
types			
Cis	ν_{MN}		460–510 ($M(NH_3)_2X_2$) 425–558 ($M(NH_3)_2X_4$)
$M(NH_3)_2X_2$ $M(NH_3)_2X_4$	δ_{NMN}		218–283 ($M(NH_3)_2X_2$) 182–240 ($M(NH_3)_2X_4$)
types	ρ_{NH_3}		746–798 ($M(NH_3)_2X_2$) 773–802 ($M(NH_3)_2X_4$)
T_d			
$M(NH_3)_2X_2$	ν_{MN}		392–430
type	ρ_{NH_3}		623–685
D_{4h}			
$M(NH_3)_4X_2$ type	ν_{MN}		488–510 (M = Pt, Pd) 420 (M = Cu)
	δ_{NMN}		160–297 (M = Pt, Pd, Cu)
	ρ_{NH_3}		778–868 (M = Pt, Pd) 713 (M = Cu)

Table 7-9 shows some force constants of the MN bond for several ammine complexes. The disagreement between various workers is readily apparent.

7.3. COMPLEXES OF 2,2′-BIPYRIDYL, 2,2′,2″-TERPYRIDYL, 1,10-PHENANTHROLINE, ANILINE, AND 8-AMINOQUINOLINE

As early as 1962 it was suggested that the metal–nitrogen absorption in complexes of 2,2′-bipyridyl and 1,10-phenanthroline[18] due to the stretching vibration would be found lower than 200 cm^{-1}. However, since this time,

Table 7-9. Force Constants for Several Ammine Complexes

Complex	k(MN),[a] mdynes/Å
$[Cr(NH_3)_6]Cl_3$	0.94[1]
$[Co(NH_3)_6]Cl_3$	1.05[1]
$[Co(ND_3)_6]Cl_3$	1.05[1]
$[Pt(NH_3)_2Cl_2]$	2.2[17] (2.4)[15]
$[Pt(NH_3)_2Br_2]$	2.2[17] (2.6)[15]
$[Pt(NH_3)_2I_2]$	2.6[15]
$[Pd(NH_3)_2Cl_2]$	1.9[17] (2.1)[15]
$[Pd(NH_3)_2Br_2]$	1.9[17]
$[Pd(NH_3)_2I_2]$	1.8[17]
$[Pt(NH_3)_6]^{4+}$	2.39[10]
$[Hg(NH_3)_2]^{2+}$	2.30[10]
$[Pt(NH_3)_4]^{2+}$	2.22[10] (1.92)[1]
$[Co(NH_3)_6]^{3+}$	1.05[1] (1.75)[10]
$[Cr(NH_3)_6]^{3+}$	1.05[1] (1.55)[10]
$[Cu(NH_3)_4]^{2+}$	1.24[10] (0.84)[1]
$[Ni(NH_3)_6]^{2+}$	0.86[10] (0.34)[1]
$[Co(NH_3)_6]^{2+}$	0.82[10] (0.33)[1]
$[Pd(NH_3)_4]Cl_2$	1.71[1] (1.97)[10]

[a]The superscripts are reference numbers.

several papers have reported on the assignments of the metal–nitrogen stretching vibrations involving these ligands and related ligands. In most cases these have been assigned in the 200–300 cm^{-1} region. A published assignment[19] for Fe(II) with 1,10-phenanthroline at 530 cm^{-1} and 2,2′-bipyridyl at 423 cm^{-1} appears to be too high in light of other transition metal(II) assignments. Tables 7-10 and 7-11 tabulate ν_{MN} assignments for these complexes, and confirmation may only be possible from transition metal isotope studies.

Recently tentative assignments for ν_{MN} stretching vibrations[20] in rare earth complexes with 2,2′-bipyridyl, 1,10-phenanthroline, and 2,2′,2″-terpyridyl have been made. These are tabulated in Tables 7-10 and 7-11. In the bipyridyl and phenanthroline complexes, the ν_{MN} vibrations appeared to vary with the metal and ligand. The bipyridyl complexes show a sharp change in ν_{MN} vibration at gadolinium and the phenanthroline at europium. There is an increase in the frequency of the ν_{MN} vibration with an increase in mass of the lanthanide. These results are similar to those found for the anhydrous rare earth nitrates for the ν_{MO} vibration.[20a] No definite assignments for ν_{MX} could be made, since no halogen-sensitive vibrations were observed. However, a recent paper makes assignments for a lanthanide–chloride stretching vibration in complexes of dibutylpyridines in the 230 cm^{-1} region.[21]

Table 7-10. Assignments for Metal–Nitrogen Vibrations in Complexes with 2,2′-Bipyridyl and 2,2′,2″-Terpyridyl

Complex		Frequency, cm^{-1}		References
		ν_{MN}	δ_{NMN}	
$Fe(bipy)_3Cl_2$		423(?)		19
$Co(bipy)_3Cl_2$		264		″
$Ni(bipy)_3Cl_2$		286		″
$Cu(bipy)_3(NO_3)_2$		297		″
$Cu(bipy)_3(ClO_4)_2$		286		26
$Cu(bipy)_2(ClO_4)_2$		319		″
$Zn(bipy)_3(NO_3)_2$		280		19
$Zn(bipy)Cl_2$		241	192	27
$Zn(bipy)Br_2$		250	190	″
$Zn(bipy)I_2$		250	194	″
$Pd(bipy)(NCS)_2$		261		28
$[Cu(OH)(bipy)]_2(NO_3)_2$		271, 246		26, 29
$[Cu(OH)(bipy)]_2(ClO_4)_2$		278, 254		26, 29
$[Cu(OH)(bipy)]_2^{2+}$ (8 compounds)		267, 278		30
$[Cr(OH)(bipy)_2]_2^{4+}$		343–378		31
$[Fe(OH)(bipy)_2]_2^{4+}$		260–294		″
$In(bipy)_{1.5}Cl_3$		238		32
$In(bipy)_{1.5}Br_3$		234		33
$In(bipy)_{1.5}I_3$		232, 226		″
$In(bipy)_{1.5}(NCO)_3$		240		″
$[\varphi_4As][In(bipy)Cl_4]\cdot MeCN$	IR	252, 239		31
	R	240		
$[In(bipy)Cl_2]PF_6\cdot H_2O$	IR	246, 231		31
	R	251, 232, 233, 195		″
$[In(bipy)Cl_2]ClO_4$		257, 252, 228		″
trans-$[Sc(bipy)_2(NCS)_2]NCS$		314		34
cis-$[Sc(bipy)_2(NCS)_2]NCS$		324, 304		″
$Sn(bipy)Cl_4$		250, 207		35
$Sn(bipy)Br_4$		260, 195		″
$Sn(bipy)I_4$		252, 198		″
$Sc(bipy)_2Cl_3$		210		20
$Y(bipy)_2Cl_3$		214		″
$La(bipy)_2Cl_3$		215		″
$Nd(bipy)_2Cl_3$		217		″
$Sm(bipy)_2Cl_3$		222, 233, 249		″
$Eu(bipy)_2Cl_3$		233, 249		″
$Gd(bipy)_2Cl_3$		224, 231, 249		″
$Dy(bipy)_2Cl_3$		234, 254		″
$Ho(bipy)_2Cl_3$		224, 235, 256		″
$Er(bipy)_2Cl_3$		243, 260		″
$Tm(bipy)_2Cl_3$		287		″
$Lu(bipy)_2Cl_3$		263		″
$Zn(terp)Cl_2$		224	167	27
$Zn(terp)Br_2$		243	167	″
$Zn(terp)I_2$		245	167	″

Table 7-11. Assignments for Metal–Nitrogen Vibrations in Complexes with 1, 10-Phenanthroline

Complex		ν_{MN}, cm^{-1}	Reference
$Fe(phen)_3Cl_2$		530	19
$Co(phen)_3Cl_2$		288	"
$Ni(phen)_3Cl_2$		299	"
$Cu(phen)_3(NO_3)_2$		300	"
$Zn(phen)_3(NO_3)_2$		288	"
trans-$[Sc(phen)_2(NCS)_2]NCS$		325	34
cis-$[Sc(phen)_2(NCS)_2]NCS$		324, 300	"
$Tl(phen)Cl_3$		290	36
$Tl(phen)Br_3$		290	"
$Tl(phen)I_3$		283	"
$[Tl(phen)_2(NO_3)_2]NO_3$		289	"
$[Tl(phen)_2Cl_2]Cl$		280	"
$[Tl(phen)_2Br_2]Br$		287	"
$[Tl(phen)_2Cl_2]NO_3$		280	"
$[Tl(phen)_2Br_2]NO_3$		287	"
$[Tl(phen)_2(ClO_4)_2]ClO_4$		294	"
$[Tl(phen)_3](ClO_4)_3$		293	"
$InCl_3(phen)_{1.5}MeCN$	IR	212	32
	R	236, 218, 189	"
$[InCl_2(phen)_2]ClO_4$	IR	236, 245	"
	R	266, 248, 221	"
$[CrOH(phen)_2]_2^{4+}$		343–378	31
$[FeOH(phen)_2]_2^{4+}$		260–274	"
$La(phen)_2Cl_3$		218, 197, 178	20
$Nd(phen)_2Cl_3$		218, 196, 175	"
$Sm(phen)_2Cl_3$		216, 199, 175	"
$Eu(phen)_2Cl_3$		235	"
$Gd(phen)_2Cl_3$		224, 235	"
$Dy(phen)_2Cl_3$		238, 247	"
$Ho(phen)_2Cl_3$		224, 235, 246	"
$Er(phen)_2Cl_3$		225, 238, 248	"
$Tm(phen)_2Cl_3$		224, 237, 248	"
$Lu(phen)_2Cl_3$		224, 238, 248	"
$Sc(phen)_2Cl_3$		275, 283	"
$Y(phen)_2Cl_3$		263, 281	"

In the case of the lanthanide–terpyridyl complexes no definite assignments[22] could be made for the ν_{MN} vibration because of the complex nature of the infrared spectra and the interference offered by the low-frequency spectra of this ligand. However, assignments were made for the ν_{MX} vibration. For the lighter members of the complexes of the type $Nd(terp)Cl_3$ the ν_{MX} is found in the region 204–228 cm^{-1}; for $Er(terp)Cl_3$ ν_{MX} is in the range 206–259 cm^{-1}, and for the complexes $RE(terp)_3Cl_3$ ν_{MX} is between 231–269 cm^{-1}. The bromides were found to show no absorption in these areas, but absorp-

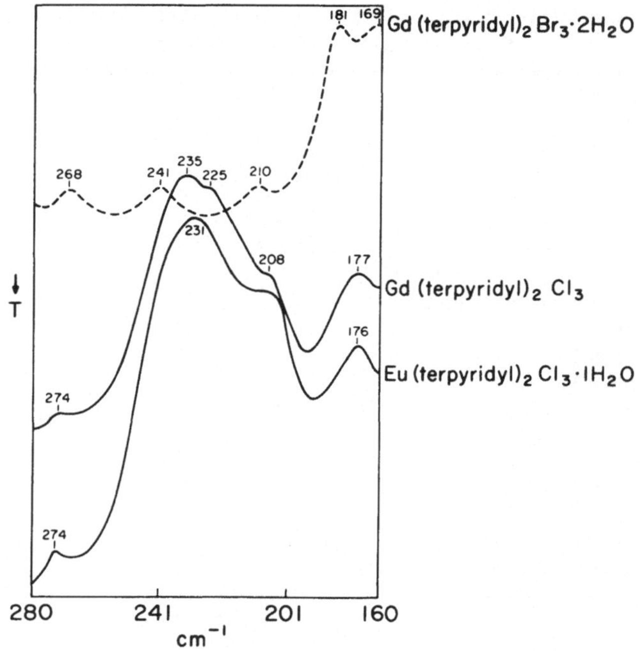

Fig. 7-2. Low-frequency spectra of several rare earth–terpyridyl complexes.

tion occurred at 140–155 cm^{-1}. Figure 7-2 shows the low frequency of several lanthanide halide complexes of terpyridyl. A recent paper has reported the preparation of $RE(terp)_3(ClO_4)_3$ with the coordination number of the lanthanides being nine.[23] These can only be made from the perchlorate. In the case of the halide salts, the halide complexing prevents the complexing of a third dipyridyl molecule. For the rare earth complexes with 8-aminoquinoline, no definite assignments could be made for ν_{MN} and ν_{MX} because of the complexity of the far-infrared spectra of these compounds.[24]

Table 7-12 shows ν_{MN} stretching vibration assignments made for transition metal(II) complexes with aniline.[25] These assignments occur at higher frequencies than comparable complexes with pyridine or substituted pyridine ligands (see Section 7-5).

7.4. METAL–IMIDAZOLE COMPLEXES AND COMPLEXES WITH SUBSTITUTED IMIDAZOLES

First-row transition metal(II) complexes with imidazole and such substituted imidazoles as 2-methylimidazole and 4- and 5-bromoimidazole

Table 7-12. Assignments for Metal–Nitrogen Vibrations in Complexes with Aniline[25]

Compound	ν_{MN}, cm^{-1}
MnAn$_2$Cl$_2$	386, 374
MnAn$_2$Cl$_2$(EtOH)$_2$	386, 370
MnAn$_2$Br$_2$	367, 350
MnAn$_2$Br$_2$(EtOH)$_2$	383, 370, 350
MnAn$_4$I$_2$	357
MnAn$_2$I$_2$	361, 355
CoAn$_2$Cl$_2$	415, 366
CoAn$_2$Br$_2$	412, 358
CoAn$_2$I$_2$	410, 349
CoAn$_2$SO$_4$	346
NiAn$_2$Cl$_2$	388, 377, 350
NiAn$_2$Cl$_2$(EtOH)$_2$	388, 377, 358
NiAn$_4$Br$_2$	363, 353
NiAn$_2$Br$_2$	386, 370, 354
NiAn$_2$Br$_2$(EtOH)$_2$	390, 362, 357
NiAn$_4$I$_2$	360, 344
NiAn$_2$I$_2$	361, 345
NiAn$_2$SO$_4$	366
CuAn$_2$Cl$_2$	430, 357
CuAn$_2$Br$_2$	424, 348
CuAn$_2$SO$_4$	385
ZnAn$_2$Cl$_2$	402, 362
ZnAn$_2$Br$_2$	400, 355
ZnAn$_2$I$_2$	396, 342
ZnAn$_2$SO$_4$	379
CdAn$_4$Cl$_2$	377, 350
CdAn$_2$Cl$_2$	377, 351
CdAn$_2$Br$_2$	365, 350
CdAn$_2$I$_2$	370, 339

have recently been prepared.[37–39] For structural representations of the imidazole and related molecules see Table 6-42. Table 7-13 compiles the available infrared data. The position of the ν_{MN} stretching vibrations also appears to be at higher frequency than that found in corresponding pyridine complexes and reflects the stronger basicity of these ligands. The stereochemistry involved in these complexes is also listed in Table 7-13. Recently, some tentative assignments for ν_{MN} and ν_{MX} for transition metal(II) complexes with pentamethylenetetrazole* have been made.[40]

* Pentamethylenetetrazole: $N\begin{smallmatrix} N=C-CH_2-CH_2 \\ | \\ N-N-CH_2-CH_2 \end{smallmatrix}CH_2$.

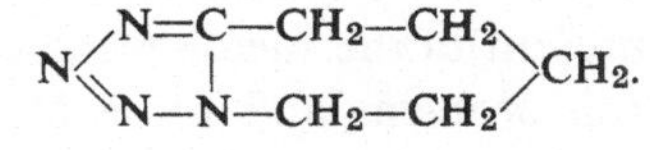

Table 7-13. Vibrational Frequencies (cm^{-1}) for Several Transition Metal(II) Complexes with Imidazole and Substituted Imidazoles[37-39]

			L*			
			Iz		MIz	BIz
Type	M	X	ν_{MN}	δ_{NMN}	ν_{MN}	ν_{MN}
LMX_2	Co	Cl	267 (O_h)			
		Br	273 ($\sim T_d$)[a]			
		I	270 ($\sim T_d$)			
L_2MX_2	Co	Cl	275 ($\sim T_d$)		275 (T_d)	250 ($\sim T_d$)
		Br	285 ($\sim T_d$)		273 (T_d)	250 ($\sim T_d$)
		I	285 ($\sim T_d$)		273 (T_d)	246 ($\sim T_d$)
	Ni	Cl			270 (O_h polymeric)	
		Br			272 (T_d)	249 (O_h)
		I			270 (T_d)	247 ($\sim T_d$)
	Cu	Cl			271 (T_d)	275 (O_h dist)
		Br			274 (T_d)	275 (O_h dist)
L_4MX_2	Ni	Br	262 (O_h)			
		I	262 (O_h)			
	Cu	Cl	286 (O_h dist)[b]		275 (O_h dist)	
		Br	288 (O_h dist)			
		NO_3	292 (O_h dist)			
L_6MX_2	Co	Cl	240, 250 (O_h)	201		256 (O_h)
		Br	239, 250 (O_h)	200		
		I	238, 245 (O_h)	200		
	Ni	Cl	262 (O_h)	213		
		Br	266 (O_h)	211		
		I	265 (O_h)	208		

* Iz = imidazole; MIz = 2-methylimidazole; BIz = bromoimidazole.
[a] Pseudotetrahedral = $\sim T_d$.
[b] No halogen-dependent bands, distorted O_h.

7.5. METAL–PYRIDINE AND RELATED COMPLEXES

A tremendous number of data are now available on the metal–nitrogen stretching (ν_{MN}) vibrations of metal complexes with pyridine and related ligands. There has been some question as to the assignment of this vibration. Bicelli[41], in 1958, predicted that the ν_{MN} vibration should occur at 150–200 cm^{-1}. In 1962, Clark[42] cited that no evidence for a ν_{MN} vibration occurred in heavy bidentate nitrogen-donor ligands above 200 cm^{-1}. An assignment for ν_{PtN} has been made at $\sim$450–480 cm^{-1}.[43] In part, this difficulty has been due to the fact that the vibration may be mixed with other vibrations and may also be weak in intensity. In addition, since the π system in the ring portion of the complexes is distorted upon complexation, some of the π component of bonding may also be present in the metal–nitrogen vibration. However, recent work has clarified this problem, and it now appears that the vibration

Table 7-14. Distribution of Normal Modes of Vibration in Compounds of Stoichiometry L_nMX_m*

Type	Stereochemistry	Symmetry group	ν_{MX}	Number infrared active	ν_{ML}	Number infrared active
MX_4	Tetrahedral	T_d	$a_1 + b_2$	1		
L_2MX_2	Tetrahedral	C_{2v}	$a_1 + b_1$	2	$a_1 + b_1$	2
MX_4	Planar	D_{4h}	$a_{1g} + b_{1g} + e_u$	1		
LMX_3	Planar	C_{2v}	$2a_1 + b_1$	3	a_1	1
L_2MX_2	*trans*-Planar	D_{2h}	$a_g + b_{3u}$	1	$a_g + b_{2u}$	1
L_2MX_2	*cis*-Planar	C_{2v}	$a_1 + b_1$	2	$a_1 + b_1$	2
L_4MX_2	*trans*-Octahedral	D_{4h}	$a_{1g} + a_{2u}$	1	$a_{1g} + b_{1g} + e_u$	1
L_4MX_2	*cis*-Octahedral	C_{2v}	$a_1 + b_1$	2	$2a_2 + b_1 + b_2$	4
L_3MX_3	*trans*-Octahedral	C_{2v}	$2a_1 + b_1$	3	$2a_1 + b_2$	3
L_3MX_3	*cis*-Octahedral	C_{3v}	$a_1 + e$	2	$a_1 + e$	2
L_2MX_2	Polymeric octahedral	C_i	$2a_g + 2a_u$	2	$a_g + a_u$	1
LMX_2	Polymeric octahedral	D_{2h}	$2a_g + 2a_u$	2	$a_g + a_u$	1
LMX_2	Tetrahedral	C_{2v}	$a_1 + b_1$	2	$a_1 + b_1$	2
LMX_3 or	Tetrahedral	C_{2v}	$2a_1 + b_1$	3	a_1	1
LMX_3^-		C_{3v}	$a_1 + e$	2	a_1	1
$(L_{1.5}MX_3)_2$		C_{3v}	$a_1 + e$	2	$a_1 + e$	2

* Taken partially from reference 44.

is located in the 200–300 cm^{-1} region. As has been the case with the bending vibration of the metal–halogen bond, much less evidence is available concerning the bending vibration of the metal–nitrogen bond in these complexes. This has been caused primarily by the lack of investigation below 200 cm^{-1} where the vibration occurs, and by the difficulty of making assignments <200 cm^{-1} because of lattice vibration overlap.

In addition to the metal–ligand vibrations, certain ligand vibrations are of interest in the low-frequency region. For example, the pyridine ligand vibrations at 605 and 405 cm^{-1} (corresponding to ring vibrations) shift toward higher frequencies upon complexation. These shifts may be considerable and are greater for the second- and third-row transition metal complexes. Furthermore, their positions may be related to the stereochemistry of the complex since the degree of shift is in the direction $T_d >$ polymeric $O_h > O_h$. The frequencies also appear to be dependent on the ionic radius of the central metal atom, increasing in frequency as the ionic radius increases.

The discussion of these complexes will be based on the number of ligands present, ranging from complexes of the type LMX_2 to L_4MX_2. Table 7-14 presents the distribution of the modes of vibrations in various complexes of the stoichiometry L_nMX_m. The number of infrared-active ν_{MN} modes and species predicted are listed in the last two columns.

LMX_2 Complexes

The LMX_2 complexes, where L is a related pyridine ligand and X is a halogen, may involve several structures. The complex may involve a distorted octahedral structure with bridging halogens and ligands and a tetrahedral symmetry with bridging ligands. Table 7-15 summarizes the data for the octahedral complexes involving the ligands 3,3′-azopyridine (apy),

Table 7-15. Metal–Nitrogen Vibrational Frequencies for LMX_2 Complexes of O_h Symmetry

LMX_2 O_h distorted (halogen and ligand bridging)	Frequency, cm^{-1}		Reference
	ν_{MN}	δ_{NMN}	
$(apy)CuCl_2$	255	191	45
$(apy)CuBr_2$	257	188	"
$(apy')CuCl_2$	252	190	"
$(apy')CuBr_2$	252	183	"
$(pz)MnCl_2$	220, 202		46
$(pz)CoCl_2$	206		"
$(pz)CuCl_2$	284		"
$(pz)CuBr_2$	284		"
$(pm)MnCl_2$	210, 195		"
$(pm)MnBr_2$	236		"
$(pm)FeCl_2$	208		"
$(pm)CoCl_2$	216		"
$(pm)CoBr_2$	221		"
$(pm)CuCl_2$	251		"
$(pm)CuBr_2$	251		"
$(pdz)MnCl_2$	189		"
$(pdz)FeCl_2$	219		"
$(pdz)CoCl_2$	199		"
$(pdz)CuCl_2$	270		"
$(pdz)CuBr_2$	275		47
$(pz)ZnCl_2$	194		"
$(pz)CdCl_2$	192		"
$(pz)CdBr_2$	186		"
$(pz)CdI_2$	186		"
$(pm)ZnCl_2$	253		"
$(pm)ZnBr_2$	250		"
$(pm)CdCl_2$	192		"
$(pm)CdBr_2$	205		"
$(pdz)CdCl_2$	191		"
$(pdz)CdBr_2$	192		"

(apy) = 3,3′-azopyridine

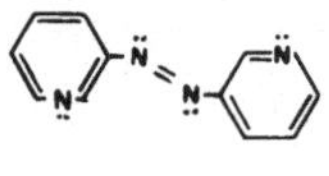

(apy′) = 4,4′-azopyridine

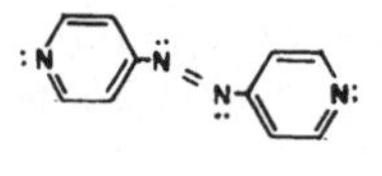

(pz) = pyrazine (see Table 6-42).
(pm) = pyrimidine (see Table 6-42).
(pdz) = pyridazine (see Table 6-42).

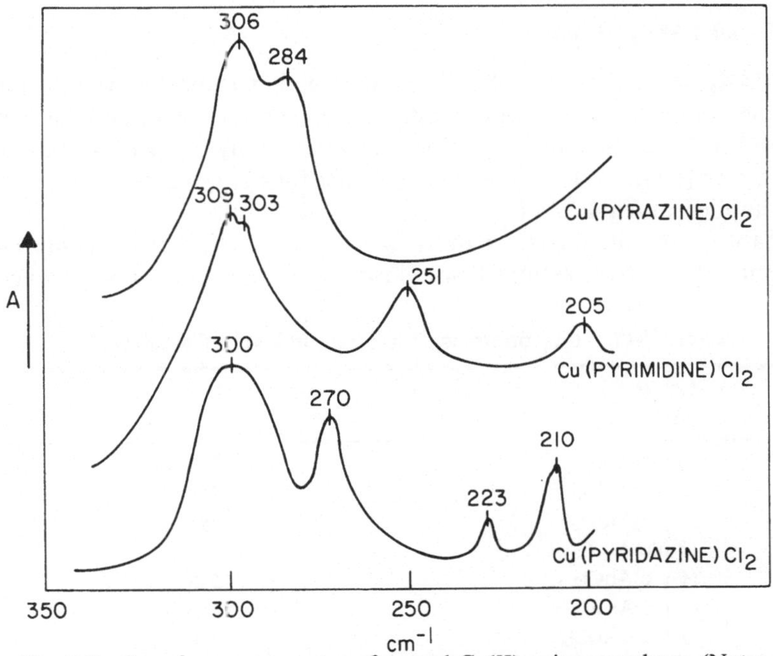

Fig. 7-3. Low-frequency spectra of several Cu(II)–azine complexes. (Note: all three ligands are free of absorption in this region.)

4,4′-azopyridine (apy′), pyrazine (py), pyrimidine (pm), and pyridazine (pdz). In the zinc, cadmium, and mercuric halide complexes many of the ν_{MN} vibrations are seen to occur below 200 cm^{-1}. Figure 7-3 shows some spectra for Cu(II)–azine complexes in the region of 200–300 cm^{-1}.

Table 7-16 summarizes the data for the tetrahedral LMX_2 complexes.

Table 7-16. ν_{MN} **Stretching Vibrations in LMX_2 Complexes in T_d Symmetry**

LMX_2 T_d (ligand bridging)	ν_{MN}, cm^{-1}	Reference
(apy)CoI_2	224	45
(apy′)$CoBr_2$	238, 211	″
(apy′)CoI_2	236, 214	″
(pz)$ZnBr_2$	232, 208	47
(pz)ZnI_2	236	″
(pz)$HgCl_2$	189	″
(pz)$HgBr_2$	208	″
(pm)$HgCl_2$	176	″
(pm)$HgBr_2$	165	″

LMX_3 and LMX_3^- Complexes

LMX_3 complexes of gold(III) halides with pyridine and substituted pyridine complexes have been investigated.[48] The ν_{AuN} vibration has been assigned in the 227–306 cm^{-1} region. Generally, only one ν_{MN} vibration is observed as predicted from the selection rules for C_{2v} symmetry. The results are compiled in Table 7-17.

Table 7-18 contains the results for several LMX_3^- anion complexes. Bradbury *et al.*[49] have assigned ν_{MN} vibrations in terms of a C_{3v} symmetry.

Table 7-17. ν_{MN} Stretching Vibrations in LMX_3 Complexes[48]

LMX_3 C_{2v}	ν_{MN}, cm^{-1}
(py)$AuCl_3$	249, 239
(α-pic)$AuCl_3$	249
(α-pic)$AuBr_3$	255
(β-pic)$AuCl_3$	234
(γ-pic)$AuCl_3$	286
(γ-pic)$AuBr_3$	286
(2,6-lut)$AuCl_3$	278
(2,6-lut)$AuBr_3$	300, 282
(3,5-lut)$AuCl_3$	289
(3,5-lut)$AuBr_3$	290
(2,4-lut)$AuCl_3$	306
(2,4-lut)$AuBr_3$	299
(4-CNpy)$AuCl_3$	249, 227
(4-CNpy)$AuBr_3$	249, 233
(py-d_5)$AuCl_3$	271, 263, 227

Table 7-18. ν_{MN} Stretching Vibrations in LMX_3^- Anion Complexes

LMX_3^- Pseudo T_d	ν_{MN}, cm^{-1}		
(C_{3v})	L = py	L = α-pic	References
$[LCoCl_3]^-$	226, 162	230, 215	49, 50
$[LCoBr_3]^-$	175, 153	204, 177, 155	〃
$[LCoI_3]^-$	161, 134		〃
$[LNiBr_3]^-$		176	〃
$[LNiI_3]^-$	164, 136		〃
$[LZnCl_3]^-$	207, 183, 161	228, 197, 150	〃
$[LZnBr_3]^-$	177, 151	174, 158	〃
$[LZnI_3]^-$	162, 138		50

$L_{1.5}MX_3$ Complexes

The complexes $L_{1.5}MX_3$, where L is bipyridyl or pyrazine, M is In(III) and Tl(III), and X is a halogen, were originally considered to have a structure $[InCl_2bipy_2]^+[InCl_4bipy]^-$. Recent far-infrared data[33] have been interpreted in terms of a nonionic structure containing bridging ligand units and involving dimeric units:

$$
\begin{array}{ccccc}
X & & \text{bipy} & & X \\
\diagdown & \diagup & & \diagdown & \diagup \\
X\!-\! & \text{In} & -\text{bipy}- & \text{In} & -X \\
\diagup & & \diagdown & \diagup & & \diagdown \\
X & & \text{bipy} & & X
\end{array}
$$

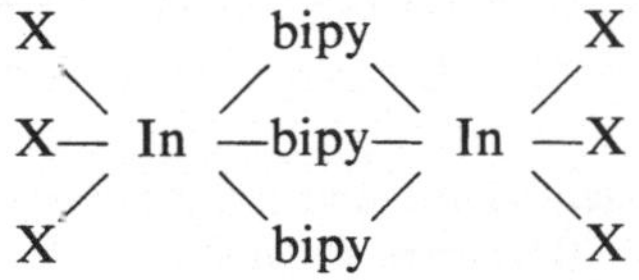

Table 7-19 records the ν_{MN} vibration for several complexes.

$L_2(OR)CuX_2$ Complexes

Table 7-20 lists ν_{MN} vibrations for several complexes of the type $L_2(OR)CuX_2$, where L is 2-aminopyridine, X is an anion, and R is H or an alkyl group. The vibration appears to be centered in the region 253–256 cm^{-1}.

Table 7-19. ν_{MN} Stretching Vibrations in $L_{1.5}MX_3$ Complexes

$L_{1.5}MX_3$	ν_{MN}, cm^{-1}		Reference
(pz)$_{1.5}$InCl$_3$	IR	265	32
	R	271	"
(pz)$_{1.5}$TlCl$_3$	IR	250, 240	"
	R	255, 237	"
(bipy)$_{1.5}$InCl$_3$		238	"
(bipy)$_{1.5}$InBr$_3$		234	33
(bipy)$_{1.5}$InI$_3$		232, 226	"
(bipy)$_{1.5}$In(NCO)$_3$		240	"

Note: See Table 7-15 for chemical abbreviations.

Table 7-20. ν_{MN} Stretching Vibrations in $L_2(OR)CuX_2$ Complexes[52]

$L_2(OR)CuX_2$	ν_{MN}, cm^{-1}
[(2-NH$_2$py)$_2$(OH)Cu]$_2$(ClO$_4$)$_2$	253
[(2-NH$_2$py)$_2$(OH)Cu]$_2$(NO$_3$)$_2$	253
[(2-NH$_2$py)$_2$(OCH$_3$)Cu]$_2$(NO$_3$)$_2$	256, 253
[(2-NH$_2$py)$_2$(OEt)Cu]$_2$(NO$_3$)$_2$	253
[(2-NH$_2$py)$_2$(OPr)Cu]$_2$(NO$_3$)$_2$	253
[(2-NH$_2$py)$_2$(OAm)Cu]$_2$(NO$_3$)$_2$	253

L_2MX_2 Complexes

The complexes L_2MX_2, where L is pyridine or a related ligand, M is a transition metal(II), and X is an anion, have been very extensively studied. These complexes may exist in various forms: in distorted polymeric octahedral structures as the Cu(II) case; in polymeric octahedral structures as for Mn(II), Fe(II), Co(II), and Ni(II); in tetrahedral structures as for Co(II), Ni(II), and Zn(II); in pseudotetrahedral structures (symmetry C_{3v}); and in planar structures as for Pt(II) and Pd(II) (symmetry for the *trans* complexes is D_{2h}; symmetry for the *cis* complexes is C_{2v}).

Table 7-21. Metal–Nitrogen Vibrational Frequencies for L_2MX_2 Complexes (Distorted Polymeric)

L_2MX_2	Frequency, cm^{-1}		
O_h (distorted, polymeric)	ν_{MN}	δ_{NMN}	Reference
$(py)_2CrCl_2$	219		44
$(py)_2CuCl_2$	268	200	"
$(py)_2CuBr_2$	269	196	"
$(\gamma\text{-pic})_2CuCl_2$	285		53
$(\beta\text{-pic})_2CuCl_2$	265		"
$(\beta\text{-pic})_2CuBr_2$	269		"
$(\gamma\text{-pic})_2CuCl_2$	258		"
$(\gamma\text{-pic})_2CuBr_2$	256		"
$(4\text{-}CH_2OHpy)_2CuCl_2$	276		"
$(4\text{-}COCH_3py)_2CuCl_2$	265		"
$(4\text{-}CONH_2py)_2CuCl_2$	264		"
$(4\text{-}CNpy)_2CuCl_2$	243		"
$(py)_2HgCl_2$	<200		44
$(py)_2HgBr_2$	<200		"
$(py)_2CdCl_2$	<200		"
$(2\text{-}Mepy)_2CuCl_2$	259	191	54
$(2\text{-}Mepy)_2CuBr_2$	268, 259	194	"
$(2,6\text{-}diMepy)_2CuCl_2$	241	154(?)	"
$(2,6\text{-}diMepy)_2CuBr_2$	244	140	"
$(2\text{-}Etpy)_2CuCl_2$	246	183	"
$(2\text{-}Etpy)_2CuBr_2$	251(?)	180	"
$(quin)_2CuCl_2$	257	151(?)	"
$(quin)_2CuBr_2$	256	140	"
$(2\text{-}Mepy)_2Cu(NO_3)_2$	247		55
$(2\text{-}Clpy)_2CuCl_2$	234		"
$(2\text{-}Clpy)_2CuBr_2$	231		"
$(2\text{-}Clpy)_2Cu(NO_3)_2$	253		"
$(2\text{-}Brpy)_2CuCl_2$	236		"
$(2\text{-}Brpy)_2CuBr_2$	239		"
$(2\text{-}Brpy)_2Cu(NO_3)_2$	240		"

(?) Obscured by other vibrations.

Distorted Polymeric Octahedral Complexes

Table 7-21 lists the ν_{MN} data and, where available, the δ_{NMN} data. Except for $(py)_2CrCl_2$ and $(py)_2HgX_2$, all of the complexes are Cu(II) complexes. The ν_{MN} vibration varies with the ligand but appears in the region of 231–285 cm^{-1}. The δ_{NMN} vibration occurs in the region of 140–200 cm^{-1}. These complexes may be considered to contain halogen-bridging bonds, with two short and two long metal–halogen bonds and the ligand located *trans* to the polymer chain. This is a manifestation of the Jahn–Teller effect, which occurs readily in Cu(II) complexes.

Polymeric Octahedral Complexes

Table 7-22 records data for ν_{MN} vibration in several complexes. Figures 7-4 and 7-5 show the low-frequency spectra of polymeric and monomeric $(py)_2CoCl_2$ and $(4\text{-Brpy})_2CoBr_2$. Table 7-14 lists the expected number of infrared-active modes for L_2MX_2 complexes existing in a polymeric octahedral structure where the halogen or anions may be considered to be bridging.

Table 7-22. ν_{MN} Stretching Vibrations in Polymeric L_2MX_2 Complexes

L_2MX_2 O_h (polymeric)	Frequency, cm^{-1} ν_{MNCS}	ν_{MN}	Reference
$(py)_2MnCl_2$		212(222)	56
$(py)_2MnBr_2$		212	56
$(py)_2Mn(NCS)_2$	254	<200	51
$(quin)_2MnCl_2$		240, 230	57
$(py)_2Fe(NCS)_2$	261	~200	51
$(py)_2FeCl_2$		222	57
$(py)_2Co(NCS)_2$	268	213	51
$(py)_2CoCl_2$		235, 243	57
$(py)_2Ni(NCS)_2$	280	229	51
$(py)_2NiCl_2$		258, 235	57
$(\beta\text{-pic})_2Ni(NCS)_2$	276	232	51
$(\beta\text{-pic})_2NiBr_2$		256, 234	57
$(\beta\text{-pic})_2NiI_2$		233	57
$(\gamma\text{-pic})_2Ni(NCS)_2$	289	225	51
$(p\text{-tol})_2Ni(NCS)_2$	279	215	51
$(p\text{-tol})_2NiCl_2$		<200	44
$(p\text{-tol})_2NiBr_2$		<200	44
$(py)_2NiBr_2$		258, 234	50
$(quin)_2Ni(NCS)_2$	291	236, 209	51
$(quin)_2NiCl_2$		258, 217	57
$(py)_2Cu(NCS)_2$	319	256, 214	51
$(quin)_2Cu(NCS)_2$	326	247	51
$(py)_2PbCl_2$		<200	44
$(py)_2CdBr_2$		<200	44
$(py)_2CdI_2$		<200	44

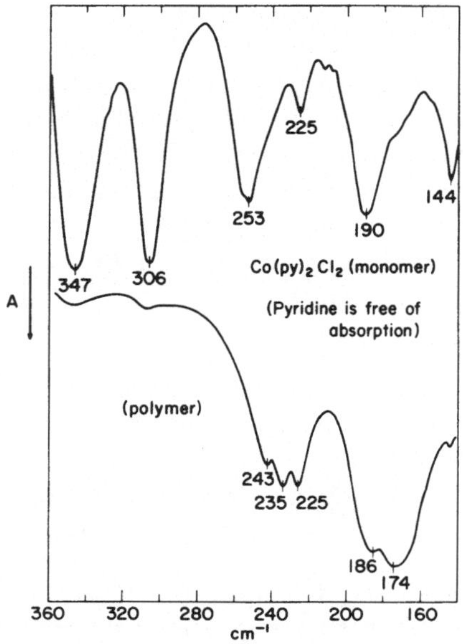

Fig. 7-4. Low-frequency spectra of $(py)_2CoCl_2$ (monomer and polymer).

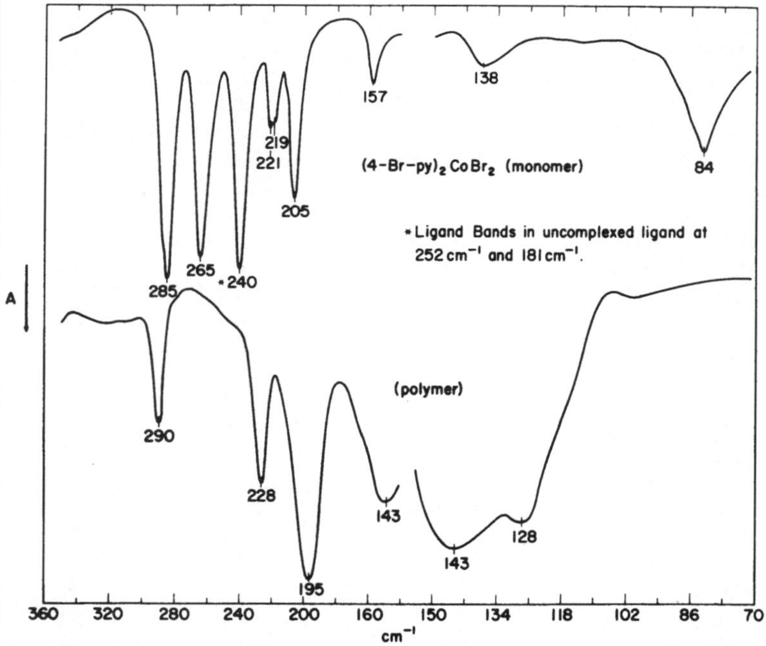

Fig. 7-5. Low-frequency spectra of $(4\text{-}Brpy)_2CoBr_2$ (monomer and polymer).

Table 7-23. ν_{MN} Stretching Vibration in L_2MX_2 Complexes of T_d Symmetry

L_2MX_2 T_d		Frequency, cm^{-1} L = pdz	py	α-pic	β-pic	γ-pic	2-Clpy	2-Brpy	4-Brpy	quin	References
M = Co	X = Cl		243–252	224–260	239	247, 234	227	216			44, 50, 56, 60
	X = Br		239–254	237,222	242	255, 248	224	215	238	249, 236	
	X = I		238–246	229–256	237	236–251	226	218	238		
	X = NCS			245(321, 287)	258(327, 278)						
M = Ni	X = Cl			227–248							44, 50
	X = Br			239						212	
	X = I		234–240	239	233–243						
	$X = BF_4$					200					
M = Zn	X = Cl	210, 202	218	224, 209	204	254, 238	210		223		44, 47, 51, 56, 58, 60, 61
	X = Br	218	219	224, 204	208	239			225		
	X = I		222	224, 214	208	250, 238					
	X = NCS		227	224	224	245, 217			215		
			(312, 268)	(283)	(301, 254)	(309, 256)					
	$X = NO_3$		250, 210								

		L = coll	lut	2-Clpy	2-Brpy	References
M = Co	X = Cl	243		222	222	50, 55
	X = Br	248, 244		222	222	
	X = I	234	244			
M = Ni	X = Cl		244			
	X = Br					
	X = I	256. 232	267, 232			

( ) ν_{MNCS}

For the transition metal (II) complexes, the ν_{MN} vibration is located in the region of 200–258 cm^{-1}. For some complexes the vibration may be lower than 200 cm^{-1}.

Tetrahedral Complexes

Table 7-23 records data for the ν_{MN} vibration for several complexes of the type L_2MX_2 in tetrahedral symmetry. For a complex of this nature with a symmetry group C_{2v} (see Table 7-14) two infrared-active ν_{MN} vibrations ought to be seen. It is not always possible to observe two frequencies because in most cases the bands are broad and not easily resolved at room temperature. The vibrations appear to occur in the range of 200–260 cm^{-1}. The ν_{MNCS} vibration is located at a higher frequency range (254–321 cm^{-1}) reflecting the stronger complexation to the inorganic ligand. Table 7-24 records data for several pseudotetrahedral complexes of the type L_2MX_2 which were assigned on the basis of a C_{3v} point group.

Planar Complexes

The complexes L_2MX_2 may also exist as planar complexes (*cis* or *trans*) when M is Pt(II) or Pd(II). For the *trans* compound, one infrared-active ν_{MN} vibration is allowed, while for the *cis* two ν_{MN} vibrations are allowed. Thus, one should be able to distinguish between a *cis* or *trans* isomer based on their low-frequency spectra. Table 7-25 records data for the *trans* (D_{2h} symmetry) L_2MX_2 complexes and the *cis* (C_{2v} symmetry) complexes. The vibrations are observed to be at higher frequencies than the ν_{MN} vibrations in tetrahedral or polymeric symmetry. In most cases, a second ν_{MN} vibration for the *cis* compound was not observed.

Burgess[59] and others[63,64] have also made assignments for ν_{MN} in L_2MX_2 complexes. However, they have not made any definite structural assignments, and, to keep the size of the tables within reason, their data have not been included.

Table 7-24. ν_{MN} Stretching Vibrations in L_2MX_2 Complexes with a Pseudo T_d Symmetry

L_2MX_2 Pseudo T_d (C_{3v})		Frequency, cm^{-1}		
M	X	L = γ-pic	L = quin	Reference
Mn^{2+}	ClO_4^-	200, 238	200	62
	BF_4^-		200	
Co^{2+}	ClO_4^-	227, 249	210	"
	BF_4^-	230, 256		"
Cu^{2+}	ClO_4^-	241, 291	220	"
	BF_4^-	241, 291		"
Zn^{2+}	ClO_4^-		220	"
	BF_4^-		200	"

Table 7-25. ν_{MN} Stretching Vibration in Planar L_2MX_2 Complexes[65]

L_2MX_2	ν_{MN}, cm^{-1}
Trans — D_{2h}	
$(py)_2PtCl_2$	284
$(py)_2PtBr_2$	300
$(py)_2PtI_2$	293
$(py\text{-}d_5)_2PtCl_2$	269
$(py)_2PdCl_2$	305
$(py)_2PdBr_2$	305
$(py)_2PdI_2$	309
Cis—C_{2v}	
$(py)_2PtCl_2$	260[a]
$(py)_2PtBr_2$	262[a]
$(py)_2PtI_2$	246[a]
$(py\text{-}d_5)_2PtCl_2$	308, 247
$(py\text{-}d_5)_2PtBr_2$	290, 253
$(py)_2PdCl_2$	266[a]
$(py)_2PdBr_2$	279[a]

[a] Second vibration not observed.

L_2MX_4, $[L_2MX_4]Y$, and $L_2M_3X_6$ Complexes

A few complexes of the type L_2MX_4, $[L_2MX_4]Y$, and $L_2M_3X_6$ have been prepared. The ν_{MN} vibrations assigned are recorded in Table 7-26.

L_3MX_3 and L_3MX_2 Complexes

The ν_{MN} vibrations for the monomeric octahedral complexes of the type L_3MX_3 and L_3MX_2 are listed in Table 7-27. For the *trans* complex three infrared-active ν_{MN} vibrations are expected, while for the *cis* complex only

Table 7-26. ν_{MN} Stretching Vibrations for L_2MX_4, $[L_2MX_4]Y$, and $L_2M_3X_6$ Complexes

Complex	ν_{MN}, cm^{-1}	Reference
L_2MX_4		
cis—D_{4h}		
$(py)_2PtCl_4$	280, 267	14
$(py)_2PtBr_4$	282, 267, 263	14
$[L_2MX_4]Y$		
trans-$[(py)_2IrCl_4]pyH$	265	44
cis-$[(py)_2IrCl_4]pyH$	262, 255	44
$L_2M_3X_6$		
$(quin)_2Mn_3Cl_6$	262	57
$(quin)_2Ni_3Cl_6$	270, 212	57

Table 7-27. ν_{MN} Stretching Vibrations for L_3MX_3 and L_3MX_2 Complexes

Complex	ν_{MN}, cm^{-1}		Reference
L_3MX_3			
O_h (monomeric)			
$(py)_3CrCl_3$	221		44
trans-$(py)_3RhCl_3$	265, 245, 230		"
cis-$(py)_3RhCl_3$	266, 245		"
trans-$(py)_3IrCl_3$	272, 264, 255		"
cis-$(py)_3IrCl_3$	270, 266		"
py_3InCl_3[a]	IR 195, 184; R 200, 189	in CH_3CN	32
	R 197	in EtOH	32
L_3MX_2			
$(py)_3Zn(NO_3)_2$	200		61
$(py)_3Cd(NO_3)_2$	160		"

[a] Structure in doubt.

two infrared-active ν_{MN} vibrations are expected (see Table 7-14). The structure of $(py)_3InCl_3$ is still in doubt.

L_4MX_2 Complexes

Complexes of the type L_4MX_2 may involve several structures, depending on whether or not the anion is coordinated. Table 7-28 summarizes the data on ν_{MN} vibrations for two zinc complexes, which are considered to be T_d having no anion bonding. Also included are complexes of Mn(II), Co(II), Ni(II), Cu(II), and Zn(II), which may involve total or partial anion coordination and are considered to involve a C_{3v} symmetry.

If the halides are completely coordinated to the cation, a monomeric octahedral symmetry ensues. Results for several of these complexes are tabulated in Table 7-28. Further a *trans* or *cis* structure is possible. Table 7-14 records the number of infrared-active ν_{MN} vibrations expected for the two isomers. Again it should be mentioned that not all the allowed vibrations may be observed.

Table 7-29 lists the frequency ranges found for ν_{MN} for the L_nMX_m complexes. Except for the LMX_3 ions, the vibration is located in the range of ~200–300 cm^{-1}. Since L varies in the various complexes so drastically, trends in the vibration for various metals were not attempted. Wherever comparisons are possible the coordination effect is noted, ν_{MN} vibrations in T_d symmetry occurring at higher frequency than those found in O_h symmetry. From the limited number of data available at present, it appears as if the second- and third-row transition metals show higher-energy metal–nitrogen vibrations than the first row. Some question still exists concerning

Table 7-28. ν_{MN} Stretching Vibrations for L_4MX_2 and $L_4MX_2^+$ Complexes

			ν_{MN}, cm^{-1}			
Complex	M^{2-}	X^-	L = py	L = γ-pic	quin	Reference
L_4MX_2						
C_{3v}	Mn	ClO_4^-	222, 233	200, 238	200	62
	Mn	BF_4^-	220		200	″
	Mn	$(F^-)BF_4^-$			215, 230	″
	Co	ClO_4	233, 240	227, 249	210	″
	Co	BF_4^-	235, 254			″
	Ni	ClO_4^-	240, 262			″
	Ni	BF_4^-	250, 266	230, 206		″
	Cu	ClO_4^-	264, 283	241, 291	220	″
	Cu	$(F^-)(BF_4^-)$			233, 236	″
	Cu	BF_4^-	264, 285	241, 291		″
	Zn	ClO_4^-			220	″
	Zn	BF_4^-			200	″
T_d	Zn	ClO_4^-	220	200		″
	Zn	BF_4^-	210	200		″
			ν_{MNCS}	ν_{MN}		
O_h	$(py)_4Mn(NCS)_2$		253	<200		51
	cis-$(py)_4Fe(NCS)_2$		266(250)	~203		″
	trans-$(py)_4Fe(NCS)_2$		267	~203		″
	$(py)_4Co(NCS)_2$		270	215, 205		″
	$(py)_4Ni(NCS)_2$		285	233		″
	$(py)_4NiI_2$			263, 234		57
	$(py)_4Cu(NCS)_2$		311	245		51
	$(coll)_4NiCl_2$			286, 226		57
	$(\beta\text{-pic})_4NiBr_2$			241		″
	$(\gamma\text{-pic})_4NiBr_2$			267, 228		″
	$(coll)_4NiBr_2$			264, 217		″
	$(\beta\text{-pic})_4NiI_2$			263, 234		″
	$(\gamma\text{-pic})_4NiI_2$			234		″
	$(lut)_4NiI_2$			266, 257, 232		″
	trans-$(py)_4CoCl_2$			217		44
	trans-$(py)_4CoBr_2$			214		″
	trans-$(py)_4NiCl_2$			236		″
	trans-$(py)_4NiBr_2$			235		″
	trans-$(py)_4NiI_2$			240, 228		″
	trans-$(\beta\text{-pic})_4NiI_2$			225, 210		″
$[L_4MX_2]^+$						
	$[(py)_4IrCl_2]Cl$			260, 255		″
	$[(py)_4RhCl_2]Cl$			246		″

Table 7-29. Frequency Ranges for the ν_{MN} Vibration in Various L_nMX_m Complexes

Complex type		ν_{MN}, cm^{-1}										
		Cr	Mn	Fe	Co	Ni	Cu	Zn	Cd	Hg	Au	In	Tl
LMX_2	(distorted O_h)		189–236	208–219	199–221		251–284	194–253	186–205				
	T_d				211–238			208–236		165–208			
LMX_3^-	(pseudo T_d—C_{3v})				134–226	136–196		138–207					
LMX_3	(C_{2v})										227–306		
$(L_{1.5}MX_3)_2$												226–271	240–250
L_2MX_2	(distorted O_h)	219					231–285			<200			
	(polymeric O_h)		<200–222	<200–222	213–235	<200–259	214–258						
	(T_d)				215–260	227–243		204–254					
				Pd		Pt							
	(planar D_{2h}—*trans*)			305–309		269–300							
	(planar C_{2v}—*cis*)			266–299		246–308							
L_2MX_4	(planar D_{4h}—*cis*)					263–282							
				Rh		Ir							
L_3MX_3	(monomeric O_h—*trans*)			230–265		255–272							
	(monomeric O_h—*cis*)			245–266		266–270							
				Mn		Co		Ni		Cu		Zn	
L_4MX_2	(C_{3v})			200–238		210–254		206–266		220–283		200–220	
	(T_d)											200–220	
				Mn		Fe		Co		Ni		Cu	
	(O_h)			<200		~203		205–215		217–286		245	

Table 7-30. Frequency Assignments in Metal–Ethylenediamine Complexes

Complex	Frequency, cm^{-1} $(NH_2)_{rock}$	ν_{MN}	δ_{NMN}	Reference
$(en)TiCl_4$		390, 350	315	69
$(en)TiBr_4$		365, 330		″
$[(en)_2CoCl_2]ClO_4$		511		70
$[(en)_2CoBr_2]ClO_4$		506		″
$[(en)_2CoCl_2]NO_3$		512		″
$[(en)_2CoBr_2]NO_3$		513		″
$(en)_2PdCl_2$	804	583, 518	290	71
$(en)_2PdBr_2$	793	578, 514	272	″
$(en)_2PdI_2$	776	571. 508	255	″
$(en)_2PtCl_2$	835	589, 546	290	″
$(en)_2PtBr_2$	825	585, 543	276	″
$(en)_2PtI_2$	813	579, 537	264	″
$(en)_3RhI_3$		570, 558		72
Raman studies				
$(en)_3ZnCl_2$ H_2O soln		423		68
D_2O soln		405		″
$(en)_3CdCl_2$ H_2O soln		400		″
D_2O soln		370		″
$(en)_3Hg(NO_3)_2$ H_2O soln		422		″
D_2O soln		400		″
Co^{III} Compounds	$\nu_{MN} + \rho_{NH_2}$	δ_{NMN}		
Cis compounds				
$[Co(en)_2Cl_2]Cl$	578, 566, 550, 518, 508	302, 270, 255		74
$[Co(en)_2Br_2]Br$	565, 535, 510, 501	281, 273, 267		″
$[Co(en)_2ClBr]Br$	568, 546, 534, 505	303, 275, 255		″
$Co(en)_2(NO_2)_2NO_3$	584, 574, 572, 553	320, 296, 278		″
$[Co(en)_2Cl(NO_2)]Cl$	585, 573, 568, 521, 515	299, 293, 260		″
$[Co(en)_2(CO_3)]Br$	580, 571, 540, 518	350, 298, 280, 273, 265		″
$Co(en)_2Cl(H_2O)SO_4$	580	289, 269		″
$Co(en)_2Cl(H_2O)Br_2$	578	299, 292, 278, 273		″
$Co(en)_2Br(H_2O)Br_2$	578	292, 281		″
$Co(en)_2(NO_2)(H_2O)(ClO_4)_2$	588	298, 289, 278		″
$Co(en)_2Cl(NCS)Cl$	573	291, 281		″
$Co(en)_2(N_3)SO_3$	577	289, 272		″
Complex		ν_{MN}		
$[Rh(MEM)_3]I_3$		586, 574, 556, 502		72
$[Rh(UDMEN)_2Cl_2]I$		575, 531		″
$[Rh(sdMEN)_2Cl_2]Cl$		498, 474		″
$[Rh(sdMEN)_2Cl_2]I$		505, 475		″
$[Rh(tMEN)_2Cl_2]Cl$		518, 490		″
$[Rh(tMEN)_2Cl_2]I$		515, 497		″
$[Rh(tetMEN)_2Cl_2]Cl$		537, 517		″

Table 7-30 (Continued)

(L)Pt(SCN)$_2$	ν_{PtN}	ν_{PtS}	Reference
L =			
en	578, 542	329, 285	73
pn	585, 540	318, 279	"
N-Me-en	571, 500	318, 270	"
N-Et-en	580, 530	314, 278	"
N-isopropyl-en	578, 531	312, 258	"
1,2-diaminoisobutane	561, 554	313, 285	"
N,N′-diMe-en	570, 540	324, 268	"
N,N′-diEt-en	562, 540	315, 268	"
N,N-diMe-en	585, 532	319, 270	"
N,N-diEt-en	570, 505	317, 279	"
N,N-diMe, N′-Et-en	556, 511	315, 258	"
N,N,N′,N′-tetraMe-en	532, 511	314, 280	"
N,N-diMe, N′,N′-diEt-en	560, 517	310, 287	"
N-φ-en	569, 520	320, 291	"

Abbreviations: en = ethylenediamine; MEM = N-methylethylenediamine; sdMEN = N,N′-dimethylethylenediamine; UDMEN = N,N′-dimethylenediamine; tMEN = N,N,N′-trimethylethylenediamine; tetMEN; = N,N,N′,N′-tetramethylethylenediamine; pn = propylenediamine; φ = phenyl.

the position of the NMN deformation, although it is expected to be in the 100–200 cm^{-1} region.

7.6. METAL–ETHYLENEDIAMINE AND RELATED COMPLEXES

Certain discrepancies have appeared in the assignments of the metal–nitrogen vibrations in metal–ethylenediamine complexes. Large frequency ranges are assigned for the ν_{MN} vibration. Powell and Sheppard[66] have assigned the vibration in the range of 500–600 cm^{-1}, and Durig[67] has made assignments for the PdN and PtN in the 430–520 cm^{-1}. Krishnan and Plane[68] assigned the vibration in the range of 370–450 cm^{-1}. The difficulty in making the assignment may be due to the coupling of the vibration with the NH_2 rocking mode and with a low-frequency skeletal mode. The metal–nitrogen bond in ethylenediamine complexes is strengthened primarily by the chelation effect, as any π interaction would be slight.

Table 7-30 tabulates assignments for ν_{MN} and ν_{NMN} for several metal–ethylenediamine and related complexes. Further studies appear in order, particularly those involving Raman, deuteration, and N^{15} isotopic substitution.

Table 7-31 shows the ν_{MN} vibrational assignment for several dialkylamido complexes of the type $M(NR_2)_X$ of Ti, V, Zr, Nb, Hf, Ta, and Th. The assignments are in the range of 528–667 cm^{-1}. The order of frequency is i-$Bu_2N > n$-$Bu_2N = n$-$Pr_2N > Et_2N > Me_2N$. This is the reverse trend that

Table 7-31. Metal–Nitrogen Stretching Frequencies for ML_4 Complexes where L = R_2N[75]

Complex	Frequency, cm^{-1}							
	Me_2N	Et_2N	n-Pr_2N	n-Bu_2N	i-Bu_2N	MeBuN	$C_5H_{10}N$	$CH_3C_5H_{10}N$
TiL_4	592	610	667	667	700	640	640	667
ZrL_4	537	577	637	637	677	606	620	601
HfL_4	533	577	637			605	620	600
VL_4	595	613						
NbL_4		587						
ThL_4		540						
R=TaL_3		585						
R′=TaL_3			630					
R″=TaL_3				634		608	625	
TaL_5	528							

R=EtN; R′=n-PrN; R″=n-BuN.

should be observed for a mass or steric effect. The trend was explained on the basis of a nitrogen–metal $p\pi$–$d\pi$ bonding.[75]

Table 7-32 tabulates data for ν_{MN} in *trans* Pt(II) and Pt(IV) complexes with CH_3NH_2. The assignments appear to be lower than the previous assignments made in Table 7-30 for Pt(II) complexes of substituted ethylenediamines. The assignments for the ν_{MN} vibrations in complexes of the type $(CH_3)_2N_{b-a}$ SiX_a are listed in Table 7-33.

Table 7-32. ν_{MN} Assignments in Pt(II) and Pt(IV) Complexes with CH_3NH_2[76]

trans Pt^{IV} compounds		*trans* Pt^{II} compounds	
R = CH_3NH_2	ν_{asym} (MN), cm^{-1}	R = CH_3NH_2	ν_{asym} (MN), cm^{-1}
$Pt(R)_2Cl_4$	514	$Pt(R)_2Cl_2$	513
$Pt(CH_3ND_2)_2Cl_4$	487	$Pt(CH_3ND_2)_2Cl_2$	485
$Pt(R)_2Br_4$	510	$Pt(R)_2Br_2$	512
$Pt(R)_2Br_2Cl_2$	511	$Pt(R)_2I_2$	508
$Pt(R)_2Cl_2(CN)_2$	523	$Pt(R)_2(NO_2)_2$	512
$Pt(R)_2Br_2(CN)_2$	526	$Pt(R)_2Cl(SCN)$	511
$Pt(R)_2Cl_2(NO_2)_2$	522	$Pt(R)_2Br(SCN)$	508
$Pt(R)_2Cl(NO_2)CN(NO_2)$	526	$Pt(R)_2I(SCN)$	507
$Pt(R)_2Br_2(CN)(NO_2)$	523	$Pt(R)_2Br(NO_2)$	491
$Pt(R)_2Br_2Cl(NO_2)$	516	$Pt(R)_2I(NO_2)$	513
$Pt(R)_2Br_2(NO_2)_2$	509	$Pt(R)_2(NO_2)(NO_3)$	511
$Pt(R)_2(NO_2)_4$	522		
$Pt(R)_2(SCN)_4$	514		
$Pt(R)_2Br_2(SCN)_2$	508		
$Pt(R)_2Br_2(SCN)(NO_2)$	519		

Table 7-33. ν_{MN} Assignments in $(CH_3)_2N_{b-a}SiX_a$ Complexes[77]

Complex	Frequency, cm^{-1}	
	ν_{asym} (MN)	ν_{sym} (MN)
$[(CH_3)_2N]SiCl_3$	721	
$[(CH_3)_2N]_2SiCl_2$	741	686
$[(CH_3)_2N]_3SiCl$	731	626
$[(CH_3)_2N]_4Si$	710	570
$[(CH_3)_2N]_2Si(CH_3)_2$	699	580
$[(CH_3)_2N]_3Si(CH_3)$	679	575
$[(CH_3)_2N]_4Ge$	578	551
$[(CH_3)_2N]_4Sn$	535	516
$[(CH_3)_2N]_4Ti$	590	532

Table 7-34. Vibration Assignments in Metal–Hydrazine Complexes

Compound	Frequency, cm^{-1}			References
	ν_{NN}	$(NH_2)_{rock}$	ν_{MN}	
N_2H_4	873	780		80
$(N_2H_4)_2MnCl_2$	960	590, 518	343	78
$(N_2H_4)_2FeCl_2$	964	607, 553	369	78
$(N_2H_4)_2CoCl_2$	974	625, 582	388	78, 81
$(N_2H_4)_2CoSO_4$			388	81
$(N_2H_4)_2NiCl_2$	978	649, 613	409	78, 81
$(N_2H_4)_2NiSO_4$			403	81
$(N_2H_4)_2CuCl_2$	985	682, 658	440	78, 81
$(N_2H_4)_2CuSO_4$			440	81
$(N_2H_4)_2ZnCl_2$	976	625, 580	385	78, 81
$(N_2H_4)_2ZnSO_4$			388	81
$(N_2H_3)_2Zn$	908	593	473, 409	82
$(N_2H_4)_2CdCl_2$	963	593, 517	388	78
$(N_2D_4)_2CdCl_2$	957	451, 400	306	78
$(N_2H_4)_2HgCl_2$	952	697		80
$(N_2H_4)HgBr_2$	976	640	440	80
$(N_2H_2)Hg_2Cl_2$	1012		599, 537, 518, 488, 475	80
$(N_2D_2)Hg_2Cl_2$	965		599, 537, 518, 488, 475	80
$(UO_2)_2(N_2H_4)(OH)_2SO_4$			590	79
$UO_2(N_2H_4)_2SO_4$	902		582	79
$Th(N_2H_4)_2(SO_4)_2$			527	79
$Ce(N_2H_4)_2(SO_4)_2$			553, 505	79
$Ni(N_2H_4)_2(SO_4)_2$			614	79

7.7. METAL–HYDRAZINE COMPLEXES

The assignments for the ν_{MN} vibration in metal–hydrazine complexes are still uncertain. In 1963, Sacconi and Sabatini[78] working with the hydrazine complexes of Mn(II), Fe(II), Co(II), Ni(II), Cu(II), Zn(II), and Cd(II) halides, assigned the ν_{MN} vibration at 339–473 cm^{-1}. However, recent assignments by Athavale *et al.*[79] for UO_2(II), Th(IV), Ce(IV), and Ni(II) hydrazine complexes were in the range of 505–614 cm^{-1}. Further research appears necessary. The ν_{NN} vibration is reported to occur in the 902–1012 cm^{-1} region and the NH_2 rock at 517–682 cm^{-1}. Table 7-34 lists the ν_{NN}, ν_{MN}, and the NH_2 rock in several metal–hydrazine complexes.

7.8. METAL COMPLEXES CONTAINING NITRILE GROUPS

In recent years several complexes involving nitrile groups have been prepared. In complexes of the type L_6MX_2, where M can be Mg^{2+}, Mn^{2+}, Fe^{2+}, Co^{2+}, Ni^{2+}, Cu^{2+}, Zn^{2+}, or Cd^{2+}, L is CH_3CN, and X is a heavy anion of the type MCl_4^{2-}, assignments for the ν_{MN} have been made in the region of 202–330 cm^{-1}. For Pd^{2+} and Pt^{2+} complexes of the type L_4MX_2, assignments were made at 420–444 cm^{-1}. The frequency order for this vibration was Mn$<$Fe$<$Co$<$Ni$<$Cu$>$Zn, in agreement with the Irving–Williams stability order. The assignments for the L_6M^{2+} entity are based on an octahedral symmetry. Assignments for a M–NCC wagging vibration were made in the region of 160–245 cm^{-1}. Table 7-35 tabulates these data.

Assignments for $SnCl_4 \cdot 2L$, where L is CH_3CN, CH_2CHCN, or φCN, are also shown in Table 7-35. These vibrations are found in the vicinity of 192–248 cm^{-1}. Other assignments were made for the Pt(II) and Pd(II) complexes of the type $MX_2 \cdot 2L$. These assignments are in the region of 90–125 cm^{-1} and appear to be much lower than the other assignments for Pd(II). Some doubt is thus cast on the ν_{MN} vibration in nitrile complexes, and more work is necessary to elucidate this frequency range.

7.9. METAL ISOTHIOCYANATE AND RELATED COMPLEXES

The thiocyanate ligand is a very interesting ligand, for it may complex through the sulfur atom, through the nitrogen atom, or through both as a bridging ligand. Infrared spectroscopy is very helpful in elucidating which type of bonding exists. Table 7-36 summarizes data for the ν_{MN} vibrations in several isothiocyanate complexes of the type $M(NCS)_6^{n-}$ and $M(NCS)_4^{n-}$ where M is a transition metal(II). Table 7-37 collects data for isothiocyanate complexes of the type $M(NCS)_6^{n-}$ where M is Nb and Ta. The bending vibrations δ_{MNCS} are assigned at higher frequency than the

Table 7-35.   Assignments for ν_{MN} and (M–NCC)$_{wag}$ Vibrations in
Metal–Nitrile Complexes

Octahedral symmetry	Frequency, cm^{-1}		Reference
	ν_{MN}	(M–NCC)$_{wag}$	
L$_6$Mg(AlCl$_4$)$_2$	330	235	83
L$_6$Mg(TlCl$_4$)$_2$	330	235	"
L$_6$Mn(TlCl$_4$)$_2$	237	160	"
L$_6$Mn(SbCl$_6$)$_2$	237	165	"
L$_6$Mn(FeCl$_4$)$_2$	238	163	"
L$_6$Mn(AlCl$_4$)$_2$	237	166	"
L$_6$Fe(GaCl$_4$)$_2$	246	185	"
L$_6$Fe(InCl$_4$)$_2$	246	185	"
L$_6$Fe(TlCl$_4$)$_2$	246	184	"
L$_6$Co(ClO$_4$)$_2$	257	202, 195	"
L$_6$Co(GaCl$_4$)$_2$	251	199	"
L$_6$Co(InCl$_4$)$_2$	252	198	"
L$_6$Ni(SbCl$_6$)$_2$	271	216	"
L$_6$Ni(GaCl$_4$)$_2$	268	220, 210	"
L$_6$Ni(FeCl$_4$)$_2$	268	215	"
L$_6$Ni(TlCl$_4$)$_2$	270	211	"
L$_6$Ni(ClO$_4$)$_2$	272	231, 208	"
L$_6$Cu(FeCl$_4$)$_2$	310	219	"
L$_6$Cu(SbCl$_6$)$_2$	315	220	"
L$_6$Cu(GaCl$_4$)$_2$	308	220	"
L$_6$Zn(FeCl$_4$)$_2$	213	182	"
L$_6$Zn(TlCl$_4$)$_2$	215	181	"
L$_6$Zn(SbCl$_6$)$_2$	215	180(?)	"
L$_6$Cd(FeCl$_4$)$_2$	204		"
L$_6$Cd(SbCl$_6$)$_2$	206		"
L$_6$Cd(TlCl$_4$)$_2$	202		"
L$_4$CuClO$_4$ ⎫	163		"
L$_4$CuGaCl$_4$ ⎪	163		"
L$_4$CuInCl$_4$ ⎬ T$_d$	163		"
L$_4$CuBF$_4$ ⎪	163		"
L$_4$AgClO$_4$ ⎭	187		"
L$_4$CuSnCl$_6$ ⎫	320	245	"
L$_4$Cu(ClO$_4$)$_2$ ⎪ Square	330	230	"
L$_4$Pd(SbCl$_6$)$_2$ ⎬ planar	440, 420	252, 235	"
L$_4$Pd(TlCl$_4$)$_2$ ⎭	444, 420	251, 235	"

Cis compounds		ν_{MX}	
SnCl$_4$·2MeCN	222, 207	366, 339, 305	84
SnCl$_4$·2CH$_2$CHCN	204, 192	360, 340, 306	"
SnCl$_4$·2C$_6$H$_5$CN	248, 220	357, 345, 339, 304	"
PtCl$_2$·2MeCN	IR 120, 100	358, 347, 330	85
	R 124	362, 354, 336	"
PtBr$_2$·2MeCN	IR 116, 95	246	"
	R 117, 104	258, 245	"

Table 7-35 (Continued)

Compound	Frequency, cm^{-1} ν_{MN}	ν_{MX}	Reference
Cis compounds			
$PtCl_2 \cdot 2C_6H_5CN$	IR 109, 104, 100	355, 344	85
	R	369, 340	"
Trans compounds			
$PdCl_2 \cdot 2MeCN$	IR 125	361	"
$PtBr_2 \cdot 2MeCN$	IR 117	231	"
	R 107	231, 211	"
$PdCl_2 \cdot 2C_6H_5CN$	IR 104	368	"
	R 92	306	"
$PtBr_2 \cdot 2C_6H_6CN$	R 90	234, 204	"

L=CH_3CN

Table 7-36. ν_{MN} **Assignments in Metal–Isothiocyanate Complexes**[86]

Complex	ν_{MN}, cm^{-1}
O_h symmetry (anion)	
$(C_9H_8N)_4[Mn(NCS)_6]$	~222
$(C_9H_8N)_4[Ni(NCS)_6]$	~237
$[(CH_3)_4N]_4[Ni(NCS)_6]$	245
$[(C_2H_5)_4N]_4[Ni(NCS)_6]$	239
$[(CH_3)_4N]_3[Fe(NCS)_6]$	298, 272, 233
$[(CH_3)_4N]_3[Cr(NCS)_6]$	364
$[(CH_3)_4N]_3[Mo(NCS)_6]$	303
T_d symmetry (anion)	
$[n\text{-}C_4H_9(C_6H_5)_3P]_2Mn(NCS)_4$	287
$[(C_2H_5)_4N]_2Fe(NCS)_4$	293
$[(CH_3)_4N]_2Co(NCS)_4$	304
$[(C_2H_5)_4N]_2Co(NCS)_4$	307
$[n\text{-}C_4H_9(C_6H_5)_3P]_2Co(NCS)_4$	303
$(cat^{2+})Co(NCS)_4$[a]	303
$[(CH_3)_4N]_2Zn(NCS)_4$	280
Distorted T_d (anion)	
$(cat^{2+})Ni(NCS)_4$	309, 281
$[(C_6H_5)_4As]_2Ni(NCS)_4$	309
$[(C_6H_5)_4As]_2Ni(NCS)_4$	294, 266, 230
$[(CH_3)_4N]_2Ni(NCS)_4$	269, 235
$[(C_2H_5)_4N]_2Cu(NCS)_4$	327, 249
$(cat^{2+})Cu(NCS)_4$	313, 294

[a] cat^{2+} = [p-xylylenebis (triphenylphosphonium)]$^{2+}$.

Table 7-37. Metal–Nitrogen Assignments in Several Metal–
Isothiocyanate Complexes[87]

Complex	Sample condition	Frequency, cm^{-1}		
		$\delta_{(MNCS)_1}$	$\delta_{(MNCS)_2}$	ν_{MN}
K[Nb(NCS)$_6$]	Mull	449–490	399	340
	Soln A	511	402	353
(C$_4$H$_9$)$_4$N[Nb(NCS)$_6$]	Mull	502		330
	Soln A	510	395	355
	Soln B	505	412	341
(C$_6$H$_5$)$_4$As[Nb(NCS)$_6$]	Mull	510		335
	Soln B	506	402	341
K$_2$[Nb(NCS)$_6$]	Mull	504		342
	Soln A	512	392	330
K[Ta(NCS)$_6$]	Mull	501	400	313
	Soln A	508	398	343
(C$_4$H$_9$)$_4$N[Ta(NCS)$_6$]	Mull	504		298
	Soln A	509	400	344
(C$_6$H$_5$)$_4$As[Ta(NCS)$_6$]	Mull	505	401	314
	Soln B	506	406	322

$\delta_1 = $ M—N—C—S; $\delta_2 = $ M—N—C—S; Soln A = acetonitrile; Soln B = 1,2-dichloroethane.

Table 7-38. Various Assignments in Metal–Isothiocyanate Complexes[88]

Complex	Symmetry (anion)	Frequency, cm^{-1}		
		ν_{CS}	ν_{MNCS}	δ_{NCS}
(C$_2$H$_5$)$_3$PH[V(NCS)$_6$]	O_h	840	310	485
K$_3$[Cr(NCS)$_6$]	O_h	820	358	474
[(C$_2$H$_5$)$_4$N]$_3$[Cr(NCS)$_6$]	O_h	820		481
[(C$_2$H$_5$)$_4$N]$_3$[Fe(NCS)$_6$]	O_h	823	270	479
NH$_4$[Cr(NCS)$_4$(NH$_3$)$_2$]	D_{4h}	827	342	467
ND$_4$[Cr(NCS)$_4$(ND$_3$)$_2$]	D_{4h}	827	342	467
pyH[Cr(NCS)$_4$(NH$_3$)$_2$]	D_{4h}	837	345	489
pyH[Cr(NCS)$_4$(ND$_3$)$_2$]	D_{4h}	837	343	488
chol[Cr(NCS)$_4$(NH$_3$)$_2$]	C_{2h}	840	346	490
chol[Cr(NCS)$_4$(ND$_3$)$_2$]	C_{2h}	840	346	481
K[Cr(NCS)$_4$(py)$_2$]	C_{2v}	840	382, 335, 322	484
(C$_2$H$_5$)$_4$N[Cr(NCS)$_4$(py)$_2$]	C_{2v}	848	382, 362, 334	488, 485, 483
K[Cr(NCS)$_4$(bipy)]	C_{2v}	847	382, 360	485, 478
(C$_2$H$_5$)$_4$N[Cr(NCS)$_4$(diars)]	C_{2v}		358, 341	486
(CH$_3$)$_3$PH[Cr(NCS)$_4$(CH$_3$)$_3$P)$_2$]	D_{4h}	846	368	480
(C$_2$H$_5$)$_4$N[Cr(NCS)$_4$((CH$_3$)$_3$P)$_2$]	D_{4h}	846	367	480
(C$_2$H$_5$)$_3$PH[Cr(NCS)$_4$((C$_2$H$_5$)$_3$P)$_2$]	D_{4h}	846	362	482
(C$_2$H$_5$)$_4$N[Cr(NCS)$_4$((C$_2$H$_5$)$_3$P)$_2$]	D_{4h}	848	357	482
(n-C$_4$H$_9$)$_3$PH[Cr(NCS)$_4$((n-C$_4$H$_9$)$_3$P)$_2$]	D_{4h}	846	362	482
C$_6$H$_5$P(CH$_3$)$_2$H[Cr(NCS)$_4$(C$_6$H$_5$P(CH$_3$)$_2$)$_2$]	D_{4h}	845	365	481
C$_6$H$_5$P(C$_2$H$_5$)$_2$H[Cr(NCS)$_4$(C$_6$H$_5$P(C$_2$H$_5$)$_2$)$_2$]	D_{4h}	855	365	484
K[Cr(NCS)$_4$(C$_6$H$_5$P(C$_2$H$_5$)$_2$)$_2$]	C_{2v}		361, 342	484
(C$_2$H$_5$)$_4$N[Cr(NCS)$_4$((C$_6$H$_5$)$_2$P(C$_2$H$_5$)$_2$]	(?)		362, 356, 348	478

chol = R$_3$PH$^+$, where R = Me, Et, n-Bu, or mixed alkyl-aryl groups.

Table 7-39. ν_{MN} and ν_{MS} in Complexes Involving Bridging NCS Groups[89]

Complex	Frequency, cm^{-1}	
	ν_{MN}	ν_{MS}
$(\varphi_3As)_2Pd(NCS)_2$	265	
$(bipy)\cdot Pd(NCS)_2)_2$	261	
$[(Et_4dien)Pd(NCS)]SCN$	268	
$(\varphi_3P)_2Pd(NCS)_2$	270	
Bridging NCS		
$MnHg(NCS)_4$	278	213
$FeHg(NCS)_4$	296, 256	216
$CoHg(NCS)_4$	303, 276	219
$NiHg(NCS)_4$	303	
$CuHg(NCS)_4$	318, 282	
$ZnHg(NCS)_4$	327, 250	217
$ZnCd(NCS)_4$	303, 276	230

ν_{MN} vibration. Table 7-38 shows infrared data for several other isothiocyanate complexes. Also listed is the ν_{CS} vibration, which is seen to move to higher frequency when the bonding is to the nitrogen atom. Table 7-39 shows the assignments for ν_{MN} and ν_{MS} in bridging thiocyanates.

Much less work has been done with the isocyanate complexes. Table 7-40 tabulates data for several of these complexes.

The ν_{MN} vibrations in the above complexes appear to be fairly well established. The ν_{MN} vibration in isothiocyanate complexes appears to range from 222 to 355 cm^{-1} depending on the metal and is at lower frequency than the ν_{MN} vibrations in the isocyanates, which occur in the range of 298–410 cm^{-1}. The ν_{CS} vibrations in the isothiocyanates are located at higher frequency than those found in the thiocyanates. The cyanide stretching region at ~2000–2100 cm^{-1} is also useful in making structure assignments. For example, in M–NCS compounds the ν_{CN} vibration occurs at lower frequency than for the M–SCN complexes.

Only a few metal–selenocyanate complexes have been prepared, and these indicate metal–selenium bonding and thus will not be reported in this chapter.

7.10. METAL–AMINO ACID COMPLEXES

These complexes have been extensively investigated in the mid-infrared region, but it has only been in recent years that the low-frequency region has been studied and at present there is some disagreement on the low-frequency assignments. One assignment is based on an ionic M–O bond with no ν_{MO} vibrations possible.[95] Essentially this would involve a monodentate amino

Table 7-40. Assignments Made in Several Metal–Isocyanate Complexes

Complex	Sample condition	Frequency, cm^{-1}					Reference
		ν_{sym}(NCO)	δ_A(M–NCO)	δ_B(M–NCO)	ν_{MN}	δ_{MN}	
Tetrahedral T_d							
$[Et_4N]_2[Mn(NCO)_4]$	Mull	1338	623		325		91
$[Et_4N]_2[Fe(NCO)_4]$	Mull	1337	619		325		″
$[Et_4N]_2[Co(NCO)_4]$	Mull	1335	620, 617		345		″
$[Et_4N]_2[Ni(NCO)_4]$	Mull	1330	619, 617		341		″
$[Et_4N]_2[Cu(NCO)_4]$	Mull	1328	619, 617, 612		338		″
$[Et_4N]_2[Zn(NCO)_4]$	Mull	1335	623		321		″
$[Et_4N]_2[Cd(NCO)_4]$	Mull	1328	621		298		″
$[\varphi_4As][Fe(NCO)_4]$	Mull	1370	626, 619		410		90
$[Et_4N]_2[Zn(NCO)_4]CH_3NO_2$	Soln			187	330, 326		92
$Si(NCO)_4$		1482	608	546	727	338	93
$Si(NCO)_4$				338			90
$Ge(NCO)_4$		1426	608	528	672	259	93
$Ge(NCO)_4$			528	259		198	90
D_{4h}							
$[Me_4N]_2[Pd(NCO)_4]$	Mull	1319	613, 604, 594		408, 384, 350		94
	Acetone		614, 595				″
O_h							
$[Me_4N]_2[Sn(NCO)_6]$	Mull	1307	667, 622		383		90

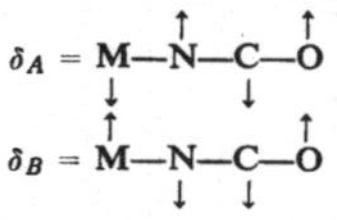

acid bonded only through the nitrogen atom. The other assignment considers the M–O bond to be highly covalent and therefore ν_{MO} vibrations are assigned.[96] This mode would involve a bidentate amino acid coordinated through both the nitrogen and the oxygen atom. Nakamoto[96] has made assignments based on the latter consideration. Lane[97] has made assignments based on Quagliano's view[95] that the M–O bond in these complexes is essentially ionic. Normal coordinate treatment of these molecules has not unambiguously decided between the two cases, and at this time the matter is not resolved. Data from both views will be presented.

Table 7-41 presents data on several metal–glycine compounds. Table 7-42 records data on several metal–amino acid complexes, where the amino acids are alanine, leucine, and valine. In the work by Walter *et*

Table 7-41. Frequency Assignments Made in Metal–Glycine Complexes

Complex	Frequency, cm^{-1}			Reference
	$(NH_2)_{rock}$	ν_{MN}	ν_{MO}	
trans-$Pt(GlH)_2Cl_2$	737	555		99
trans-$Pt(GlH)_2Br_2$	732	553		″
trans-$Pt(GlH)_2I_2$	730	559		″
trans-$Pt(NH_3)_2(Gl)_2$	798	555, 513		″
cis-$Pt(GlH)_2Cl_2$	755	552		″
$K[PtCl_2(Gl)]$	755	550	388	″
β-cis-$[Pt(Gl)_2]$	772	553	398	″
α-cis-$[Pt(Gl)_2]$	768	554	405, 400	″
trans-$[Pt(Gl)_2]$	792	549	411	″
$Pt(GlH)Cl(Gl)$	786	538	407	″
trans-$Pd(Gl)_2$	771	550	420	96
trans-$Cu(Gl)_2$	644	439	333	″
trans-$Ni(Gl)_2$	630	439	290	″
cis-$Cu(Gl)_2 \cdot H_2O$	669	460	335, 285	″
cis-$Cu(ND_2CH_2COO)_2 \cdot D_2O$	515	455	333, 280	″
cis-$Cu(NH_2CD_2COO)_2 \cdot H_2O$	661	450	335, 283	″
cis-$Cu(ND_2CD_2COO)_2 \cdot D_2O$	505	450	332, 280	″
$Ni(Gl)_2 \cdot 2H_2O$	630	435	287	100
$KNi(NHCH_2COO)(Gl)$		486	285	″
$K_2Ni(NHCH_2COO)_2$		482		″
trans-$[Pt(Gl)_2]$		418		97
cis-$[Pt(Gl)_2]$		406, 389[a]		″
trans-$[Pd(Gl)_2]$		423		″
cis-$[Pd(Gl)_2]$		413, 373[a]		″
trans-$Cu(Gl)_2 \cdot 2H_2O$	639	332		″
cis-$Cu(Gl)_2 \cdot 2H_2O$	663	332, 275[a]		″

Abbreviations: GlH = Glycine (NH_2CH_2COOH); Gl = Glycino $(NH_2CH_2COO)^-$.
[a] High-frequency band is asymmetric stretch, low-frequency band is symmetric stretch.

Table 7-42. Frequency Assignments Made in Metal–Amino Acid Complexes

Complex	Frequency, cm^{-1}			Reference
	ν_{MN}	ν_{MO}	$(NH_2)_{rock}$	
R = DL-α-alanine $(CH_3–CHNH_2–CO_2)^-$				
CoR_2	322			98
NiR_2	328			″
CuR_2	335			″
ZnR_2	305			″
CdR_2	282			″
PdR_2	419			″
PtR_2	422			″
$Ni(NH_2CH_2CH_2CO_2)_2·2H_2O$	366	285	655	100
$KNi(NHCH_2CH_2CO_2)(NH_2CH_2CH_2CO_2)$	380		660	″
$K_2(NHCH_2CH_2CO_2)_2$	380			″
R = DL-leucine $((CH_3)_2CH–CH_2–CHNH_2CO_2)^-$				
CoR_2	312			101
NiR_2	323			″
CuR_2	400			″
ZnR_2	314			″
CdR_2	290			″
PdR_2	385			″
PtR_2	409			″
R = DL-valine $((CH_3)_2–CH–CHNH_2CO_2)^-$				
$NiR_2·2H_2O$	365	638		102
CuR_2	394	640		″
PdR_2	385	749		″
PtR_2	399	833		″
R = DL-isoleucino $((C_2H_5)(CH_3)CH–CH(NH_2)(CO_2))^-$				
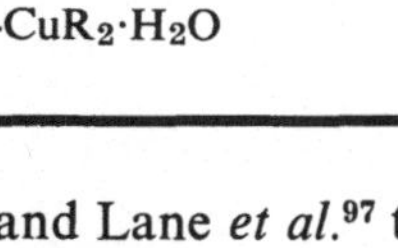				
trans-PtR_2	419_{asym}		831	103
trans-PdR_2	406_{asym}		721	″
cis-PdR_2	429_{asym}		729, 702	″
	356_{sym}			″
cis-$CuR_2·H_2O$	431_{asym}		633	″
	357_{sym}			″

al.[98] and Lane *et al.*[97] the values obtained for the force constants, based on ν_{MN} assignments only, varied in the order Pt(II)>Pd(II)>Cu(II)>Ni(II).

7.11. METAL–NITRO COMPLEXES

The nitro group is another interesting ligand similar to the thiocyanate ligand. It may coordinate with metals through the nitrogen atom, forming nitro complexes, or through the oxygen atom, forming nitrito complexes, and it may act as a bridging ligand. Low-frequency data are available for the nitro and nitrito complexes.

A strong possibility for π bonding between the metal and the nitrogen atom exists in the nitro complexes. This is reflected by the weaker force constant for $Co(NO_2)_6^{3-}$ as compared to $Co(NH_3)_6^{3+}$ and a decrease in the NO force constant in going from the isolated NO_2^- ion to the $Co(NO_2)_6^{3-}$ compound. Shimanouchi has shown that considerable coupling of lattice modes with the low-frequency bands occurs below 300 cm^{-1}, and thus these vibrations are not to be considered pure.

For the octahedral complexes of the type $M(NO_2)_6^{n-}$, where M can be Fe, Co, Ni, Cu, Rh, or Ir, the ν_{MN} stretching vibration appears to be centered at 286–457 cm^{-1}, the NO_2 rocking mode is at 254–300 cm^{-1}, and the NO_2

Table 7-43. Frequency Assignments in Several Metal–Nitro Complexes

Complex	Frequency, cm^{-1}					Reference
	ν_{MN}	$(NO_2)_{rock}$	Skeleton modes	Lattice modes		
$K_3[Co(NO_2)_6]$	416, 410	293, 285	195, 154	132, 124, 106		104
$Rb_3[Co(NO_2)_6]$	413	287	191, 141	109, 74		″
$Cs_3[Co(NO_2)_6]$	409	281	186, 136	94, 63		″
$K_2CaNi(NO_2)_6$	286	254	191, 175	136, 104		″
$Na_2BaNi(NO_2)_6$	292	256	200, 162	130, 81		″
	δ_{NO_2}	$(NO_2)_{wag}$	ν_{MN}	$(NO_2)_{rock}$	δ_{NMN}	
$K_3[Co(NO_2)_6]$	833	626	415	292		105
$K_3[Rh(NO_2)_6]$ R(soln)	847, 832		304, 277		150	″
IR(solid)	833	627	386	283		″
$K_3[Ir(NO_2)_6]$ R(soln)	846, 835		319, 300		140	″
IR(solid)	830	657	390	300		″
$Cs_3[Co(NO_2)_6]$		625	415			106
$K_2Na[Co(NO_2)_6]$		630	417			″
$CaK_2[Ni(NO_2)_6]$			412			″
$PbNa_2[Fe(NO_2)_6]$		640	445			″
$PbK_2[Cu(NO_2)_6]$			450			″
$(NH_4)_3[Rh(NO_2)_6]$		630	386			″
$K_3[Ir(NO_2)_6]$		650	390			″
$K_2[Pt(NO_2)_6]$		625	394			″
$Na_3[Ir(NO_2)_6]$	850	643	413, 373			105
$Na_3[Rh(NO_2)_6]$	850	619	425, 362			″
$Na_3[Co(NO_2)_6]$	851, 832	612	450, 369			″
$Na_3[Co(NO_2)_6]$	845, 831	623	451, 373	277, 250	216, 196, 182, 145, 124	104a

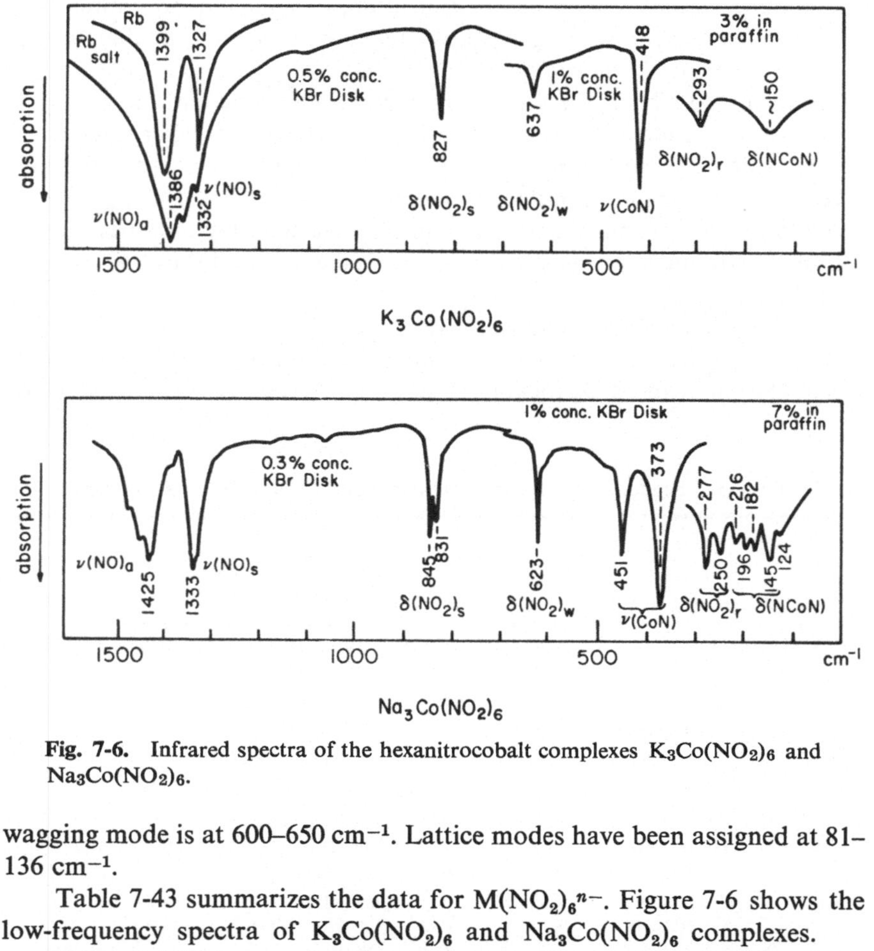

Fig. 7-6. Infrared spectra of the hexanitrocobalt complexes $K_3Co(NO_2)_6$ and $Na_3Co(NO_2)_6$.

wagging mode is at 600–650 cm^{-1}. Lattice modes have been assigned at 81–136 cm^{-1}.

Table 7-43 summarizes the data for $M(NO_2)_6{}^{n-}$. Figure 7-6 shows the low-frequency spectra of $K_3Co(NO_2)_6$ and $Na_3Co(NO_2)_6$ complexes.

Table 7–44 lists some low-frequency data for ammine–nitro complexes of Co, Rh, and Pt. The ν_{MN} stretching vibration has been assigned at 317–404 cm^{-1} and is at a lower frequency than the ν_{MN} in metal–ammine complexes.

The complexes $[Pt(NO_2)_4]^{2-}$ have been studied by Nakamoto.[107] He assigned the ν_{PtN} stretching vibration at 307–450 cm^{-1}, the wagging vibration at 613–636 cm^{-1}, and the rocking vibration at 115 cm^{-1}. Durig[108] observed the spectrum of $M_2Pt(NO_2)_4$ and $M_2Pd(NO_2)_4$ and assigned the ν_{PtN} and ν_{PdN} of a_{1g} symmetry in the Raman at 322 cm^{-1} and 298 cm^{-1}. The other Raman-active band occurred at 322 cm^{-1} in $K_2[Pt(NO_2)_4]$ and 298 cm^{-1} for $K_2[Pd(NO_2)_4]$. The infrared-active modes were observed at 375 and 355 cm^{-1} in the platinum compound and 355 cm^{-1} in the palladium compound.

Table 7-44. Vibrational Assignments in Metal–Ammine–Nitro Complexes[105]

Complex	Frequency, cm^{-1}	
	ν_{MNO_2}	ν_{MNH_3}
$[Co(NH_3)_5NO_2]Cl_2$	375	497, 433
$[Co(NH_3)_5NO_2](NO_3)_2$	357	488, 457
cis-$[Co(NH_3)_4(NO_2)_2]Cl$	326	518, 483, 461
trans-$[Co(NH_3)_4(NO_2)_2]Cl$	343	500
1, 2, 4-$[Co(NH_3)_3(NO_2)_3]$[a]	355, 341	512, 483, 471
1, 2, 3-$[Co(NH_3)_3(NO_2)_3]$[b]	404	468
$[Rh(NH_3)_5NO_2]Cl_2$	350	474, 450
cis-$[Pd(NH_3)_2(NO_2)_2]$		496, 463
cis-$[Pt(NH_3)_2(NO_2)_2]$	357, 330	515
trans-$[Pt(NH_3)_2(NO_2)_2]$	317	518
$[Pt(NH_3)_3(NO_2)]_2[PtCl_4]$	328	508

[a] Symmetry considered C_{2v}.
[b] Symmetry considered C_{3v}.

Table 7-45 lists the data for $M(NO_2)_4^{2-}$ and $M(NO_2)_x(X_2)_y^{2-}$, where M is Pt or Pd and X is a halogen.

When the NO_2 group becomes complexed, four low-frequency vibrations of the NO_2 portion result in addition to the internal vibrations of the nitro group. These are the out-of-plane wag, the in-plane rock, a metal–nitrogen stretch, and a torsion of the nitro group. The mid-infrared vibrations of the nitro group are useful and consist of

$$\nu_{asym}(NO_2) \quad 1363\text{–}1497\ cm^{-1}$$
$$\nu_{sym}(NO_2) \quad 1300\text{–}1373\ cm^{-1}$$
$$\delta_{NO_2} \quad 798\text{–}849\ cm^{-1}$$

These may be contrasted to the M–ONO vibration in nitrito complexes

$$\nu_{asym}(NO_2) \quad 1440\text{–}1505\ cm^{-1}$$
$$\nu_{sym}(NO_2) \quad 995\text{–}1066\ cm^{-1}$$

Table 7-45. ν_{MN} Assignments in Several Metal–Nitro Complexes[105]

Complex	ν_{MNO_2}, cm^{-1}
$K_2[Pt(NO_2)_4]$	377–552
$K_2[Pt(NO_2)_3Cl]$	370, 341
cis-$K_2[Pt(NO_3)_2Cl_2]$	369
trans-$K_2[Pt(NO_3)_2Cl_2]$	396, 365
trans-$K_2[Pt(NO_3)_2Br_2]$	384, 362
$K_2[Pt(NO_2)Cl_3]$	375
$K_2[Pd(NO_2)_4]$	363, 330
cis-$K_2[Pd(NO_2)_2Cl_2]$	363
trans-$K_2[Ir(NO_2)_4Cl_2]$	404, 374

δ_{NO_2}	825–850 cm^{-1}
ν_{MONO}	340–355 cm^{-1}

7.12. METAL–AZIDE COMPLEXES

Several azide complexes have now been prepared. Some of the early spectroscopic studies were made by Dehnicke and co-workers.[109] More recently, Forster and Horrocks[110] and Schmidke and Garthoff[111] have studied complexes of the type $M(N_3)_4^{2-}$, $M(N_3)_6^{2-}$, $M(N_3)_4^{-}$, and $M(N_3)_6^{3-}$. Table 7-46 records various vibrations involving the nitrogen atom. The assignments of the $M(N_3)_4^{2-}$ complexes were based on a D_{2d} symmetry for which M is Co or Zn. The $Sn(N_3)_6^{2-}$ complex was based on a D_{3d} symmetry, and the $M(N_3)_6^{3-}$ and the $M(N_3)_4^{2-}$, and $M(N_3)_4^{-}$ complexes were assigned on the basis of O_h and D_{4h} structures respectively. The assignments for the $\nu_1(N_3)$ vibration are higher in the $Co(N_3)_4^{2-}$, $Zn(N_3)_4^{2-}$, and $Sn(N_3)_6^{2-}$ complexes than in the $M(N_3)_6^{3-}$ complexes, where M is Ru(III), Rh(III), Ir(III), or Pt(IV), and in the $Au(N_3)_4^{-}$ complexes. The $\nu_2(N_3)$ and the ν_{MN_3} assignments are similar in both classes of compounds. The ν_{MN_3} assignments made for MX_aN_3 were M is Cr(VI), Ti(IV), V(V), or Mo(V) appear to range from 495 to 565 cm^{-1} and are higher than the previous assignments. Only a limited number of metal–ligand bending data are available.

7.13. METAL–NITROSYL COMPLEXES

Several metal–nitrosyl complexes have been investigated by spectroscopic techniques. The complexes demonstrate a high degree of π bonding, and it may be greater than that shown by the corresponding carbonyl bonds. The assignments for the ν_{NO} vibrations are well known and have been the subject of discussion in several reviews. The vibrations ν_{MNO} and δ_{MNO} are less well established. The confusion is apparently based on the fact that these two vibrations are similar in energy, and it has not been determined which vibration is of the highest energy. It is evident that additional studies such as normal coordinate treatments, ^{15}N and ^{13}C isotope studies, and Raman polarization analysis of metal–nitrosyl complexes are needed to firmly establish these assignments. Metal isotope studies might also be helpful.

Table 7-47 includes various assignments for vibrations of several ruthenium nitrosyl complexes. The assignments are in agreement except for the results for the ν_{RuNO} vibration. Fairey and Irving[114] assign this vibration at $\sim$570 cm^{-1}, while Durig *et al.*[115] assign it at higher frequencies. No assignments for the δ_{MNO} are made by Fairey and Irving. Durig's assignments are in the 298–398 cm^{-1} region, although some recent data place the δ_{MNO} vibration at higher frequency.

Table 7-46. Vibrational Assignments Made for Several Metal–Azide Complexes

Complex		Frequency, cm^{-1} $\nu_1(N_3)_{sym}$	$\nu_2(N_3)_{def}$	$\nu_{3(MN)}$	Medium	References
$[(CH_3)_4N]_2Na[Fe(N_3)_6]$		1280	633, 604	325	KI	111
$[n\text{-}Bu_4N]_3[Ru(N_3)_6]$		1285	670, 636, 591	364	KI	″
$[\varphi_4As]_3[Rh(N_3)_6]$		1288	582	373, 365	KBr	″
$[n\text{-}Bu_4N]_3[Rh(N_3)_6]$		1291, 1262	681, 592, 530	380	KBr	″
$[n\text{-}Bu_4N]_3[Ir(N_3)_6]$		1290	680, 581, 530	372, 352	KI	″
$[Et_4N]_2[Pt(N_3)_6]$		1268	693, 580	402	KBr	″
$[(cetyl)Me_3N]_2[Pt(N_3)_6]$		1270, 1260	693, 580	410, 400	KBr	″
$[(cetyl)Me_3N]_2[Pd(N_3)_4]$		1278	670, 582	408, 385	KBr	″
$[\varphi_4As]_2[Pt(N_3)_4]$		1277	575	387	KBr	″
$[(cetyl)Me_3N]_2[Pt(N_3)_4]$		1288, 1278	686, 578	408, 394	KBr	″
$[\varphi_4As][Au(N_3)_4]$		1261	693, 579	433, 426	KBr	″
$[Et_4N][Au(N_3)_4]$		1262	696, 573	437	KBr	″
$[(C_2H_5)_4N]_2[Co(N_3)_4]$	IR solid	1338, 1280	642, 610	368		110
	IR soln	1338, 1280[c]	640, 611[b]	371[a]		″
$[(C_2H_5)_4N]_2[Zn(N_3)_4]$	IR solid	1342, 1290	649, 615	351		″
	IR soln	1341, 1288[c]	649, 615[b]	350[a]		″
	R solid			362		″
	R soln			360		″
$[(C_2H_5)_4N]_2[Sn(N_3)_6]$	IR solid	1340, 1288	659, 601	390, 330		″
	IR soln	1342, 1289[c]	662, 602[b]	391, 335[a]		″
	R solid			335		″
					δ_{MN}	
CrO_2ClN_3		1223	637	495		109
$TiCl_3N_3$		1230	659	565	286	109
$VOCl_2N_3$		1249	698	560	272	109
$MoCl_4N_3$		1260, 1230, 1215	683, 614	552	333	109
VCl_4N_3		1190	703	556		112
$SnCl_3N_3$		1238, 1227	696	556	272	112, 113

[a] Nitromethane. [b] Acetone. [c] Acetonitrile.

Table 7-47. Vibrational Assignments for Several Ruthenium–Nitrosyl Complexes

Complex	Frequency, cm^{-1}							Reference
	ν_{NO}	$(NH_3)_{rock}$	ν_{MNO}	ν_{MNH_3}	δ_{MNO}	δ_{NMN}	ν_{MX}	
$[Ru(NH_3)_4NOCl]Cl_2$	1880	858	608	483		270	328	114, 115
$[Ru(NH_3)_4NOBr]Br_2$	1870	845	591	472	298	264	222	114, 115
$[Ru(NH_3)_4NOI]I_2$	1862	825	572	466	390	255	173	114, 115
$[Ru(NH_3)_5NO]Cl_3$	1903, 1865	845	602	474	395	270		115
$[Ru(NH_3)_4NOOH]Cl_2$	1845	853	571	500, 480				114
	1834	848	628	495, 473	395	267		115
$[Ru(NH_3)_4NOOH]Br_2$	1850	846	570	499, 473				114
	1840	827	624	495, 468	395	261		115
$[Ru(NH_3)_4NOOH]I_2$	1855	834, 816	568	496, 463				114
	1840	828	624	497, 469	398	258		115
$[Ru(ND_3)_4NOOD]I_2$	1830	660	621	460, 444		235		115
$Ag_2[RuNOCl_4OH]$	1883		600		358		321	115
$Ag_2[RuNOCl_4OD]$	1880, 1873		585		355		318	115
				ν_{MNO_2}				
$Na_2[Ru(NO)(NO_2)_4OH]\cdot 2H_2O$	1893		638	490		317		115
$Na_2[Ru(NO)(NO_2)_4OD]\cdot 2D_2O$	1900		631	480		318		115
$Ag_2[Ru(NO)(NO_2)_4OH]$	1883		625	470		325		115
$Ag_2[Ru(NO)(NO_2)_4OD]$	1890		630	470		315		115

Table 7-48. ***Trans*** **and** ***Cis*** **Effects in Nitrosylruthenium Complexes[115]**

Complex	Frequency, cm^{-1}	
	ν_{RuNO}	ν_{RuOH}
trans effect		
$K_2RuNO(CN)_5$	634	
$[RuNO(NH_3)_4OH]Cl_2$	628	
$[RuNO(NH_3)_4Cl]Cl_2$	608	
$[RuNO(NH_3)_4NH_3]Cl_3$	602	
$[RuNO(NH_3)_4Br]Br_2$	591	
$[RuNO(NH_3)_4I]I_2$	572	
cis effect		
$Na_2RuNO(NO_2)_4OH \cdot 2H_2O$	638	568
$[RuNO(NH_3)_4OH]Cl_2$	628	565
$Ag_2RuNOCl_4OH$	600	519

Durig's work established the *trans*-labelizing effect of the NO ligand.[115] In complexes of the type $RuNOX_5^{2-}$ (where X is Cl, Br, I, or CN) the ν_{RuX} vibration for the group *trans* to the NO group occurs at a frequency 30–40 cm^{-1} lower than the corresponding group in the *cis* position. Similar results have occurred for complexes of the type $[RuNO(NH_3)_4X]^{n+}$. Table 7-48 summarizes these data. The *trans*-directing influence is apparent. The frequency of the ν_{RuNO} vibration, in the *trans* position to the X group, decreases in the order CN>OH>Cl>NH_3>Br>I. The influence of *cis* ligands is also observed in these complexes, although it is less than that of the *trans* ligands. Previously, with square planar complexes, the *cis* effect has been considered to be negligible and has been neglected. Table 7-48 also includes these results. Figure 7-7 shows the spectra of several nitrosylruthenium complexes.

Nitrogen vibrations for complexes of the type $[M(NO)X_5]^{2-}$ have been assigned and are tabulated in Table 7-49. The assignments for ν_{MNO} range from 552 to 613 cm^{-1} when M is ruthenium to 585 to 623 cm^{-1} when M is osmium. Table 7-50 summarizes assignments for some pentammine nitrosyl cobalt ion complexes $[Co(NO)(NH_3)_5]^{2+}$. Durig[119] makes assignments for ν_{MNO} in the range 573–644 cm^{-1} and δ_{MNO} at $\sim$578 cm^{-1}. The shifting toward lower frequency of the ν_{MNO} vibration with ^{15}N substitution can be noted.

Table 7-51 includes data for $[M(NO)(CN)_5)^{n-}$, where M is Fe, Mn, and Cr.

7.14. SUMMARY

It is apparent that further research is necessary to establish the position of the metal–nitrogen vibrations in a number of complexes. Of considerable interest would be further Raman studies (particularly polarizability

Table 7-49. Vibrational Assignments for Metal–Nitrosyl–Pentahalide $[M(NO)X_5]^{2-}$ Complexes

Complex	Frequency, cm^{-1}						References
	ν_{NO}	ν_{MNO}	$(H_2O)_{rock}$	$(H_2O)_{wag}$	ν_{MX}	Lattice	
$Na_2[Ru(NO)Cl_5]\cdot 3H_2O$	1908	586	550				116
$K_2[Ru(NO)Cl_5]$	1916, 1905	598, 580			334, 228	143, 96	116, 117
$(NH_4)_2[Ru(NO)Cl_5]$	1909, 1900	600, 583					116
$Rb_2[Ru(NO)Cl_5]$	1902, 1891	602, 587			330, 288	107, 80	116, 117
$Rb_2[Ru(NO)Cl_5]\cdot 2H_2O$	1909	584, 575	604	483			116
$Cs_2[Ru(NO)Cl_5]$	1878	591, 582			322, 279	<70	116, 117
$Cs_2[Ru(NO)Cl_5]\cdot 2H_2O$	1905, 1897	586, 576	620	483			117
$[(C_2H_5)_4N]_2[Ru(NO)Cl_5]$	1830	613, 588			315, 281	<70	″
$Na_2[Ru(NO)Br_5]\cdot 3H_2O$	1890	575	540				116
$K_2[Ru(NO)Br_5]$	1881	583			257, 221	81	117
$(NH_4)_2[Ru(NO)Br_5]$	1878	573					″
$Rb_2[Ru(NO)Br_5]$	1871	570			255, 220	70	″
$Rb_2[Ru(NO)Br_5]\cdot 2H_2O$	1894	586	560	450			″
$Cs_2[Ru(NO)Br_5]$	1853	583, 571			251, 222	<70	″
$Cs_2[Ru(NO)Br_5]\cdot 2H_2O$	1892, 1887	586, 580	540	560			″
$[(CH_3)_4N]_2[Ru(NO)Br_5]$	1840, 1821	610, 577			247, 217, 175	75	″
$[(C_2H_5)N]_2[Ru(NO)Br_5]$	1832	609, 523			245, 222	<70	″
$Na_2[Ru(NO)I_5]\cdot 3H_2O$	1850	551	510	445			″
$K_2[Ru(NO)I_5]$	1879, 1857	597, 571, 553			214, 172	105	″
$(NH_4)_2[Ru(NO)I_5]$	1910, 1898, 1848	586, 573, 554					″
$Rb_2[Ru(NO)I_5]$	1903, 1890, 1855	583, 571, 553					″
	1843						
$Cs_2[Ru(NO)I_5]$	1875, 1860, 1855	586, 570, 552			210, 169	<70	″
$[(C_2H_5)N]_2[Ru(NO)I_5]$		605, 578, 559			208, 172	<70	″
			δ_{MNO}				
$K_2[Os(NO)Cl_5]$	1865	617, 593			320, 298		″
$K_2[Os(NO)Br_5]$		623, 585			230, 217		″
$K_2[Ir(NO)Br_5]$	1973, 1955	552, 534			236, 225		″
$K_2[Ru(^{14}NO)Cl_5]$	1915, 1904	606	588		336, 327, 386		118
$K_2[Ru(^{15}NO)Cl_5]$	1874, 1865	600	588, 572				″
$K_2[Ru(^{14}NO)Cl_5]$	1888, 1880	606	606		257, 221		″
$K_2[Ru(^{15}NO)Cl_5]$	1850, 1843	605	572, 557				116

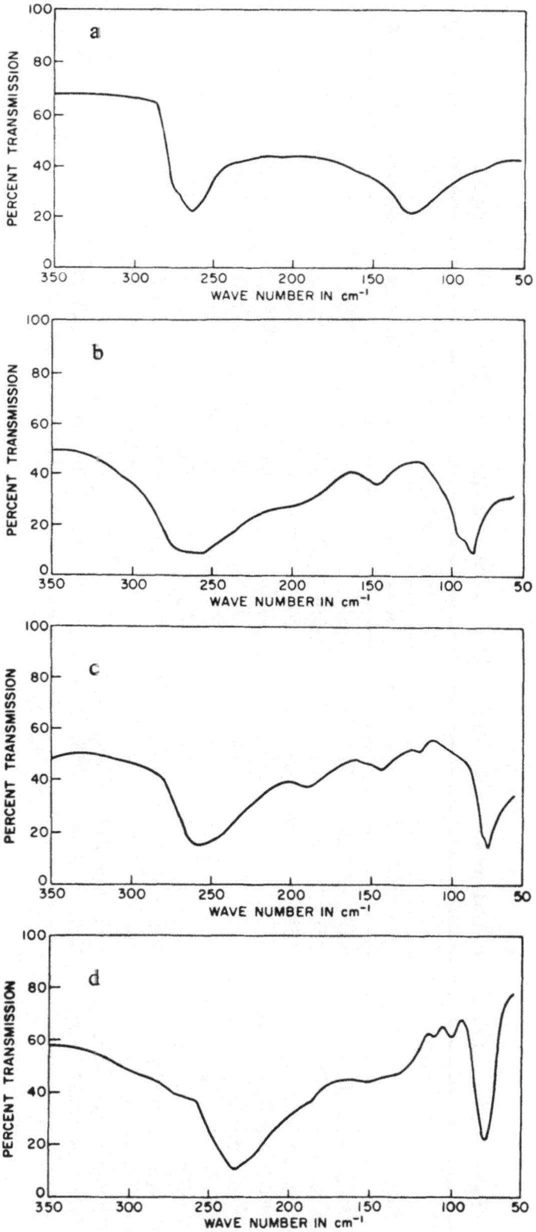

Fig. 7-7. Far-infrared spectra of (a) $[RuNO(NH_3)_4OH]Cl_2$; (b) $[RuNO(NH_3)_4OH]Br_2$; (c) $[RuNO(NH_3)_4OH]I_2$; (d) $[RuNO(ND_3)_4OD]I_2$.

Table 7-50. Vibrational Assignments for Metal–Nitrosyl–Pentammine $[M(NO)(NH_3)_5]^{2+}$ Complexes

Complex	Frequency, cm^{-1}								Reference
	$\nu_{(NO)}$	$(NH_3)_{def}$	$(NH_3)_{rock}$	ν_{MNO}	δ_{MNO}	ν_{MNH_3}	$\delta_{H_3NMNH_3}$	$\delta_{O_2NMNO_2}$	
$[Co(^{14}NO)(NH_3)_5]Cl_2$	1610	1292	812	644	578	469, 442	290		119
$[Co(^{15}NO)(NH_3)_5]Cl_2$	1588	1293	811	630	580	466, 443	290		"
$[Co(^{14}NO)(ND_3)_5]Cl_2$	1610	990	697	640	580	452, 416	262		"
$[Co(^{14}NO)(NH_3)_5]_2(NO_3)_4$			830	577		485, 455	310	247	"
$[Co(^{15}NO)(NH_3)_5]_2(NO_3)_4$			830	566		487, 452	311	245	"
$[Co(^{14}NO)(ND_3)_5]_2(NO_3)_4$		1003	710	575		465, 430	275	230	"
$[Co(^{14}NO)(NH_3)_5]_2Br_4$		1300	835	573		490, 446	315	243	"
$[Cr(^{14}NO)(NH_3)_5]Cl_2$	1685	1281	791, 776, 734	573	535	464, 455, 440	306, 264		118
$[Cr(^{15}NO)(NH_3)_5]Cl_2$	1649	1281	786, 774, 736	573	526	464, 455, 438	306, 264		"
$[Cr(^{14}NO)(NH_3)_5](ClO_4)_2$	1727	1299	764, 732, 700	577	531	462, 421	264, 255		"
$[Cr(^{15}NO)(NH_3)_5](ClO_4)_2$	1692	1300	758, 733	574	521	462, 418	263, 256		"

Table 7-51. Vibrational Assignments for Metal–Nitrosyl–Cyano $[M(NO)(CN)_5]^{n-}$ Complexes

	Frequency, cm^{-1}				
Complex	ν_{NO}	ν_{MN}	δ_{MNO}	Bending and lattice modes	Reference
$Na_2[Fe(NO)(CN)_5]\cdot 2H_2O$	1935	658, 647		189, 145, 138, 117	116
$Cs_2[Fe(NO)(CN)_5]$	1929, 1910	663, 658, 651		139	″
$[(CH_3)_4N]_2[Fe(NO)(CN)_5]$	1908, 1883	660, 656		141, 91	″
$[(C_2H_5)_4N]_2[Fe(NO)(CN)_5]\cdot H_2O$	1880, 1845	662, 653		149, 99	″
$K_3[Mn(NO)(CN)_5]$	1700	655		174, 130	″
$K_3[Cr(NO)(CN)_5]$	1630	616		169, 126	″
$[(CH_3)_4N]_3[Cr(NO)(CN)_5]$	1625	607		161, 101	″
$K_3[Cr(^{14}NO)(CN)_5]\cdot H_2O$	1643	620	610		118
$K_3[Cr(^{15}NO)(CN)_5]\cdot H_2O$	1610	617	600		″

data), low-temperature and high-pressure investigations, ^{15}N isotopic studies, and transition metal isotope studies.

A consideration of the factors involved in determining the position of a metal–nitrogen vibration is not as clear-cut as the similar consideration made for the metal–halogen vibration in Section 6.12. This is because of the increased complexity of the metal–halide molecule when it becomes coordinated to one or more organic ligands. The effects of the ligand may be operating in the opposite direction from the effects caused by other factors. Thus, it becomes extremely difficult to screen and weigh the various factors under consideration. At the present writing, it would appear that the ν_{MX} vibration, demonstrating more sensitivity to these factors, is of more diagnostic value than the ν_{MN} vibration. However, for thorough analysis of the far-infrared region, the assignments of both vibrations is necessary.

Oxidation Number

Perhaps the best example of the oxidation effect is demonstrated by the cobalt (III) hexammines, which show a ν_{MN} vibration $>150\ cm^{-1}$ higher than the Co^{2+} complexes (see Table 7-8).

Stereochemistry and Coordination Number

Table 7-14 illustrates the usefulness of assigning the ν_{MX} and the ν_{MN} vibrations in the far-infrared region for determination of the stereochemistry of a coordination complex, providing the expected bands are all resolved. The effect of stereochemistry and coordination number on the position of the ν_{MN} vibration, although not as easily discernable as in the case of the metal halides, is apparent. Table 7-52 records comparisons of $M(NH_3)_6Cl_2$ complexes versus $M(NH_3)_2Cl_2$ complexes.

Basicity of Ligand

The effects of the basicity of the ligand upon the position of the ν_{MN} vibration are masked by other factors such as coordination number, stereochemistry, and chelation. Table 7-53 tabulates some comparisons made for

Table 7-52. Effects of Stereochemistry and Coordination Number on the Position of the ν_{MN} Position

O_h symmetry, CN=6	ν_{MN}, cm^{-1}	T_d symmetry, CN=4	ν_{MN}, cm^{-1}
$Zn(NH_3)_6Cl_2$	300	$Zn(NH_3)_2Cl_2$	421
$Zn(NH_3)_6Br_2$	294	$Zn(NH_3)_2Br_2$	414
$Zn(NH_3)_6I_2$	282	$Zn(NH_3)_2I_2$	397
$Co(NH_3)_6I_2$	312	α-$Co(NH_3)_2I_2$	430

Table 7-53. Ligand Effect on the Position of the ν_{MN} Vibration

Compound	ν_{MN}, cm^{-1}
$Co(NO)(NH_3)_5Cl_2$	644 (ν_{MNO})
$Co(An)_2Cl_2$	414, 366
$Co(N_2H_4)_2Cl_2$	388
$(Et_4N)_2Co(N_3)_4$	368
$(Et_4N)_2Co(NCO)_4$	345
$Co(NH_3)_6Cl_2$	318
$Co(phen)_3Cl_2$	288
$Co(Iz)_2Cl_2$	275
$Co(bipy)_3Cl_2$	264
$Co(CH_3CN)_6Cl_2$	252
$Co(py)_2Cl_2$	243, 235

Co^{2+} complexes for various ligands, and it is observed that a ligand effect is present.

The effects of the mass of the central atom and the effect of the counterion are less apparent, probably because of the complexity of the compounds and the interplay of the various factors.

Distinguishing between a bridging nitrogen ligand and a nonbridging ligand (terminal) is less possible than was the case for the halide ligand. This is probably due to the fact that the bridging of a nitrogen ligand involves a bridge through the entire molecule, and no dramatic change occurs in the position of the ν_{MN} vibration as compared to the terminal ligand.

As more definite assignments for metal–nitrogen vibrations are made, it may be possible to further enlarge on these comparisons.

BIBLIOGRAPHY

1. L. Sacconi, A. Sabatini, and D. Gans, *Inorg. Chem.* **3,** 1772 (1964).
2. W. P. Griffith, *J. Chem. Soc.*, 899 (1966).
3. T. Shimanouchi and I. Nakagawa, *Inorg. Chem.* **3,** 1805 (1964).
4. I. Nakagawa and T. Shimanouchi, *Spectrochim. Acta* **22,** 759 (1966).
5. K. Nakamoto, P. J. McCarthy, J. Fujita, R. A. Condrate, and G. T. Behnke, *Inorg. Chem.* **4,** 36 (1965).
6. T. E. Haas and J. R. Hall, *Spectrochim. Acta* **22,** 988 (1966).
7. J. H. Terasse, H. Poulet, and J. P. Mathieu, *Spectrochim. Acta* **20,** 305 (1964).
8. N. Tanaka, M. Kamada, J. Fujita, and E. Kymo, *Bull. Chem. Soc. Japan* **37,** 222 (1964).
9. G. Blyholder and S. Vergez, *J. Phys. Chem.* **67,** 2149 (1963).
10. J. Hiraishi, I. Nakagawa, and T. Shimanouchi, *Spectrochim. Acta* **24A,** 819 (1968).
11. T. Shimanouchi and I. Nakagawa, *Spectrochim. Acta* **18,** 89 (1962).
12. R. J. H. Clark and C. S. Williams, *J. Chem. Soc.* (A), 1425 (1966).
13. C. H. Perry, D. P. Athans, E. F. Young, J. R. Durig, and B. R. Mitchell, *Spectrochim. Acta* **23A,** 1137 (1967).

14. D. M. Adams and P. J. Chandler, *J. Chem. Soc.* (A), 1009 (1967).
15. P. J. Hendra, *Spectrochim. Acta* **23A,** 1275 (1967).
16. R. Layton, J. W. Sink, and J. R. Durig, *J. Inorg. Nucl. Chem.* **28,** 1965 (1966).
17. P. J. Hendra and N. Sadasivan, *Spectrochim. Acta* **21,** 1271 (1965).
18. R. J. H. Clark, *J. Chem. Soc.*, 1377 (1962).
19. R. G. Inskeep, *J. Inorg. Nucl. Chem.* **24,** 763 (1962).
20. J. R. Ferraro, L. J. Basile, and D. L. Kovacic, *Inorg. Chem.* **5,** 391 (1966).
20a. A. Walker and J. R. Ferraro, *J. Chem. Phys.* **43,** 2689 (1965).
21. F. A. Hart and J. E. Newbery, *J. Inorg. Nucl. Chem.* **31,** 1725 (1969).
22. L. J. Basile, D. L. Gronert, and J. R. Ferraro, *Spectrochim. Acta* **24A,** 707 (1968).
23. D. A. Durham, G. H. Frost, and F. A. Hart, *J. Inorg. Nucl. Chem.* **31,** 833 (1969).
24. L. J. Basile, D. L. Kovacic, and J. R. Ferraro, *Inorg. Chem.* **6,** 406 (1967).
25. I. S. Ahuja, D. H. Brown, R. H. Nuttall, and D. W. A. Sharp, *J. Inorg. Nucl. Chem.* **27,** 1105 (1965).
26. W. R. McWhinnie, *J. Inorg. Nucl. Chem.* **27,** 1063 (1965).
27. C. Postmus, J. R. Ferraro, and W. Wozniak, *Inorg. Chem.* **6,** 2030 (1967).
28. R. N. Keller, N. B. Johnson, and L. L. Westmoreland, *J. Am. Chem. Soc.* **90,** 2729 (1968).
29. W. R. McWhinnie, *J. Chem. Soc.*, 2959 (1964).
30. J. R. Ferraro and W. R. Walker, *Inorg. Chem.* **4,** 1382 (1965).
31. J. R. Ferraro, R. Driver, W. R. Walker, and W. Wozniak, *Inorg. Chem.* **6,** 1586 (1967).
32. R. A. Walton, *J. Chem. Soc.* (A), 61 (1965).
33. A. J. Carty, S. J. Patel, and D. G. Tuck, *Can. J. Chem.* to be published.
34. N. P. Crawford and G. A. Melson, *J. Chem. Soc.*, 427 (1969).
35. M. F. Farona and J. G. Grasselli, *Inorg. Chem.* **6,** 1675 (1967).
36. F. Ya. Kol'ba, N. G. Yaroslavskii, L. V. Konovalov, A. U. Barsukov, and V. E. Mironov, *Zh. Neorgan. Khim.* **13,** 153 (1968).
37. C. E. Taylor and A. E. Underhill, *J. Chem. Soc.* (A), 368 (1969).
38. W. J. Eilbeck, F. Holmes, C. E. Taylor, and A. E. Underhill, *J. Chem. Soc.* (A), 128 (1968).
39. W. J. Eilbeck, F. Holmes, C. E. Taylor, and A. E. Underhill, *J. Chem. Soc.* (A), 1189 (1968).
40. D. M. Bowers and A. I. Popov, *Inorg. Chem.* **7,** 1595 (1968).
41. L. P. Bicelli, *Ann. Chim. Rome* **48,** 749 (1958).
42. R. J. H. Clark, *J. Chem. Soc.*, 1377 (1962).
43. A. D. Allen and T. Theopanides, *Can. J. Chem.* **42,** 1551 (1964).
44. R. J. H. Clark and C. S. William, *Inorg. Chem.* **4,** 350 (1965).
45. P. J. Beadle, M. Goldstein, D. M. L. Goodgame, and R. Grzeskowiak, *Inorg. Chem.* **8,** 1490 (1969).
46. J. R. Ferraro, J. Zipper, and W. Wozniak, *Appl. Spectry.* **23,** 160 (1969).
47. J. R. Ferraro, W. Wozniak, and G. Roch, *Ric. Sci.* **38,** 433 (1968).
48. L. Cattalini, R. J. H. Clark, A. Orio, and C. P. Poon, *Inorg. Chem. Acta* **2,** 62 (1968).
49. J. Bradbury, K. P. Forest, R. H. Nuttall, and D. W. A. Sharp, *Spectrochim. Acta* **23A,** 2701 (1967).
50. D. H. Brown, K. P. Forrest, R. H. Nuttall, and D. W. A. Sharp, *J. Chem. Soc.* (A), 2146 (1968).
51. R. J. H. Clark and C. S. Williams, *Spectrochim. Acta* **22,** 1081 (1966).
52. W. R. McWhinnie, *J. Inorg. Nucl. Chem.* **27,** 1063 (1965).
53. P. T. T. Wong and D. G. Brewer, *Can. J. Chem.* **46,** 139 (1968).

54. M. Goldstein, E. R. Mooney, A. Anderson, and H. A. Gebbie, *Spectrochim. Acta* **21,** 105 (1965).
55. W. R. McWhinnie, *J. Inorg. Nucl. Chem.* **27,** 2573 (1965).
56. R. J. H. Clark and C. S. Williams, *Chem. Ind. London,* 1317 (1964).
57. J. R. Allan, D. H. Brown, R. H. Nuttall, and D. W. A. Sharp, *J. Inorg. Nucl. Chem.* **27,** 1305 (1965).
58. C. Postmus, J. R. Ferraro, and W. Wozniak, *Inorg. Chem.* **6,** 2030 (1967).
59. J. Burgess, *Spectrochim. Acta* **24A,** 277 (1968).
60. N. S. Gill and H. J. Kingdon, *Australian. J. Chem.* **19,** 2197 (1966).
61. T. J. Doellette and H. M. Haendler, *Inorg. Chem.* **8,** 1777 (1969).
62. D. H. Brown, R. H. Nuttall, J. McAvoy, and D. W. A. Sharp, *J. Chem. Soc.,* 892 (1966).
63. C. W. Frank and L. B. Rogers, *Inorg. Chem.* **5,** 615 (1966).
64. W. E. Hatfield and K. Whyman, in *Transition Metal Chemistry* (R. L. Carlin, Ed.) Marcell Dekker, New York (1969).
65. J. R. Durig, B. R. Mitchell, D. W. Sink, J. N. Willis, and A. S. Wilson, *Spectrochim. Acta* **23A,** 1121 (1967).
66. D. B. Powell and N. Sheppard, *Spectrochim. Acta* **17,** 68 (1961).
67. J. R. Durig, R. Layton, D. W. Sink, and B. R. Mitchell, *Spectrochim. Acta* **21,** 1367 (1965).
68. K. Krishnan and R. A. Plane, *Inorg. Chem.* **5,** 852 (1966).
69. D. Horley and E. G. Torrible, *Can. J. Chem.* **43,** 3201 (1965).
70. N. Tanaka, N. Sato, and J. Fujita, *Spectrochim. Acta* **22,** 577 (1966).
71. G. W. Watt and D. S. Klett, *Inorg. Chem.* **5,** 1278 (1966).
72. G. W. Watt and P. W. Alexander, *J. Am. Chem. Soc.* **89,** 1814 (1967).
73. R. J. Mureinik and W. Robb, *Spectrochim. Acta* **24A,** 837 (1968).
74. M. N. Hughes and W. R. McWhinnie, *J. Inorg. Nucl. Chem.* **28,** 1659 (1966).
75. D. C. Bradley and M. H. Gitlitz, *J. Chem. Soc.* (A), 980 (1969).
76. Yu. Ya. Kharitonov, I. K. Dymina, and T. N. Leonova, *Zh. Neorgan. Khim.* **12,** 830 (1967).
77. H. Buerger and W. Sawodny, *Inorg. Nucl. Chem. Letters* **2,** 209 (1966).
78. L. Sacconi and A. Sabatini, *J. Inorg. Nucl. Chem.* **25,** 1389 (1963).
79. V. T. Athavale and C. S. Padmanabha Iyer, *J. Inorg. Nucl. Chem.* **29,** 1003 (1967).
80. K. Broderson, *Z. Anorg. Allgem. Chem.* **290,** 24 (1957).
81. M. S. Barvinkov and I. S. Bukhareva, *Zh. Fiz. Khim.* **41,** 525 (1967).
82. J. Gaubeau and U. Kull, *Z. Anorg. Allgem. Chem.* **316,** 182 (1962).
83. J. Reedijk and W. L. Groenwald, *Rec. Trav. Chim.* **87,** 1079 (1968).
84. M. F. Farona and J. G. Grasselli, *Inorg. Chem.* **6,** 1675 (1967).
85. R. A. Walton, *Can. J. Chem.* **46,** 2347 (1968).
86. D. Forster and D. M. L. Goodgame, *Inorg. Chem.* **4,** 715 (1965).
87. G. F. Knox and T. M. Brown, *Inorg. Chem.* **8,** 1401 (1969).
88. M. A. Bennett, R. J. H. Clark, and A. D. J. Goodwin, *Inorg. Chem.* **6,** 1625 (1967).
89. R. N. Keller, N. B. Johnson, and L. L. Westmoreland, *J. Am. Chem. Soc.* **90,** 2729 (1968).
90. D. Forster and D. M. L. Goodgame, *J. Chem. Soc.,* 262 (1968).
91. A. Sabatini and I. Bertini, *Inorg. Chem.* **4,** 959 (1965).
92. D. Forster and W. De W. Horrocks, *Inorg. Chem.* **6,** 339 (1967).
93. F. A. Miller and G. Carlson, *Spectrochim. Acta* **17,** 977 (1961).
94. D. Forster and D. M. L. Goodgame, *J. Chem. Soc.,* 1286 (1965).

95. A. J. Saraceno, I. Nakagawa, S. Mizushima, C. Curran, and J. V. Quagliano, *J. Am. Chem. Soc.* **80,** 5018 (1958).
96. R. A. Condrate and K. Nakamoto, *J. Chem. Phys.* **42,** 2590 (1965).
97. T. J. Lane, J. A. Durkin, and R. J. Hooper, *Spectrochim. Acta* **20,** 1013 (1964).
98. J. F. Jackowitz, J. A. Durkin, and J. G. Walter, *Spectrochim. Acta* **23A,** 67 (1967).
99. J. A. Kieft and K. Nakamoto, *J. Inorg. Nucl. Chem.* **29,** 2561 (1967).
100. G. W. Watt and J. F. Knifton, *Inorg. Chem.* **6,** 1010 (1967).
101. J. F. Jackowitz and J. L. Walter, *Spectrochim. Acta* **22,** 1393 (1966).
102. I. Nakagawa, R. J. Hooper, J. L. Walter, and T. J. Lane, *Spectrochim. Acta* **21,** 1 (1965).
103. R. J. Hooper, T. J. Lane, and J. L. Walter, *Inorg. Chem.* **3,** 1568 (1964).
104. I. Nakagawa and T. Shimanouchi, *Spectrochim. Acta* **23A,** 2099 (1967).
104a. I. Nakagawa, T. Shimanouchi, and Y. Yamasaki, *Inorg. Chem.* **3,** 772 (1964).
105. M. LePostelloc, J. P. Mathieu, and H. Poulet, *J. Chim. Phys.* **60,** 319 (1963).
106. Y. Puget and C. Duval, *Compt. Rend.* **250,** 4141 (1960).
107. K. Nakamoto, J. Fujita, and H. Murata, *J. Am. Chem. Soc.* **80,** 4817 (1958).
108. J. R. Durig and D. W. Wertz, *Appl. Spectry.* **22,** 627 (1968).
109. K. Dehnicke and J. Straehle *Z. Anorg. Allgem. Chem.* **339,** 171 (1965).
110. D. Forster and W. De W. Horrocks, *Inorg. Chem.* **5,** 1510 (1966).
111. H. H. Schmidtke and D. Garthoff, *J. Am. Chem. Soc.* **89,** 1317 (1967).
112. J. Straehle and K. Dehnicke, *Z. Anorg. Allgem. Chem.* **338,** 287 (1965).
113. K. Dehnicke, *J. Inorg. Nucl. Chem.* **27,** 809 (1965).
114. M. B. Fairey and R. J. Irving, *Spectrochim. Acta* **26,** 359 (1966).
115. E. E. Mercer, W. A. McAllister, and J. R. Durig, *Inorg. Chem.* **5,** 1881 (1966).
116. N. M. Sinitsyn and K. I. Petrov, *Zh. Strukt. Khim.* **9,** 45 (1968).
117. P. Gans, A. Sabatini, and L. Sacconi, *Inorg. Chem.* **5,** 1877 (1966).
118. E. Miki, *Bull. Chem. Soc. Japan* **41,** 1835 (1968).
119. E. E. Mercer, W. A. McAllister, and J. R. Durig, *Inorg. Chem.* **6,** 1816 (1967).

Review Articles and General Bibliography

1. D. M. Adams, *Metal-Ligand and Related Vibrations*, St. Martin's Press, New York (1968).
2. J. R. Durig and D. W. Wertz, *Appl. Spectry.* **22,** 627 (1968).
3. D. W. James and M. J. Nolan, in *Progress in Inorganic Chemistry, Vol. 9* (F. A. Cotton, Ed.), Interscience Publishers, New York (1968) pp. 195–275.
4. R. E. Hester, *Coord. Chem. Rev.* **2,** 319 (1967).
5. J. W. Brasch, Y. Mikawa and R. J. Jakobsen, in *Applied Spectroscopy Reviews, Vol. I*, (E. G. Brame, Ed.) M. Dekker, Inc., New York (1968) pp. 187–235.
6. J. R. Ferraro, in *Far-Infrared Properties of Solids* (S. S. Mitra and S. Nudelman, Eds.), Plenum Press, New York (1970).
7. P. J. Hendra and P. M. Stratton, *Chem. Rev.* **69,** 325 (1969).
8. D. M. Adams, in *Molecular Spectroscopy* (P. Hipple, Ed.), Elsevier Publishing Co., London (1968).

Chapter 8

MISCELLANEOUS METAL-LIGAND VIBRATIONS

8.1. INTRODUCTION

The tendency to form predominantly covalent bonds between metals and donor atoms of ligands is greatest when the donor atom is a nonmetal from Group V through VII in the periodic table. The recent interest in the far-infrared region has centered on the search for vibrations involving metal–ligand donor atoms. Figure 8-1 shows a number of such vibrations. Chapters 5 through 7 have discussed the vibrations involving metal–oxygen, metal–nitrogen, and metal–halide atoms. For the most part, these vibrations are fairly well characterized. However, the remaining vibrations listed in Fig. 8-1 are less well known. This has been primarily due to a lack of spectroscopic studies, until recently, of compounds involving these linkages. Recent syntheses of compounds containing metal–sulfur, metal–selenium, metal–tellurium, metal–phosphorus, metal–arsenic, and metal–metal bonds and subsequent spectroscopic studies have given some preliminary information

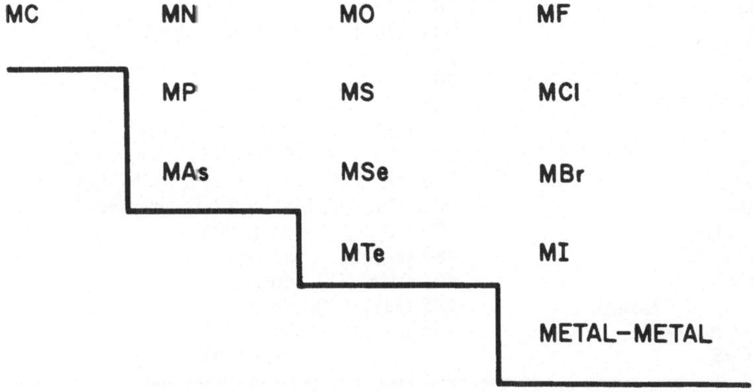

Fig. 8-1. Metal–ligand vibrations.

concerning the stretching vibrations of these bonding atoms. Little data is presently available on the bending vibrations involving these atoms. Furthermore, these vibrations would not be the easiest to study. It is difficult to prepare compounds in which the donor atom is replaced by an atom of the same group in the periodic table but of heavier mass and observe the shift of the metal–ligand vibration toward lower frequency as has been done with the metal–halide vibrations. Many of the vibrations discussed in this chapter would be expected to show only a small dipole moment change and thus be of weak to medium intensity in the infrared region. By contrast, the vibrations would be expected to be of stronger intensity in the Raman, and it would appear that they could be better studied by the Raman method. However, to date, Raman studies are practically nil. Additional complications in infrared studies may result from the coupling of these vibrations with bending vibrations involving other atoms in the molecule, lattice vibrations, and ligand vibrations.

Since we are not discussing organometallic compounds in this book, no attempt at a lengthy discussion of the metal–carbon vibrations will be made in this chapter. Instead, a summary of the latest available information on these vibrations will be presented.

8.2. METAL–SULFUR VIBRATIONS

Of the vibrations discussed in this chapter, only the metal–sulfur vibrations appear to be fairly well characterized. Although the preliminary assign-

Table 8-1. Low-Frequency Absorptions (cm^{-1}) in Various Inorganics Containing Sulfur[1-7]

ZnS	350, 274[a]
CdS	304, 242[a]
HgS	340, 276, 123, 87, 38 (cinnabar)
CaS	255
SrS	208
BaS	180
As_4S_4	370, 362, 337, 173, 109 (realgar)
As_2S_3	379, 350, 345, 310, 295, 185, 140 (orpiment)
Sb_2S_3	330, 273, 235, 174, 143, 110, 66, 57 (stibnite)
Bi_2S_3	282, 235, 205, 188, 66, 61 (bismuthate)
$(NH_4)_3VS$	470 (asym), 404 (sym)
M_2MoS_4[b]	480 (asym), 460 (sym)
M_2WS_4[b]	466 (asym)(?), 466 (sym)
$(NH_4)_2MoO_2S_2$	485 (asym)(?), 485 (sym)
$(NH_4)_2WO_2S_2$	465 (asym), 480 (sym)
ReO_3S	504 (sym)

[a] See Table 9-5 in Chapter 9 for values of other lattice frequencies.
[b] M = alkali or alkaline earth metal.

ments made for most of the metal–sulfur vibrations are still considered to be tentative, accumulation of additional data will tend to substantiate and adjust some of these assignments. Tables 8-1 to 8-4 tabulate the available data for inorganic compounds and complexes involving monodentate, bidentate, and bridging sulfur ligands.

Certain conclusions based on these data may be made:

(1) Several binary metal sulfides are observed to absorb radiation at 300 cm^{-1} and lower. The absorption appears to be mass dependent on the cation, similar to any lattice vibration.

(2) Several sulfur-containing minerals have low-frequency absorptions, which have been used to identify them.

(3) Table 8-2 summarizes data for sulfur ligands acting as monodentate ligands. Wherever comparisons are possible, the ν_{MS} stretching vibration occurs at a lower frequency for the complexes ML_4X_2 (CN = 6) than for the complexes ML_2X_2 (CN = 4).

(4) The ν_{PtS} and ν_{PdS} stretching vibrations are found consistently at higher frequency than for metal–ligand vibrations involving the first-row transition metals. This is consistent with similar findings for ν_{PtF} and ν_{PdF} stretching vibrations (Chapter 6).

(5) In monodentate thiocyanato complexes the ν_{MS} stretching vibration is at 183–308 cm^{-1}.

(6) Table 8-3 summarizes the data for bidentate and bridging sulfur complexes. The complexes involving xanthates, dithiolates, 1,2-dithiolenes, and other bidentate sulfur ligands appear to absorb radiation in the 267–435 cm^{-1} region and may absorb it at higher frequency than the monodentate complexes. The bridging complexes absorb radiation at 213–230 cm^{-1}.

(7) The metal–sulfur vibration is of weak to medium intensity in the infrared (see Fig. 8-2). It is mass dependent and sensitive to coordination number, type of bonding (e.g., monodentate, bidentate, and bridging), and the *d* electronic configuration of the cation (see Table 8-2).

(8) Only limited δ_{MS} bending vibrations have been assigned. These are for the $M(tu)_4X_2$ complexes, where tu is thiourea:

$$\left.\begin{array}{l}\delta_{MSC}\text{—}156\text{–}182\ cm^{-1}\\ \delta_{SMS}\text{—}\ 83\text{–}112\ cm^{-1}\end{array}\right\}\quad M{=}Ni^{2+},\ Fe^{2+},\ Co^{2+},\ Mn^{2+}$$

$$\delta_{MSC}\text{—}145\ cm^{-1}\qquad M{=}Pb^{2+}$$

For the complexes $M(tu)_2X_2$:

$$\delta_{MSC}\text{—}158\text{–}161\ cm^{-1}\ ;\ M{=}Co^{2+}$$
$$\delta_{MSC}\text{—}139\text{–}146\ cm^{-1}\ ;\ M{=}Zn^{2+}$$
$$\delta_{MSC}\text{—}139\text{–}147\ cm^{-1}\ ;\ M{=}Cd^{2+}$$

Table 8-2. Low-Frequency Absorptions (cm^{-1}) in Complexes Containing Monodentate Sulfur Ligands

	ML_2X_2					ML_4X_2			
M^{2+}	L=tu[8,9]	L=ta[10]	L=$(CH_3)_2S$[12,10a]	L=BBS,BPS[11]	L=$(CH_3)_3AsS$[13]	L=$(CH_3)_3PS$[14]	L=$(CH_3)_3AsS$[13]	L=tu[8,9]	L=ta[10]
Co	242–278	238–255			285–293	311–330	294, 312	205–227	
Zn	227–266	218–237			276–286		287		
Cd	185–231	189–252						220	173
Ni		201, 220						220	237–294
Mn								205–277	254, 257
Pt			311–315 *trans* 305–310 *cis*	315–395				238–277	256, 269
Pd			307–316 *trans*					248–277	
Fe		230						205–227	
Pb								170	

	$R_2M(SCN)_4$[15,16] R = $(CH_3)_4N^+$ or $(C_2H_5)_4N^+$	$K_2M(SCN)_4$[15,16]	$KM(SCN)_4$[15,16] and $K_2M(SCN)_4$	R_3MSSnR_3[17]	$M_2Cl_4L_2$[10a] L = SEt_2, SPr_2	MX_4L_2[10a] *trans* L = SMe_2
Pt	226, 280		183, 293		327–358	311
Pd	207, 289, 298		256, 300			
Au	224, 277, 301		274, 308			
Hg		210	255, 288			
Ge				404		
Pb				336(asym), 278(sym)		
Sn				376(asym), 330(sym)		

Abbreviations: tu = thiourea; ta = thioacetamide; BBS = but-3-enyl butylsulfide ($RS(CH_2)_2CH{:}CH_2$); BPS = *n*-butylpent-4-enyl sulfide ($RS(CH_2)_3CH{:}CH_2$).

Table 8-3. Low-Frequency Absorptions (cm^{-1}) in Complexes Containing Bidentate and Bridging Sulfur Ligands

M	ML_2: L = dtp[18]	ML_2: L = dt[18,20]	ML_n $n = 2,3$: L = ethyl or methyl xanthate[19]	MX_2L: L = $\varphi SC_2H_4S\varphi$[21]	
Ni^{2+}	226, 327	333–435	351–383		Bidentate sulfur ligand
Pt^{2+}	221, 312	352–401 ($R = \varphi$)	312–362	308–331	
Pd^{2+}		310–405 ($R = \varphi, CH_3$)	323–347	324–350	
Hg^{2+}	239, 280				
Pb^{2+}	237, 291				
Cr^{3+}	313		340–376		
Rh^{3+}	293				
In^{3+}	286				
Bi^{3+}	271				
Co^{3+}			340–361		
Sn^{2+}		303–343[22] 333–392(sym)[a] 267–363(asym)[b]			

M	$MHg(SCN)_4$[16]	
Mn^{2+}	213	Bridging sulfur ligand
Fe^{2+}	216	
Co^{2+}	219	
Zn^{2+}	217	
Co^{2+} in $CoCd(SCN)_4$	230	
	$M_2X_4(SR_2)_2$[10a] X = Cl, Br R = Et, Pr, *n*-Bu	
Pt	380–422	Bridging sulfur ligand

Abbreviations: dtp = O,O′-diethyldithiophosphate, dt = dithiolate ion ($R_2C_2S_2^-$).

[a] ML_2 compounds where L = dithiastannacyclopentane.[23]

[b] ML_2 compounds where L = dithiastannacyclohexane.[23]

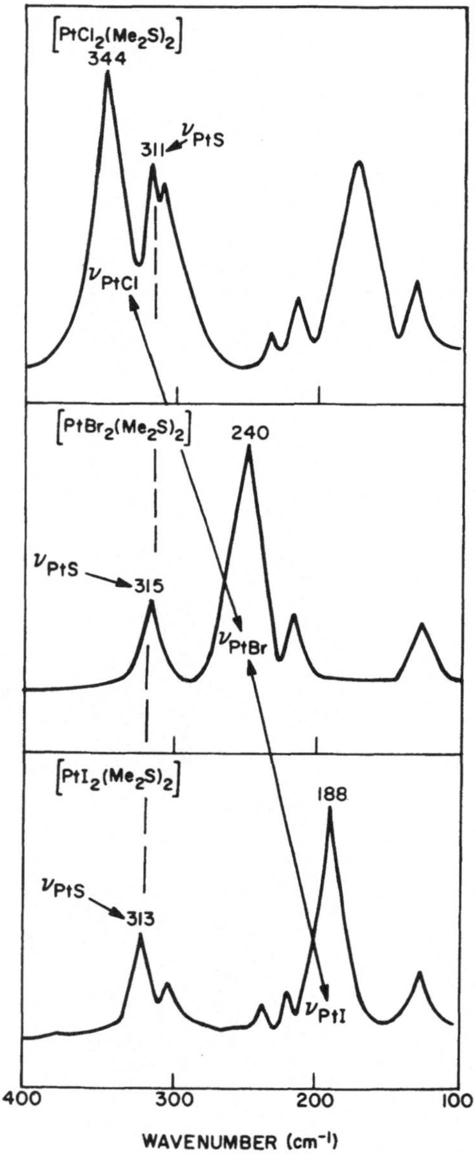

Fig. 8-2. Metal–sulfur vibration in $PtX_2(Me_2S)_2$ complexes. (Courtesy of the Chemical Society, Burlington House, London. From Allkins and Hendra.[12])

Table 8-4. Raman Frequencies for Several Sulfur-Containing Complexes of Tellurium[24]

Complex	ν_{TeS}, cm^{-1}
$Te(etu)_2I_2$	229p
$Te(tmtu)_2Cl_2$	240p, 233p
$Te(tmtu)_2Br_2$	232p
$Te(tmtu)_2I_2$	231p
$Te(etu)^{2+}$	235p
$Te(tu)_4{}^{2+}$	234p
$Te(tu)_2{}^{2+}$	264

Abbreviations: tu = thiourea; etu = ethylenethiourea; tmtu = tetramethylthiourea; p = polarized.

(9) Very few Raman data are available.

A very recent study of the metal complexes of 1,2-ditholene has been made.[24a] The complexes may be denoted as follows: $[M–S_4]_n{}^z$, $[M–S_6]^z$, and $[M(L)_x–S_{2,4})^z$, where L may be π-C_5H_5, NO, φ_3P, or O and may be illustrated by structures 1 to 3.

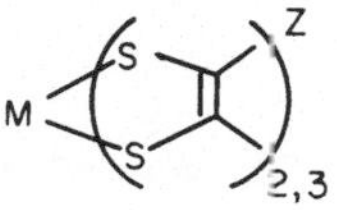

(1)

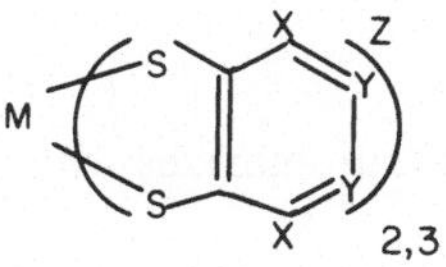

(2)

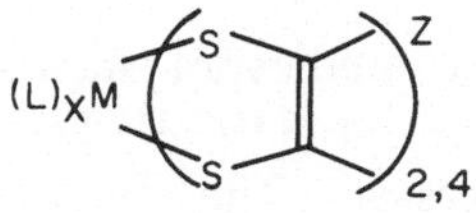

(3)

Table 8-5 summarizes the infrared data for several planar and dimeric dithiolene complexes.

Table 8-5. Infrared Data for Several Planar and Dimeric Dithiolene Complexes

Complex	ν_{MS}, cm^{-1}
$[NiS_4C_4Me_4]^0$	435, 333
$[PtS_4C_4Me_4]^0$	405, 310
$[NiS_4C_4\varphi_4]^0$	408, 354
$[PdS_4C_4\varphi_4]^0$	401, 352
$[PtS_4C_4\varphi_4]^0$	403, 373

Table 8-6. Infrared Data for Tris-Dithiolene Complexes

Complex[a]	ν_{MS}, cm^{-1}
$[VS_6C_6H_6]^0$	385, 361
$[VS_6C_6H_6]^-$	392, 363
$[VS_6C_6H_6]^{2-}$	367, 350
$[VS_6C_6\varphi_6]^0$	406, 346
$[VS_6C_6\varphi_6]^-$	398, 349
$[CrS_6C_6\varphi_6]^0$	421, 356
$[MoS_6C_6H_6]^0$	380, 354
$[MoS_6C_6\varphi_6]^0$	403, 356
$[MoS_6C_6Me_6]^0$	402, 306
$[WS_6C_6H_6]^0$	369, 329
$[WS_6C_6\varphi_6]^0$	403, 359
$[WS_6C_6Me_6]^0$	398, 300
$[ReS_6C_6H_6]^0$	338, 333
$[ReS_6C_6H_6]^-$	361, 333
$[ReS_6C_6\varphi_6]^0$	373, 359
$[ReS_6C_6\varphi_6]^-$	361, 350

[a] For the anionic complexes, φ_4As^+ or Et_4N^+ were the cations.

Table 8-6 summarizes the infrared data for several tris-dithiolene complexes.

8.3. METAL–SELENIUM VIBRATIONS

Very few spectroscopic studies of metal–selenium compounds are available at the present time.[1,2,10a,12,15,16,21] This is, therefore, an area where considerable research is necessary. The vibration would be expected to be weak in intensity and to occur below the metal–sulfur vibration. A factor contributing to the difficulty of making assignments for this vibration is the coupling with lattice and bending modes, which overlap with the ν_{MSe} stretching vibration.

Table 8-7 summarizes the results of investigations concerning the metal–selenium vibration. In the limited systems studied thus far, the following tentative conclusions may be drawn:

(1) The ν_{MSe} stretching vibration in complexes involving dimethylselenide are found at lower frequency than those for other complexes. This is the reverse of the results found for the ν_{MS} stretching vibration with complexes involving dimethylsulfide.

(2) The various metal–selenium vibrations are found at lower frequencies than those of metal–sulfur vibrations, as expected.

(3) A trend exists which indicates that complexes involving a coordina-

Table 8-7. Low-Frequency Stretching Vibrations ν_{MSe} (cm^{-1}) in Various Pd and Pt Complexes

M	$[ML_2]^{2+}$ L = su CN = 4	$MX_2(Me_2Se)_2$ X = halogen CN = 4 trans	cis	$M_2Cl_4L_2$ L = $SeEt_2$, $SePr_2$	$[ML_4]^{2+}$ L = su CN = 6	$[M(SeCN)_6]^{2+}$	$MX_2{\cdot}L$ L = $(\varphi Se)_2C_3H_6$ CN = 4	$(MX_2Me_2Se)_2$	Remarks	
Pd	270[12a]	219[12]	193, 152[12]	282–290[10a]	235–253[12a]	200–222[16]			Monodentate	IR
Pt	295[12a]	220–233[12]			214–231[12a]	240[15]				
Pd							296–314[21]		Bidentate	
Pt							285–296[21]			
Pd								224[12]	Bridged	
Pt								243[12]		
Pd					170–186p[12a]					R
Pt		172–176[12]			182–202p[12a]					

Note: Lattice vibrations[1,2]—ZnSe, 215 cm^{-1}; CdSe, 185 cm^{-1}. See Table 9-5 in Chapter 9 for other values of lattice frequencies for these crystals.
Abbreviations: su = selenourea; φ = phenyl.

tion number of 4 have vibrations at higher frequencies than those which have a coordination number of 6.

(4) For the most part, only palladium and platinum complexes have been studied.

(5) Very few bending vibrations have been assigned (δ_{MSec} at 150–169 cm^{-1} and δ_{SeMSe} at 94–108 cm^{-1} in selenourea complexes).

(6) Only limited Raman studies have been made thus far.[12,12a]

(7) Very little differentiation may be made between complexes containing monodentate, bidentate, and bridging ligands at present because of the limited number of data available.

(8) The symmetrical stretch in complexes of the type $[ML_4]^{2+}$ (L is selenourea), which have a square planar symmetry, found only in the Raman, is located below 200 cm^{-1}. The asymmetric stretch found in the infrared is located above 200 cm^{-1}.

8.4. METAL–TELLURIUM VIBRATIONS

Even less research has been accomplished with metal–tellurium vibrations than with metal–selenium vibrations.[1,2,12] Table 8-8 summarizes the results for these vibrations, and it is observed that a few tentative conclusions may be reached:

(1) Little Raman work has been done.

(2) Only Pt and Pd complexes have been studied.

(3) No bending vibration data is available at present.

(4) Hendra[12] has indicated that the stretching vibrations increase in the order, $\nu_{PtS} > \nu_{PtSe} \approx \nu_{PtTe}$.

(5) The symmetrical vibration of ν_{PtTe} is located in Raman at ~169 cm^{-1}, while the asymmetric ν_{PtTe} is at 245 cm^{-1}.

Table 8-8. Summary of Spectroscopic Data on Metal–Tellurium Vibrations[12]

	Frequency, cm^{-1}				
	MX_2Y_2 X = halogen Y = Me_2Te		$M_2X_4L_2$ L = $TeEt_2$, $TePr_2$	$(MX_2{\cdot}Me_2Te)_2$[a]	
M	*trans*	*cis*	*trans*	X = halogen	Remarks
Pt	245 (asym)	187, 156	177–197[10a]	183	
	227 (asym)			158 or 228	IR
Pt	169 (sym)				R

Note: Lattice vibrations[1,2] found at ZnTe, 179 cm^{-1}; CdTe, 141 cm^{-1}. See Table 9-5 for other 9 lattice frequencies.

[a] Bridged molecules.

Table 8-9. Summary of Infrared Data for the ν_{MP} Vibration, cm^{-1}

M	$M_2X_4(PZ_3)_2$[27†]	$(M(PEt_3)_4)(MBr_4)$[28]	$MX_2(PZ_3)_2$[25–27‡] Planar (D_{4h}) *cis*	*trans*	T_d	$MX_2(PZ_3)_2$[25§]	$MH_3(P(CH_3)_3)$[29]	$MX_4(P\varphi_3)$[30]
Pt	373–380,[27a] 361–458[10a,27a]	389	362–443	346–428				
Pd	365–381,[27a] 355–444[10a-27a]		341–390	343–409				
Au				348–424		347–391		
Ni				343–426[31]	168–196[31]			
Ti								434–463
Ga							326	
Zn					153–166[32]			
Cd					133–136[32]			
Hg					91–137[32]			
Co								
Mo								
W								

Table 8-9 (Continued)

M	MX_4R[30]	$MX_2(P\varphi_3)_2$[33] C_{2v}	$(Et_4N)(\varphi_3PMX_3)$[33] C_{3v}	$M(CO)_4(P(CH_3)_3)_2$[34] *cis*	$M(PF_3)_4$[35,36] (zero-valent)	$M(CO)_n(PF_3)_m$[37] $n=1,2,3;\ m=3,2,1$
Pt						
Pd						
Au						
Ni		148–190	177		195, 219	209–262
Ti	453, 414					
Ga						
Zn		128–166	134, 146			
Cd						
Hg						
Co		142–187	167			
Mo				200–214		
W				189, 205		

X = Halogen.
†Z = Me, Et, Pr, ethoxy, chlorine.
‡Z = Me, Et, Pr, Bu, φ, *p*-tolyl, *p*-anisyl, (diphenylphosphino)methane, (diphenylphosphino)ethane, chlorine.
§Z = Me, –C≡C–*t*-butyl, –C≡C–C_6H_5.
φ = Phenyl.
R = $\varphi_2P\cdot C_2H_4\cdot P\varphi_2$.
Raman data[38]: for $M[P(OMe)_3]_4$, ν_{MP} = 178 cm^{-1}, for M = Ni.

8.5 METAL-PHOSPHORUS VIBRATION

A renewed interest in phosphorus chemistry has occurred recently. A number of new phosphine and substituted phosphine complexes have been synthesized. Some infrared and Raman studies have been made,[10a 25–38] although much too little information is presently available. Some assignments for a metal–phosphorus (ν_{MP}) stretching vibration have been made. However, few if any confirmatory x-ray data on structure are available. Certain tentative conclusions based on a limited number of data can be made.

(1) The vibration ν_{MP} is of weak to medium intensity and therefore may be difficult to locate in the low-frequency region, particularly where the ligand may also show absorption.

(2) The vibration appears to behave normally; that is, it is apparently dependent on the oxidation state of the metal. The ν_{MP} vibrations for zero-valent metals are located at lower frequencies than those for divalent compounds.

(3) Compounds in a T_d environment show ν_{MP} vibrations at a higher frequencies than those for compounds in a C_{3v} and C_{2v} structure. In other words, symmetry plays a role in determining the position of the vibration, much the same as Clark demonstrated for the ν_{MX} vibration.

(4) Boorman and Carty[31] have presented data from a series of nickel (II) halide–phosphine complexes to indicate that one may distinguish between a T_d or D_{4h} structure on the position of the ν_{NiP} vibration; the D_{4h} complexes show ν_{NiP} vibrations at 335–424 cm^{-1} while the T_d complexes have the ν_{NiP} vibration occurring at 168–196 cm^{-1}. The low ν_{NiP} frequencies in T_d complexes are accounted for in terms of less π bonding in the T_d complexes as compared with the D_{4h} complexes.

Table 8-10. Summary of Infrared Data for the ν_{MAs} Vibration (cm^{-1})

M	$MX(AsMe_3)$[24]	$MX_2(AsR_3)$[26,27,39]	$MX_4(AsEt_3)_2$[39]	$M_2X_4(AsR_3)_2$[27,27a]
Pt		237–302 (R = Me, Et, Pr)	321–328	272–287 (R = Me)
Pd		268–272 (R = Me, Et, Pr)		282–287 (R = Me)
		cis	*trans*	*trans*[10a]
Pt		303 (R = Et) (276)	305 (R = Et) (265)	318 (R = Et) 334 (R = Pr)
Pt		242–276 (R = Me)		
Pd				342 (R = Et) 332 (R = Pr)
Au	265–268			

Note: Raman data[40]: for $(NiCO)_3 \cdot AsEt_3$, ν_{MAs} = 207 cm^{-1}.
( ) = Data uncertain. x = Halogen.

Table 8-11. Infrared Data Concerning ν_{MM} Stretching Vibrations

Compound	Metal bond	Frequency, cm^{-1}
$Pt(NH_3)_4PtCl_4$[41]	(Pt–Pt)	180–195
$Pd(NH_3)_4PdCl_4$[41]	(Pd–Pd)	179–195
$\varphi_3Sn–SnEt_3$[53]	(Sn–Sn)	208
$Et_3Sn–SnBu_3$[53]	(Sn–Sn)	197
$\varphi_3Sn–SnMe_3$[53]	(Sn–Sn)	194
$(Et_4N)_3Pt(SnX_3)_5$[42,43] (X = halogen)	(Sn–Pt)	193–210
$(Et_4N)_4[Rh_2X_2(SnX_3)_4]$[41,43] (X = halogen)	(Sn–Rh)	209–217
$(CH_3)_3Sn-Ge\varphi_3$[53]	(Sn–Ge)	225
$(C_2H_5)_3Sn-Ge\varphi_3$[53]	(Sn–Ge)	230
$\varphi_3Sn–GeBu_3$[53]	(Sn–Ge)	235
$(CH_3)Cl_2SnMn(CO)_5$ and $Cl_3SnMn(CO)_5$[53]	(Sn–Mn)	170–201
$R_3*Sn–Mn(CO)_5$[53]	(Sn–Mn)	174–182
$(CH_3)_3SnMn(CO)_5$[53]	(Sn–Mn)	179
$(CH_3)_2ClSnMn(CO)_5$[53]	(Sn–Mn)	197
$(CH_3)_2BrSnMn(CO)_5$[53]	(Sn–Mn)	191
$(CH_3)_2ISnMn(CO)_5$[53]	(Sn–Mn)	179
π-$C_5H_5(CO)_3Cr–Sn(CH_3)_3$[50]	(Sn–Cr)	183
π-$C_5H_5(CO)_3Mo–Sn\varphi$[53]	(Sn–Mo)	169
π-$C_5H_5(CO)_3Mo–Sn(CH_3)_3$[50,53]	(Sn–Mo)	170[50], 172[53]
π-$C_5H_5(CO)_3Mo–SnCl_3$[51]	(Sn–Mo)	190
$R*_3Sn–Mo(CO)_3R'$[53]	(Sn–Mo)	169–172
$R*_3Sn–Fe(CO)_2R'$[53]	(Sn–Fe)	174–185
$X_2Sn[Fe(CO)_2(\pi$-$C_5H_5)_6]$[51]		
X = Cl	(Sn–Fe)	233
Br	(Sn–Fe)	235, 201
I	(Sn–Fe)	232, 196
$Me_3SnFe(CO)_2\pi–C_5H_5$[53]	(Sn–Fe)	185
$\varphi_3Sn–Fe(CO)_2\pi–C_5H_5$[53]	(Sn–Fe)	174
$R_3*Sn–Co(CO)_4$[53]	(Sn–Co)	176
$Cl_2Sn[Co(CO)_4]_2$[51]	(Sn–Co)	213, 172
$Br_2Sn[Co(CO)_4]_2$[51]	(Sn–Co)	210
$I_2Sn[Co(CO)_4]_2$[51]	(Sn–Co)	190
π-$C_5H_5(CO)_3W–Sn(CH_3)_3$[50]	(Sn–W)	166
π-$C_5H_5(CO)_3W–Ge(CH_3)_3$[50]	(Ge–W)	169 ± 4
π-$C_5H_5(CO)_3Mo–Ge(CH_3)_3$[50]	(Ge–Mo)	178
π-$C_5H_5(CO)_3Cr–Ge(CH_3)_3$[50]	(Ge–Cr)	191
$Fe(CO)_4(HgCl_2)$[43]*cis*	(Fe–Hg)	219
$Fe(CO)_4Hg$[43]	(Fe–Hg)	196
$Fe(CO)_4Cd$[43]	(Fe–Cd)	266, 216
$M_6X_8Y_6^{2-}$ ions[52]		
M = Mo; X = halogen (Cl or Br)	(Mo–Mo)	210–260
M = W; Y = halogen (Cl, Br, or I)	(W–W)	150(?)

$R^* = \varphi$, Me; $R' = \pi$-C_5H_5.

Table 8-12. Raman Data Concerning ν_{MM} Stretching Vibrations

Compound	Metal bond	Frequency, cm^{-1}
Hg_2^{2+}(aqueous)[45,45a]	(Hg–Hg)	169
in HgCl	(Hg–Hg)	169
in Hg_2Br_2	(Hg–Hg)	136
in Hg_2I_2	(Hg–Hg)	113
$Hg[Co(CO)_4]$[46]	(Hg–Co)	152(p)
Cd_2^{2+} in $Cd_2(AlCl_4)_2$ melt[47]	(Cd–Cd)	183
$Cd[Co(CO)_4]_2$[46]	(Cd–Co)	161(p)
$Sn_2(CH_3)_6$[48]	(Sn–Sn)	190
$Ge_2(CH_3)_6$[48]	(Ge–Ge)	273(p)
$Si_2(CH_3)_6$[48]	(Si–Si)	404(p)
$Ge[Co(CO)_4]_2$[48]	(Ge–Co)	273
$Re_2(CO)_{10}$[49]	(Re–Re)	120
π-$C_5H_5(CO_3)_3Mo$–$Ge(CH_3)_3$[50]	(Mo–Ge)	184
π-$C_5H_5(CO_3)_3W$–$Ge(CH_3)_3$[50]	(W–Ge)	177
π-$C_5H_5(CO_3)_3Cr$–$Sn(CH_3)_3$[50]	(Cr–Sn)	174 $\pm$ 4
π-$C_5H_5(CO_3)_3Mo$–$Sn(CH_3)_3$[50]	(Mo–Sn)	166
π-$C_5H_5(CO_3)_3W$–$Sn(CH_3)_3$[50]	(W–Sn)	158
$Cl_2Sn[Co(CO)_4]_2$[51]	(Sn–Co)	214, 172
$Br_2Sn[Co(CO)_4]_2$[51]	(Sn–Co)	168
$Cl_3SnRh(nbd)_2$[51]	(Sn–Rh)	165
$Cl_3SnIr(cod)_2$[51]	(Sn–Ir)	165
Me_3Sn–$Sn(CO)_5$[53]	(Sn–Sn)	175
φ_3Sn–$Mn(CO)_5$[53]	(Sn–Mn)	174
φ_6Sn_2[53]	(Sn–Sn)	208
$Sn[Sn(\varphi_3)_4$[53]	(Sn–Sn)	207
Me_6Sn_2[53]	(Sn–Sn)	190

Abbreviations: nbd = norbornadiene; cod = cycloocta-1,5-diene.

(5) Additional Raman data are necessary.

(6) No bending (δ_{MP}) vibrations have been assigned to date.

Table 8-9 summarizes the data.

8.6. METAL–ARSENIC VIBRATIONS

Very few spectroscopic studies involving compounds with metal–arsenic links have been made to date.[10a,24,26,27,39,40] The meager number of data available make all conclusions very tentative. Table 8-10 summarizes the data. The ν_{PtAs} and ν_{PdAs} stretching vibrations appear to be found in the range 242–342 cm^{-1}. No bending assignments have been made and few Raman data exist. It would be expected that this vibration will be, in most cases, weak to medium in intensity in the infrared, since a small dipole moment change would be expected for the metal–arsenic bond.

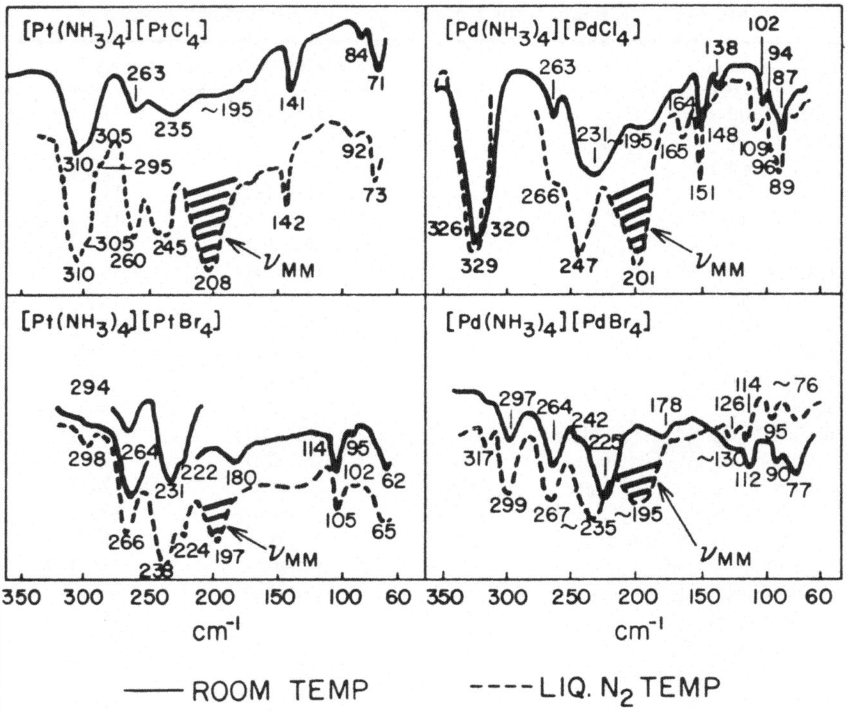

Fig. 8-3. Metal–metal vibration in $[M(NH_3)_4][MX_4]$, where M is Pt or Pd and X is a halogen. (Courtesy of Pergamon Press, New York. From Hiraishi *et al.*[41])

8.7. METAL–METAL VIBRATIONS

In 1934, Woodward[44] showed that the diatomic mercurous ion existed in aqueous solution, and thus provided the first evidence for a metal–metal bond. Most of the progress in this field is recent, and vibrational studies of metal–metal vibrations are minimal.[41–53] Virtually none of the assignments made for metal–metal stretching vibrations can be considered to be thoroughly substantiated. The region of 100–220 cm^{-1} has been suggested as the metal–metal stretching region, although the vibration would be expected to be mass dependent and dependent on the strength of the bond. Tables 8-11 and 8-12 summarize the available data, but it should be emphasized that these assignments are only tentative, as insufficient data are available. Figure 8-3 illustrates the spectra of several $[M(NH_3)_4]$ $[MX_4]$ complexes,[41] (M is Pd and Pt), where metal–metal bonds are said to exist. The vibration at about 200 cm^{-1} is considered as the metal–metal stretching vibration, and liquid

N_2 temperatures were necessary to bring out the band. Considerable work appears to have been done in assigning the Sn–M vibration. A recent discussion concerning the Raman intensities of metal–metal vibrations in various compounds has been published by Quicksall and Spiro.[53a,53b]

8.8. SUMMARY OF THE ν_{MC} AND δ_{MCX} VIBRATIONS*

a) Octahedral carbonyls $M(CO)_6$
- ν_{MCO} 366–428 cm^{-1}
- δ_{MCO} 468–787 cm^{-1}

b) (π-arene) $M(CO)_3$ compounds
- ν_{MCO} 304–489 cm^{-1}
- δ_{MCO} 500–682 cm^{-1}

c) Neutral binary metal carbonyls and hydrocarbonyls
- ν_{MCO} 357–437 cm^{-1}
- δ_{MCO} 460–752 cm^{-1}

d) Binary metal carbonyls $M(CO)_n$, $n=4$, 5, or 6
- ν_{MCO} 357–556 cm^{-1}
- δ_{MCO} 300–785 cm^{-1}

e) Linear dicyanides $M(CN)_2$ and $M(CN)_4$
- ν_{MCN} 360–452 cm^{-1}
- δ_{MCN} 250–354 cm^{-1}

f)† π-bonded M–C compounds

$\nu_{M\pi}$ monoolefin	—C_2H_4, M = Pt	383–427
	—allyl, M = Pd	361–370 (sym)
		369–403 (asym)
diolefin	—nonconjugated, M = Pt	290–310
	conjugated	407–415
benzene ($M\varphi_2$ type),	where M = Cr, Mo, W, or V	330–460

g)† Sigma-bonded M–C compounds

C = CH_2		400–700
C = Et		400–700
C = vinyl, allyl, acetylenic		400–700
C = φ		200–500
R_3MX type	M = Se	623–688 (sym)
		685–805 (asym)
	M = Sn	510–557
	M = Pt	570–590

* D. M. Adam's book[54] contains a thorough discussion of these vibrations through 1965.

† Vibrations listed in f) and g) above are discussed by Nakamoto.[55]

$(R_2MO)_x$ type	M = Ge	550–630
R_5M type	M = Sb	516 (equatorial)
		456 (axial)

BIBLIOGRAPHY

1. A. Mitsuishi, H. Yoshinaga, and S. Fujita, *J. Phys. Soc. Japan* **13,** 1235 (1958).
2. F. A. Miller, G. L. Carlson, F. F. Bentley, and W. H. Jones, *Spectrochim. Acta* **16,** 135 (1960).
3. C. Karr, Jr. and J. J. Kovach, *Appl. Spectry.* **23,** 219 (1969).
4. G. Gattow, A. Franke and A. Müller, *Naturwiss.* **52,** 428 (1965).
5. A. Müller and B. Krebs, *Z. Anorg. Allgem. Chem.* **344,** 56 (1966).
6. M. J. F. Leroy and C. Kaufman, *Bull. Soc. Chim. France*, 3090 (1966).
7. S. S. Mitra, in *Optical Properties of Solids* (S. Nudelman and S. S. Mitra, Eds.), Plenum Press, New York (1969).
8. C. D. Flint and M. Goodgame, *J. Chem. Soc.* (A), 744 (1966).
9. D. M. Adams and J. B. Cornell, *J. Chem. Soc.* (A), 884 (1967).
10. C. D. Flint and M. Goodgame, *J. Chem. Soc.* (A), 2178 (1968).
10a. D. M. Adams and P. S. Chandler, *J. Chem. Soc.* (A), 588 (1969).
11. D. C. Goodall, *J. Chem. Soc.* (A), 887 (1968).
12. J. R. Allkins and P. J. Hendra, *J. Chem. Soc.* (A), 1325 (1967).
12a. P. J. Hendra and Z. Jovic, *Spectrochim. Acta* **24A,** 1713 (1968).
13. A. M. Brodie, S. H. Hunter, G. A. Rodley, and C. J. Wilkins, *J. Chem. Soc.* (A), 987 (1968).
14. A. M. Brodie, S. H. Hunter, G. A. Rodley, and C. J. Wilkins, *J. Chem. Soc.* (A), 2039 (1968).
15. A. Sabatini and I. Bertini, *Inorg. Chem.* **4,** 959 (1965).
16. D. Forster and D. M. L. Goodgame, *Inorg. Chem.* **4,** 715 (1965).
17. M. Schumann and M. Schmidt, *J. Organometallic Chem.* **3,** 485 (1965).
18. D. M. Adams and J. B. Cornell, *J. Chem. Soc.* (A), 1299 (1968).
19. G. W. Watt and B. J. McCormick, *Spectrochim. Acta* **21,** 753 (1965).
20. G. N. Schrauzer and V. P. Mayweg, *J. Am. Chem. Soc.* **87,** 1483 (1965).
21. J. Pluscec and A. D. Westland, *J. Chem. Soc.*, 5371 (1965).
22. R. C. Poller and S. A. Stillman, *J. Chem. Soc.* (A), 1024 (1966).
23. A. Finch, R. C. Poller and D. Steel, *Trans. Faraday Soc.* **61,** 2628 (1965).
24. P. J. Hendra and Z. Jovic, *J. Chem. Soc.* (A), 735 (1967); 911 (1968).
24a. J. A. McCleverly, in: *Progress in Inorganic Chemistry*, Vol. 10, Interscience, New York (1968), p. 49.
25. G. E. Coates and C. Parker, *J. Chem. Soc.*, 421 (1963).
26. M. S. Taylor, A. L. O'Dell and H. A. Raethel, *Spectrochim. Acta* **24A,** 1855 (1968).
27. R. J. Goodfellow, J. C. Evans, P. L. Goggen, and D. A. Duddell, *J. Chem. Soc.* (A), 1604 (1968).
27a. R. J. Goodfellow, P. L. Goggin and L. M. Venanzi, J. *Chem. Soc.* (A), 1897 (1967).
28. D. M. Adams and P. J. Chandler, *Chem. Comm.*, 68 (1966).
29. N. N. Greenwood, E. J. F. Ross, and A. Storr, *J. Chem. Soc.*, 1400 (1965).
30. A. D. Westland and L. Westland, *Can. J. Chem.* **43,** 426 (1965).
31. P. M. Boorman and J. A. Carty, *Inorg. Nucl. Chem. Letters* **4,** 101 (1968).
32. G. B. Deacon and J. H. S. Green, *Chem. Comm.*, 629 (1966); *Spectrochim. Acta* **24A,** 1921 (1968).

33. J. Bradbury, K. P. Forest, R. H. Nuttall, and D. W. A. Sharp, *Spectrochim. Acta* **23A,** 2701 (1967).
34. N. F. Curtis, R. W. Hay, and Y. M. Curtis, *J. Chem. Soc.* (A), 182 (1968).
35. L. A. Woodward and J. R. Hall, *Nature* **181,** 831 (1958).
36. L. A. Woodward and J. R. Hall, *Spectrochim. Acta* **16,** 654 (1960).
37. A. Loutellier and M. Bigorgne, *Bull. Soc. Chim. France*, 3186 (1965).
38. M. Bigorgne, *Compt. Rend.* **250,** 3484 (1960).
39. D. M. Adams, P. J. Chandler, and R. G. Churchill, *J. Chem. Soc.* (A), 1272 (1967).
40. G. Bouquet and M. Bigorgne, *Bull. Soc. Chim. France*, 433 (1966).
41. J. Hiraishi, I. Nakagawa, and T. Shimanouchi, *Spectrochim. Acta* **24A,** 819 (1968).
42. D. M. Adams and P. J. Chandler, *Chem. Ind.*, 269 (1965).
43. D. M. Adams, D. J. Covic, and R. D. W. Kermitt, *Nature* **205,** 589 (1965).
44. N. A. D. Carey and H. C. Clark, *Inorg. Chem.* **7,** 94 (1968).
45. L. A. Woodward, *Phil. Mag.* **18,** 823 (1934).
45a. J. R. Durig, K. K. Lau, G. Nagarajan, M. Walker, and J. Bragin, *J. Chem. Phys.* **50,** 2130 (1969).
46. H. Stammreich, K. Kawai, O. Sala, and P. Krumholz, *J. Chem. Phys.* **35,** 2175 (1916).
47. J. D. Corbett, *Inorg. Chem.* **1,** 700 (1962).
48. M. P. Brown, E. Cartmell, and G. W. B. Fowles, *J. Chem. Soc.*, 506 (1960).
49. F. A. Cotton and R. M. Wing, *Inorg. Chem.* **4,** 1328 (1965).
50. D. J. Cardin, S. A. Keppie, and M. F. Lappert, *Inorg. Nucl. Chem. Letters* **4,** 365 (1968).
51. D. M. Adams, J. N. Crosby, and R. D. W. Kemmitt, *J. Chem. Soc.*, 3056 (1968).
52. F. A. Cotton, R. M. Wing, and R. A. Zimmerman, *Inorg. Chem.* **6,** 11 (1967).
53. N. A. D. Carey and H. C. Clark, *Chem. Comm.*, 292 (1967).
53a. C. O. Quicksall and T. G. Spiro, *Raman Newsletter*, No. 7, 4 (1969).
53b. T. G. Spiro, in *Progress in Inorganic Chemistry*, Vol. 11 (S. J. Lippard, Ed.), Interscience, New York (1970), pp. 1–51.
54. D. M. Adams, *Metal–Ligand and Related Vibrations*, St. Martin's Press, New York (1968),
55. K. N. Nakamoto, in *Characterization of Organometallic Compounds* (M. Tsutsuki, Ed.), Interscience Publishers (1969).

Chapter 9

OTHER LOW-FREQUENCY VIBRATIONS

Table 1-3 in Chapter 1 listed various types of data that could be obtained in the low-frequency region of the infrared. Several of these are worthy of further comment, and this chapter will discuss some of them.

9.1. ION-PAIR VIBRATIONS

Evans and Lo[1] indicated that certain low-frequency absorptions found for hydrogen-bonded hydrogen dihalide ions could be assigned as ion-aggregate vibrations. Table 9-1 summarizes these data. The assignments were based on a comparison of chloride and bromide spectra, on the established existence of ion aggregates in those solutions, and on the evidence[2] that ion-pair vibrations should have shorter interionic distances than lattice modes of the solid salts.

Further evidence for ion-pair vibrations has come from Edgell and co-workers[3,3a] and French and Wood[3b] in studies of alkali metal salts in various solvents. The vibrations are broad, of medium intensity, and are cation, anion, and solvent dependent. Table 9-2 presents the frequency of vibrations found in various solvents. The lithium salts show the vibration in the 373–425 cm^{-1} region; the sodium salts in the 175–203 cm^{-1} region; and the

Table 9-1. Absorptions in Far-Infrared Region for Salts in Benzene Solution and Assigned to Ion-Aggregate Vibrations

Salt	Frequency, cm^{-1}
$(n\text{-}C_4H_9)_4N{\cdot}Cl$	120 ± 3
$(n\text{-}C_5H_{11})_4N{\cdot}Cl$	119 ± 3
$(n\text{-}C_4H_9)_4N{\cdot}Br$	80 ± 4
$(n\text{-}C_5H_{11})_4N{\cdot}Br$	80 ± 4
$(n\text{-}C_4H_9)_4N{\cdot}ClHCl$	102 ± 5
$(n\text{-}C_4H_9)_4N{\cdot}ClDCl$	102 ± 5
$(n\text{-}C_4H_9)_4N{\cdot}BrHBr$	73 ± 5

Table 9-2. Summary of Interionic Frequencies of Alkali Metal Ions[3,3a]

Salt	Frequency, cm^{-1} THF	DMSO-d_6	Piperidine	Pyridine or benzene
Li salts				
$Co(CO)_4^-$	407			
	413			
NO_3^-	407	425		
$B\varphi_4^-$	410			
	412			
Cl^-	387	425		
Br^-	378	424		
I^-	373	424		
Na salts				
$Co(CO)_4^-$	190	199	183	180
	192			
$B\varphi_4^-$	198	203		175
I^-	184	194		
$RCr_2(CO)_{10}^-$		200		
$Cr_2(CO)_{10}^{2-}$		200		
NO_3^-		200		
K salts				
$Co(CO)_4^-$	150			
	142			
$B\varphi_4^-$				133
NH_4^+ salts				
$B\varphi_4^-$				198
$(n\text{-}C_4H_9)_4N^+$ salts				
Cl^-				120
Br^-				180

potassium salts at 133–142 cm^{-1}. Edgell studied the effects of pressure on LiCl in tetrahydrofuran from 0 to 20 kbars. Blue shifts of the order of 1.5 cm^{-1} per kilobar of applied pressure occurred, characteristic of the behavior of lattice vibrations with pressure. The results were interpreted in terms of a model in which the alkali ion vibrates in a cage under the influence of electrostatic and repulsive forces. Two types of cages are found: in tetrahydrofuran solution the cage also involves the anion, while in dimethylsulfoxide the cage involves only the solvent molecules. The studies involving these interionic vibrations are important, as they may lead to obtaining direct information regarding short-range forces, structure, and dynamics of ions in electrolytic solutions.

9.2. LOW-FREQUENCY VIBRATIONS OF MOLECULES TRAPPED IN CLATHRATES

The low-frequency vibrations involving gas molecules trapped in a caged structure of a clathrate have recently been investigated.[4,5] Burgiel *et al.*[4] studied the far-infrared spectra of twelve different gases trapped in cages of β-quinol clathrates. Table 9-3 shows some of the results obtained for both nonpolar monatomic and diatomic gas molecules and for polar diatomic gases. For the nonpolar monatomic and diatomic gases one low-frequency vibration was observed, and this was assigned to a translation motion ("rattling") of the molecule in the cage. For the polar diatomic gases two bands were observed and assigned to a hindered rotation and a translation

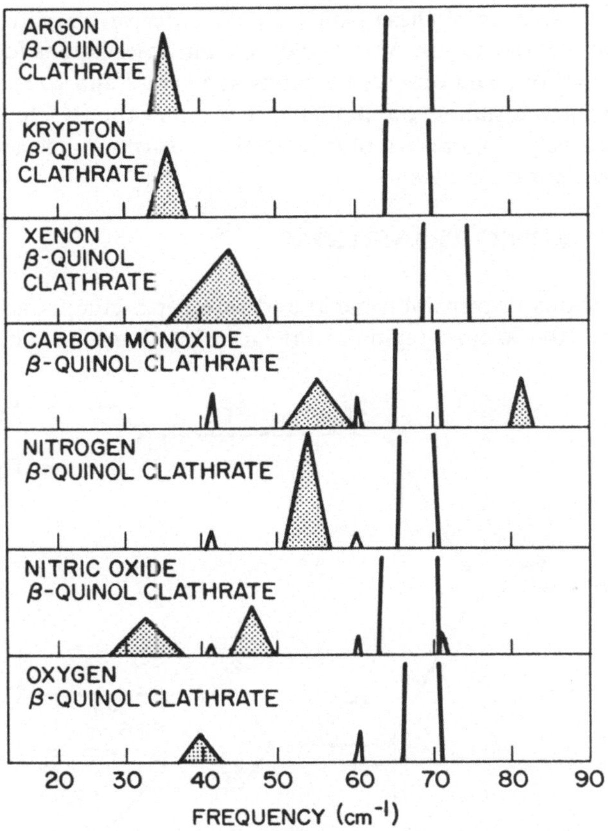

Fig. 9-1. Absorption spectra of several gas molecules trapped in β-quinol clathrate from 15–90 cm^{-1}. (Courtesy of the American Institute of Physics, New York. From Burgiel *et al.*[4])

Table 9-3. Far-Infrared Absorptions of Gaseous Molecules Trapped in β-Quinol Clathrates

Gas	Translations, cm^{-1}	Hindered rotations, cm^{-1}
A	35	
Kr	35	
Xe	42	
CO	55.2	81.5
N_2	53.5	
NO	46.5	33.0
O_2	40.0	
HCl	52	20

mode. The intensities of these bands are an order of magnitude less than electrical dipole transitions. Allen[5] made similar observations for HCl gas in a β-quinol clathrate and observed a bands at 52 cm^{-1} and 20 cm^{-1} which he assigned to the translation and hindered rotation of the HCl molecule in the cage, respectively. Figure 9-1 illustrates the absorption spectra of several gases in β-quinol clathrates.

9.3. ELECTRONIC TRANSITIONS

Absorptions in transition metal and rare earth compounds have been found in the 10–100 cm^{-1} region of the far infrared and have been assigned

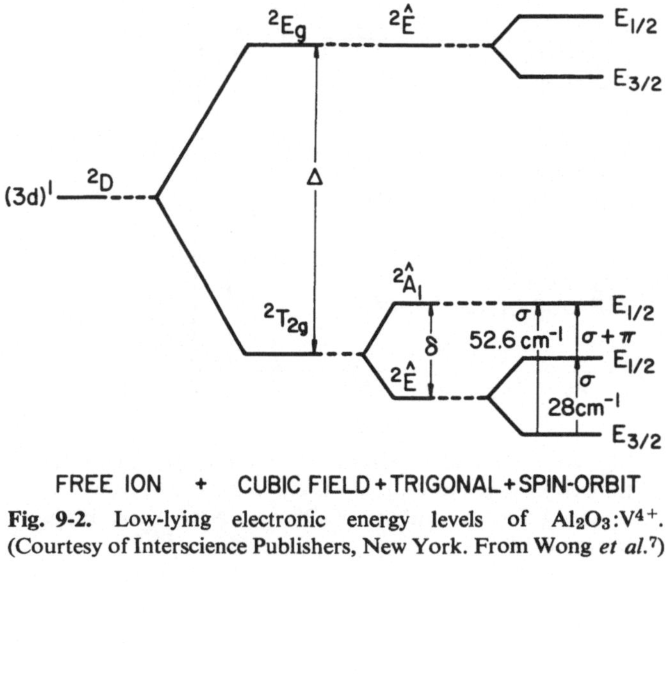

Fig. 9-2. Low-lying electronic energy levels of $Al_2O_3:V^{4+}$. (Courtesy of Interscience Publishers, New York. From Wong *et al.*[7])

to electronic transitions.[6–14] In some cases they may couple to vibrational modes and are called vibronic modes.[6]

A few examples will serve to illustrate the value of the far-infrared region for the observation of these vibrations.

(1) Two far-infrared bands are observed at 28.0 cm^{-1} and 52.6 cm^{-1} in vanadium-doped corundum.[7] These have been assigned to the transitions from the ground state to the first two excited states of $Al_2O_3:V^{4+}$. Figure 9-2 illustrates the electronic configuration of $Al_2O_3:V^{4+}$. Similar results were obtained for Al_2O_3: Co^{2+} with an absorption occurring at 110 cm^{-1}. FeF_2 has an absorption at 52.7 cm^{-1}, MnF_2 at 10 cm^{-1}, MnO at 27.5 cm^{-1}, and NiO at 36 cm^{-1}.[8,9,12] MnF_2 doped with rare earth ions also shows far-infrared absorptions.[10] Many more examples exist.

Table 9-4. Far-Infrared Absorptions for Several Garnets and Er_2O_3

Compound	Frequency, cm^{-1}
YbIG[a]	14.0
ErIG	10.0
SmIG	33.5
HoIG	38.5
Er_2O_3	39.3, 40.9, 75.7, 80.6, 90.1, 99.0

[a] Ytterbium iron garnet, etc.

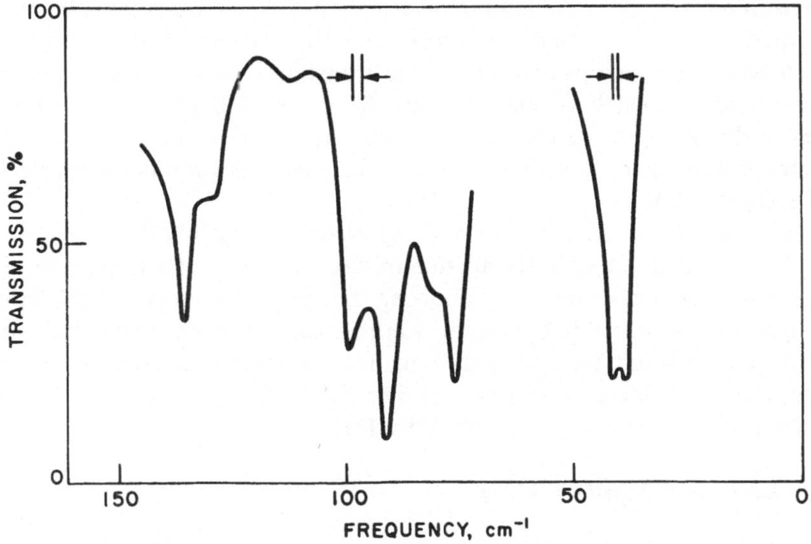

Fig. 9-3. Far-infrared spectrum of Er_2O_3 at 4.2°K. (Courtesy of American Institute of Physics, New York. From Bloor *et al.*[13])

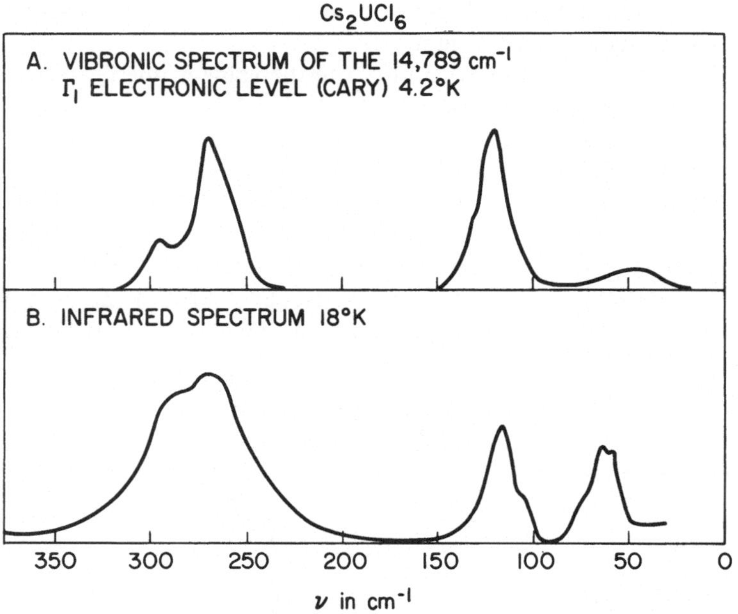

Fig. 9-4. Vibronic spectrum of Cs_2UCl_6. (Courtesy of Interscience Publishers, New York. From Satten and Stafsudd.[14])

(2) Observations with rare earth compounds also exist. Rare earth garnets with a chemical formula of $3RE_2O_3 \cdot 5Fe_2O_3$ show absorptions between 14 and 39 cm^{-1}.[11,12] Table 9-4 records the absorptions of several garnets which have been assigned to electronic transitions occurring between the lowest-lying rare earth ion energy levels. Results with Er_2O_3[13] are also tabulated in the table, the results agreeing with those obtained from the optical spectra and neutron inelastic scattering data. Figure 9-3 shows the spectrum of Er_2O_3 at 4.2°K.

Figure 9-4 illustrates the vibronic spectrum of Cs_2UCl_6.[14]

The far-infrared absorptions due to electronic transitions may be distinguished from lattice bands by allowing the sample to warm and obtaining the spectrum at room temperature. Lattice bands persist, while electronic transitions wash out because of the equalization of population of lower and upper electronic levels. For a further discussion of electronic transitions in the far-infrared region, see the book by Hadni.[14a]

9.4. LATTICE VIBRATIONS

The motions of groups in a solid with respect to other groups are termed external or lattice modes. The groups may be ions or molecules, giv-

Table 9-5. Lattice Frequencies (cm^{-1}) in Various Solids[a]

1. Space Group ($Fm3m$–O_h^5); NaCl Structure; $\Gamma = f_{1u}$(IR)

	ν_{TO}[b]	ν_{LO}[b]	References
LiH	590	1120	15, 17
LiF	306	659	15, 17
LiCl	191	398	15, 18
LiBr	159	325	15, 18
LiI	144		15
NaF	244	418	15, 17
NaCl	164	264	15, 17
NaBr	134	209	15, 19
NaI	117	176	15, 19
KF	190	326	15, 18
KCl	146	214	15, 19
KBr	113	168	15, 19
KI	101	139	15, 19
RbF	156	286	15, 18
RbCl	116	173	15, 19
RbBr	88	167	15, 19
RbI	75	103	15, 19
CsF	127		15
AgCl	106	146	15, 16
AgBr	79	138	15, 16
AgI	108		15, 16
MgO	401	718	15, 17
NiO	401	580	15, 20
CoO	349	542	15, 20
PbS	70		21
PbTe	31		51

2. Space Group ($Pm3m$–O_h^1); CsCl Structure; $\Gamma = f_{1u}$(IR)

	ν_{TO}	ν_{LO}	References
CsCl	99	165	15, 19
CsBr	73	112	15, 19
CsI	62	85	15, 19
TlCl	63	158	15, 19
TlBr	43	101	15, 19
TlI	52		15, 19

3. Space Group ($Fd3m$–O_h^7); Diamond Structure; $\Gamma = f_{2g}$(R)

	ν_{TO}	Reference
C	1330	22
Si	520	23
Ge	301	23

Table 9-5 (Continued)

4. Space Group ($P6_3mc$–C_{6v}^4); Wurtzite Structure

$\Gamma =$	a_1(IR,R) +	e_1(IR,R) +	$2e_2$(R) +	$2b_1$(IA)	References
AlN		668			24
BeO		678			25
ZnO	377[c]	407	437	101	15, 26
SiC	786	794			27
CdS	228[c]	235	252	44	15, 28
CdSe	166[c]	172			15, 29

5. Space Group ($F\bar{4}3m$–T_d^2); Sphalerite Structure; $\Gamma = f_2(\mathrm{IR} + \mathrm{R})$

	ν_{TO}	ν_{LO}	References
BN	1065	1340	15, 29
SiC	794		27
GaP	367	403	15, 30
GaAs	269	292	15, 30
GaSb	231	240	15, 30
InP	304	345	15, 30
InAs	217	243	15, 30
InSb	185	197	15, 31
AlSb	319	340	15, 30
CdTe	141	168	15, 33
ZnS	274	350	15, 32
ZnSe	215		15, 30
ZnTe	179	306	15, 30

6. Space Group ($P4/mmm$–D_{4h}^{14}); Rutile Structure

$\Gamma =$	a_{1g} + (R)	b_{1g} + (R)	b_{2g} + (R)	e_g + (R)	a_{2u} + (IR)	$3e_u$ + (IR)	a_{2g} + (IA)	$2b_{1u}$ (IA)	References
MgF_2	410	92	515	295	399	410	247		34
MnF_2	341	61	476	247					35
FeF_2	340	73	496	257	440	320	200		36
CoF_2					360	208	196		36
NiF_2					375	286	228		36
ZnF_2	350	70	522	253	294	244	173		34
TiO_2	612	143	826	447	167	388	183		35

7. Space Group ($Fm3m$–O_h^5); Fluorite Structure

$\Gamma =$	f_{2g}(R) +	f_{1u}(IR)	References
CaF_2	322	257	37, 39
SrF_2	286	217	38
BaF_2	241	184	39
CdF_2		202	41
$SiMg_2$		267	40
$GeMg_2$		207	40
$SnMg_2$		187	40
PbF_2		102	41

Table 9-5 (Continued)

7. (Continued)				
	$\Gamma =$	f_{2g}(R) +	f_{1u}(IR)	References
EuF_2			194	42
ThO_2			283	42
UO_2		455	283	42, 43

8. Space Group ($R\bar{3}c$–D^6_{3d}); Corundum Structure

$\Gamma =$	$2a_{1g}$ +	$5e_g$ +	$2a_{2u}$ +	$4e_u$ +	$3a_{2g}$ +	$2a_{2u}$	Reference
	(R)	(R)	(IR)	(IR)	(IA)	(IA)	
Al_2O_3			635, 543, 400	569, 442, 335			44
Cr_2O_3			613, 538	532, 444, 417			45

[a] The data in this table have been obtained in various ways—laser Raman experiments, transmission through films, diffuse neutron scattering, Kramers–Kronig fits of specular curves.
[b] ν_{TO} = transverse lattice mode, where the displaced vector of ions is perpendicular to the propagation of the light wave.
ν_{LO} = longitudinal lattice mode, where the displaced vector of ions is parallel to the propagation of the light wave.
[c] This is ν_{TO}·ν_{LO} for ZnO is at 595 cm^{-1}, for CdS it is at 306 cm^{-1}; for CdSe it is at 211 cm^{-1}.

ing rise to ionic lattice vibrations or molecular lattice vibrations. These modes are to be contrasted with the vibrations arising from oscillation of atoms within a group, which are called internal modes. The forces between groups are considerably less than those found within a group, and therefore, the lattice modes would be expected to occur at lower frequency. However, for compounds of light mass the lattice mode may be of higher frequency (e.g., the longitudinal vibration in BN is at 1340 cm^{-1}). The lattice modes may be translatory or rotatory in nature; the rotatory type are often referred to as librational modes. They may be acoustical or optical. Only the optical lattice modes are observed in the infrared spectrum, since no dipole moment change occurs for the acoustical modes. However, it is possible that acoustical modes may take part in combination tones. Considerably more research has been devoted to the study of ionic lattice modes than to molecular lattice modes. Table 9-5 compiles the frequencies for several lattice vibrations in common crystals. For a more thorough discussion on this subject see Mitra and Gielisse.[46]

9.5. MISCELLANEOUS VIBRATIONS

Discussions of rotations and inversion-type vibrations (restricted or nonrestricted) in the far infrared are to be found in many books and review articles. For a discussion of low-lying torsional vibrations, see Miller.[47] For a review on low-frequency hydrogen-bonded vibrations see Brasch *et al.*[48] A discussion of low-frequency water-librational modes was presented in Chapter 5.

BIBLIOGRAPHY

1. J. C. Evans and G. Y.-S. Lo, *J. Phys. Chem.* **69,** 3223 (1965); **70,** 11 (1966); **70,** 20 (1966); **71,** 3942 (1967).
2. H. K. Bodenseh and J. B. Ramsey, *J. Phys. Chem.* **69,** 543 (1965).
3. W. F. Edgell, J. Lyford, R. Wright, W. Risen, and A. Watts, *J. Am. Chem. Soc.* **92,** 2240 (1970).

3a. W. F. Edgell and N. Pauuwe, *Chem. Commun.*, 284 (1969).

3b. M. J. French and J. L. Wood, *J. Chem. Phys.* **49,** 2358 (1968).

4. J. C. Burgiel, H. Meyer, and P. L. Richards, *J. Chem. Phys.* **43,** 4291 (1965).
5. S. J. Allen, *J. Chem. Phys.* **44,** 394 (1966).
6. M. Wagner, in *Optical Properties of Ions in Crystals* (H. W. Crosswhite and H. W. Moos, Eds.), Interscience Publishers, New York (1967).
7. J. Y. Wong, M. J. Berggren, and A. L. Schawlow, in *Optical Properties of Ions in Crystals* (H. M. Crosswhite and H. W. Moos, Eds.), Interscience Publishers New York (1967).
8. R. C. Ohlman and M. Tinkham, *Phys. Rev.* **123,** 425 (1961).
9. M. Tinkham, *Proc. Roy. Soc. London* **A236,** 535 (1956).
10. R W. Alexander, Jr. and A. J. Sievers, in *Optical Properties of Ions in Crystals* (H. M. Crosswhite and H. W. Moos, Eds.), Interscience Publishers, New York (1967).
11. A. J. Sievers and M. Tinkham, *Phys. Rev.* **129,** 1995 (1963).
12. M. Tinkham, in *Far-Infrared Properties of Solids*, Plenum Press, New York (1970).
13. D. Bloor, E. Ellis, D. H. Martin, and A. Wadham, *J. Appl. Phys.* **39,** 971 (1968).
14. R. A. Satten and O. M. Stafsudd, in *Optical Properties of Ions in Crystals* (H. M. Crosswhite and H. W. Moos, Eds.), Interscience Publishers, New York (1967).

14a. A. Hadni, *Essentials of Modern Physics Applied to the Study of the Infrared*, Pergamon Press, New York (1967).

15. S. S. Mitra, in *Optical Properties of Solids* (S. Nudelman and S. S. Mitra, Eds.), Plenum Press, New York (1969).
16. G. L. Bottger and K. L. Geddes, *J. Chem. Phys.* **46,** 3000 (1967).
17. E. Burstein, in *Lattice Dynamics* (R. F. Wallis, Ed.), Pergamon Press, New York (1964).
18. M. Hass, *J. Phys. Chem. Solids* **24,** 1159 (1963).
19. G. O. Jones, D. H. Martin, P. A. Mawer, and C. H. Perry, *Proc. Roy. Soc.* **261A,** 10 (1961).
20. P. J. Gielisse, J. N. Plendl, L. C. Mansur, R. Marshall, S. S. Mitra, R. Mykolajewycz, and A. Smakula, *J. Appl. Phys.* **36,** 2446 (1965).
21. R. Geick, *Phys. Letters* **10,** 51 (1964).
22. S. S. Mitra, in *Solid State Physics, Vol. 13* (F. Seitz and D. Turnbull, Eds.), Academic Press, Inc., New York (1962), p. 1.
23. J. H. Parker, D. W. Feldman, and M. Ashkin, *Phys. Rev.* **155,** 712 (1967).
24. A. T. Collins, E. C. Lightowlers, and P. J. Dean, *Phys. Rev.* **158,** 833 (1967).
25. R. M. Brugger, K. A. Strong, and J. M. Carpenter, *J. Phys. Chem. Solids* **28,** 249 (1967).
26. T. C. Damen, S. P. S. Porto, and B. Tell, *Phys. Rev.* **142,** 570 (1966).

27. M. Tsuboi, *J. Chem. Phys.* **40,** 1326 (1964).
28. B. Tell, T. C. Damen, and S. P. S. Porto, *Phys. Rev.* **144,** 771 (1966).
29. R. Geick, C. H. Perry, and G. Rupprecht, *Phys. Rev.* **146,** 543 (1966).
30. S. S. Mitra and R. Marshall, *J. Chem. Phys.* **41,** 3158 (1964).
31. D. L. Stierwalt, *J. Phys. Soc. Japan Suppl.* **21,** 58 (1966).
32. T. Deutsch, *Rpt. Int. Conf. Phys. Semiconductors*, Inst. Physics and Physical Soc. London (1962), p. 505.
33. A. Mitsuishi, *J. Phys. Soc. Japan* **16,** 533 (1961).
34. A. S. Barker, *Phys. Rev.* **136A,** 290 (1964).
35. S. P. S. Porto, P. A. Fluery, and T. C. Damen, *Phys. Rev.* **154,** 522 (1967).
36. M. Balkanski, P. Moch, and G. Parisot, *J. Chem. Phys.* **44,** 940 (1966).
37. D. R. Bosomworth, *Phys. Rev.* **157,** 709 (1967).
38. R. K. Chang, B. Lacina, and P. J. Pershan, *Phys. Rev. Letters* **17,** 755 (1966).
39. D. Cribier, B. Farnoux, and B. Jacrot, *Phys. Letters* **1,** 187 (1962).
40. D. McWilliams and D. W. Lynch, *Phys. Rev.* **130,** 2248 (1963).
41. J. D. Axe, J. W. Gaglianello, and J. E. Scardefield, *Phys. Rev.* **139A,** 211 (1965).
42. J. D. Axe and G. D. Pettit, *Phys. Rev.* **151,** 676 (1966).
43. P. G. Marlow, J. P. Russell, and J. R. Hardy, *Phil. Mag.* **14,** 409 (1966).
44. A. S. Barker, *Phys. Rev.* **132,** 1474 (1963).
45. D. R. Dennecke and D. W. Lynch, *Phys. Rev.* **138A,** 530 (1965).
46. S. S. Mitra and P. J. Gielisse, AFCRL-65-395, June (1965).
47. F. A. Miller, in *Molecular Spectroscopy* (P. W. Hipple, Ed.), Institute of Petroleum, London (1969).
48. J. W. Brasch, Y. Mikawa, and R. J. Jakobsen, *Applied Spectroscopy Reviews, Vol. I* (E. C. Brame, Ed.), Marcell Dekker Co., New York (1968).

Appendix 1

SELECTION RULES AND CORRELATION CHARTS

For the determination of the solid state selection rules the site symmetries for 230 space groups are included. For the development of correlation charts for various solids, correlation tables are also included. A description of their use is given in Appendix 2.

Site Symmetries for the 230 Space Groups

Space group		
Hermann–Mauguin symbols[1]	Schoenflies symbols[2]	Site symmetry[2]
$P\bar{1}$	C_i^1	$8C_i$
Pm	C_s^1	$2C_s$
Cm	C_s^3	$C_s(2)$
$P2$	C_2^1	$4C_2$
$C2$	C_2^3	$2C_2(2)$
$P2/m$	C_{2h}^1	$8C_{2h}$; $4C_2(2)$; $2C_s(2)$
$P2_1/m$	C_{2h}^2	$4C_i(2)$; $C_s(2)$
$C2/m$	C_{2h}^3	$4C_{2h}(2)$; $2C_i(4)$; $2C_2(4)$; $C_s(4)$
$P2/c$	C_{2h}^4	$4C_i(2)$, $2C_2(2)$
$P2_1/c$	C_{2h}^5	$4C_i(2)$
$C2/c$	C_{2h}^6	$4C_i(4)$; $C_2(4)$
$Pmm2$	C_{2v}^1	$4C_{2v}$; $4C_s(2)$
$Pmc2_1$	C_{2v}^2	$2C_s(2)$
$Pcc2$	C_{2v}^3	$4C_2(2)$
$Pma2$	C_{2v}^4	$2C_2(2)$; $C_s(2)$
$Pnc2$	C_{2v}^6	$2C_2(2)$
$Pmn2_1$	C_{2v}^7	$C_s(2)$
$Pba2$	C_{2v}^8	$2C_2(2)$
$Pnn2$	C_{2v}^{10}	$2C_2(2)$
$Cmm2$	C_{2v}^{11}	$2C_{2v}(2)$; $C_2(4)$; $2C_s(4)$
$Cmc2_1$	C_{2v}^{12}	$C_s(4)$
$Ccc2$	C_{2v}^{13}	$3C_2(4)$
$Amm2$	C_{2v}^{14}	$2C_{2v}(2)$; $3C_s(4)$

APPENDIX 1 (Continued)

Space group		
Hermann–Mauguin symbols[1]	Schoenflies symbols[2]	Site symmetry[2]
*Abm*2	C_{2v}^{15}	$2C_2(4)$; $C_s(4)$
*Ama*2	C_{2v}^{16}	$C_2(4)$; $C_s(4)$
*Aba*2	C_{2v}^{17}	$C_2(4)$
*Fmm*2	C_{2v}^{18}	$C_{2v}(4)$, $C_2(8)$; $2C_s(8)$
*Fdd*2	C_{2v}^{19}	$C_2(8)$
*Imm*2	C_{2v}^{20}	$2C_{2v}(2)$; $2C_s(4)$
*Iba*2	C_{2v}^{21}	$2C_2(4)$
*Ima*2	C_{2v}^{22}	$C_2(4)$; $C_s(4)$
*P*222	D_2^1	$8D_2$; $12C_2(2)$
$P222_1$	D_2^2	$4C_2(2)$
$P2_12_12$	D_2^3	$2C_2(2)$
$C222_1$	D_2^5	$2C_2(4)$
*C*222	D_2^6	$4D_2(2)$; $7C_2(4)$
*F*222	D_2^7	$4D_2(4)$; $6C_2(8)$
*I*222	D_2^8	$4D_2(2)$; $6C_2(4)$
$I2_12_12_1$	D_2^9	$3C_2(4)$
Pmmm	D_{2h}^1	$8D_{2h}$; $12C_{2v}(2)$; $6C_s(4)$
Pnnn	D_{2h}^2	$4D_{2h}(2)$; $2C_i(4)$; $6C_2(4)$
Pccm	D_{2h}^3	$4C_{2h}(2)$; $4D_2(2)$; $8C_2(4)$; $C_s(4)$
Pban	D_{2h}^4	$4D_2(2)$; $2C_i(4)$; $6C_2(4)$
Pmma	D_{2h}^5	$4C_{2h}(2)$; $2C_{2v}(2)$; $2C_2(4)$; $3C_s(4)$
Pnna	D_{2h}^6	$2C_i(4)$; $2C_2(4)$
Pmna	D_{2h}^7	$4C_{2h}(2)$; $3C_2(4)$; $C_s(4)$
Pcca	D_{2h}^8	$2C_i(4)$; $3C_2(4)$
Pbam	D_{2h}^9	$4C_{2h}(2)$; $2C_2(4)$; $2C_s(4)$
Pccn	D_{2h}^{10}	$2C_i(4)$; $2C_2(4)$
Pbcm	D_{2h}^{11}	$2C_i(4)$; $C_2(4)$; $C_s(4)$
Pnnm	D_{2h}^{12}	$4C_{2h}(2)$; $2C_2(4)$; $C_s(4)$
Pmmn	D_{2h}^{13}	$2C_{2v}(2)$; $2C_i(4)$; $2C_s(4)$
Pbcn	D_{2h}^{14}	$2C_i(4)$; $C_2(4)$
Pbca	D_{2h}^{15}	$2C_i(4)$
Pnma	D_{2h}^{16}	$2C_i(4)$; $C_s(4)$
Cmcm	D_{2h}^{17}	$2C_{2h}(4)$; $C_{2v}(4)$; $C_i(8)$; $C_2(8)$; $2C_s(8)$
Cmca	D_{2h}^{18}	$2C_{2h}(4)$; $C_i(8)$; $2C_2(8)$; $C_s(8)$
Cmmm	D_{2h}^{19}	$4D_{2h}(2)$; $2C_{2h}(4)$; $6C_{2v}(4)$; $C_2(8)$; $4C_s(8)$
Cccm	D_{2h}^{20}	$2D_2(4)$; $4C_{2h}(4)$; $5C_2(8)$; $C_s(8)$
Cmma	D_{2h}^{21}	$2D_2(4)$; $4C_{2h}(4)$; $C_{2v}(4)$; $5C_2(8)$; $2C_s(8)$
Ccca	D_{2h}^{22}	$2D_2(4)$; $2C_i(8)$; $4C_2(8)$
Fmmm	D_{2h}^{23}	$2D_{2h}(4)$; $3C_{2h}(8)$; $D_2(8)$; $3C_{2v}(8)$; $3C_2(16)$; $3C_s(16)$
Fddd	D_{2h}^{24}	$2D_2(8)$; $2C_i(16)$; $3C_2(16)$

APPENDIX 1 (Continued)

Space group		
Hermann–Mauguin symbols[1]	Schoenflies symbols[2]	Site symmetry[2]
Immm	D_{2h}^{25}	$4D_{2h}(2)$; $6C_{2v}(4)$; $C_i(8)$; $3C_s(8)$
Ibam	D_{2h}^{26}	$2D_2(4)$; $2C_{2h}(4)$; $C_i(8)$; $4C_2(8)$; $C_s(8)$
Ibca	D_{2h}^{27}	$2C_i(8)$; $3C_2(8)$
Imma	D_{2h}^{28}	$4C_{2h}(4)$; $C_{2v}(4)$; $2C_2(8)$; $2C_s(8)$
$P\bar{4}$	S_4^1	$4S_4$; $3C_2(2)$
$I\bar{4}$	S_4^2	$4S_4(2)$; $2C_2(4)$
$P\bar{4}2m$	D_{2d}^1	$4D_{2d}$; $2D_2(2)$; $2C_{2v}(2)$; $5C_2(4)$; $C_s(4)$
$P\bar{4}2c$	D_{2d}^2	$4D_2(2)$; $2S_4(2)$; $7C_2(4)$
$P\bar{4}2_1m$	D_{2d}^3	$2S_4(2)$; $C_{2v}(2)$; $C_2(4)$; $C_s(4)$
$P\bar{4}2_1c$	D_{2d}^4	$2S_4(2)$; $2C_2(4)$
$P\bar{4}m2$	D_{2d}^5	$4D_{2d}(2)$; $3C_{2v}(4)$; $2C_2(8)$; $2C_s(8)$
$P\bar{4}c2$	D_{2d}^6	$2D_2(4)$; $2S_4(4)$; $5C_2(8)$
$P\bar{4}b2$	D_{2d}^7	$2S_4(4)$; $2D_2(4)$; $4C_2(8)$
$P\bar{4}n2$	D_{2d}^8	$2S_4(4)$; $2D_2(4)$; $4C_2(8)$
$I\bar{4}m2$	D_{2d}^9	$4D_{2d}(4)$; $2C_{2v}(8)$; $2C_2(16)$; $C_s(16)$
$I\bar{4}c2$	D_{2d}^{10}	$2S_4(8)$; $2D_2(8)$; $4C_2(16)$
$I\bar{4}2m$	D_{2d}^{11}	$2D_{2d}(2)$; $D_2(4)$; $S_4(4)$; $C_{2v}(4)$; $3C_2(8)$; $C_s(8)$
$I\bar{4}2d$	D_{2d}^{12}	$2S_4(4)$; $2C_2(8)$
*P*4	C_4^1	$2C_4$; $C_2(2)$
$P4_2$	C_4^3	$3C_2(2)$
*I*4	C_4^5	$C_4(2)$; $C_2(4)$
$I4_1$	C_4^6	$C_2(4)$
$P4/m$	C_{4h}^1	$4C_{4h}$; $2C_{2h}(2)$; $2C_4(2)$; $C_2(4)$; $2C_s(4)$
$P4_2/m$	C_{4h}^2	$4C_{2h}(2)$; $2S_4(2)$; $3C_2(4)$; $C_s(4)$
$P4/n$	C_{4h}^3	$2S_4(2)$; $C_4(2)$; $2C_i(4)$; $C_2(4)$
$P4_2/n$	C_{4h}^4	$2S_4(2)$; $2C_i(4)$; $2C_2(4)$
$I4/m$	C_{4h}^5	$2C_{4h}(2)$; $C_{2h}(4)$; $S_4(4)$; $C_4(4)$; $C_i(8)$; $C_2(8)$; $C_s(8)$
$I4_1/m$	C_{4h}^6	$2S_4(4)$; $2C_i(8)$; $C_2(8)$
P4mm	C_{4v}^1	$2C_{4v}$; $C_{2v}(2)$; $3C_s(4)$
P4bm	C_{4v}^2	$C_4(2)$; $C_{2v}(2)$; $C_s(4)$
$P4_2cm$	C_{4v}^3	$2C_{2v}(2)$; $C_2(4)$; $C_s(4)$
$P4_2nm$	C_{4v}^4	$C_{2v}(2)$; $C_2(4)$; $C_s(4)$
P4cc	C_{4v}^5	$2C_4(2)$; $C_2(4)$
P4nc	C_{4v}^6	$C_4(2)$; $C_2(4)$
$P4_2mc$	C_{4v}^7	$3C_{2v}(2)$; $2C_s(4)$
$P4_2bc$	C_{4v}^8	$2C_2(4)$
I4mm	C_{4v}^9	$C_{4v}(2)$; $C_{2v}(4)$; $2C_s(8)$
I4cm	C_{4v}^{10}	$C_4(4)$; $C_{2v}(4)$; $C_s(8)$
$I4_1md$	C_{4v}^{11}	$C_{2v}(4)$; $C_s(8)$

APPENDIX 1 (Continued)

Space group		
Hermann–Mauguin symbols[1]	Schoenflies symbols[2]	Site symmetry[2]
$I4_1cd$	C_{4v}^{12}	$C_2(8)$
$P422$	D_4^1	$4D_4$; $2D_2(2)$; $2C_4(2)$; $7C_2(4)$
$P42_12$	D_4^2	$2D_2(2)$; $C_4(2)$; $3C_2(4)$
$P4_122$	D_4^3	$3C_2(4)$
$P4_12_12$	D_4^4	$C_2(4)$
$P4_222$	D_4^5	$6D_2(2)$; $9C_2(4)$
$P4_22_12$	D_4^6	$2D_2(2)$; $4C_2(4)$
$P4_322$	D_4^7	$3C_2(4)$
$P4_32_12$	D_4^8	$C_2(4)$
$I422$	D_4^9	$2D_4(2)$; $2D_2(4)$; $C_4(4)$; $5C_2(8)$
$I4_122$	D_4^{10}	$2D_2(4)$; $4C_2(8)$
$P4/mmm$	D_{4h}^1	$4D_{4h}$; $2D_{2h}(2)$; $2C_{4v}(2)$; $7C_{2v}(4)$; $5C_5(8)$
$P4/mcc$	D_{4h}^2	$2D_4(2)$; $2C_{4h}(2)$; $D_2(4)$; $C_{2h}(4)$; $2C_4(4)$; $4C_2(8)$; $C_s(8)$
$P4/nbm$	D_{4h}^3	$2D_4(2)$; $2D_{2d}(2)$; $2C_{2h}(4)$; $C_4(4)$; $C_{2v}(4)$; $4C_2(8)$; $C_s(8)$
$P4/nnc$	D_{4h}^4	$2D_4(2)$; $D_2(4)$; $S_4(4)$; $C_4(4)$; $C_i(8)$; $4C_2(8)$
$P4/mbm$	D_{4h}^5	$2C_{4h}(2)$; $2D_{2h}(2)$; $C_4(4)$; $3C_{2v}(4)$; $3C_s(8)$
$P4/mnc$	D_{4h}^6	$2C_{4h}(2)$; $C_{2h}(4)$; $D_2(4)$; $C_4(4)$; $2C_2(8)$; $C_s(8)$
$P4/nmm$	D_{4h}^7	$2D_{2d}(2)$; $C_{4v}(2)$; $2C_{2h}(4)$; $C_{2v}(4)$; $2C_2(8)$; $2C_s(8)$
$P4/ncc$	D_{4h}^8	$D_2(4)$; $S_4(4)$; $C_4(4)$; $C_i(8)$; $2C_2(8)$
$P4_2/mmc$	D_{4h}^9	$4D_{2h}(2)$; $2D_{2d}(2)$; $7C_{2v}(4)$; $C_2(8)$; $3C_s(8)$
$P4_2/mcm$	D_{4h}^{10}	$2D_{2h}(2)$; $2D_{2d}(2)$; $D_2(4)$; $C_{4h}(4)$; $4C_{2v}(4)$; $3C_2(8)$; $2C_s(8)$
$P4_2/nbc$	D_{4h}^{11}	$3D_2(4)$; $S_4(4)$; $C_i(8)$; $5C_2(8)$
$P4_2/nnm$	D_{4h}^{12}	$2D_{2d}(2)$; $2D_2(4)$; $2C_{2h}(4)$; $C_{2v}(4)$; $5C_2(8)$; $C_s(8)$
$P4_2/mbc$	D_{4h}^{13}	$2C_{2h}(4)$; $S_4(4)$; $D_2(4)$; $3C_2(8)$; $C_s(8)$
$P4_2/mnm$	D_{4h}^{14}	$2D_{2h}(2)$; $C_{2h}(4)$; $S_4(4)$; $3C_{2v}(4)$; $C_2(8)$; $2C_s(8)$
$P4_2/nmc$	D_{4h}^{15}	$2D_{2d}(2)$; $2C_{2v}(4)$; $C_i(8)$; $C_2(8)$; $C_s(8)$
$P4_2/ncm$	D_{4h}^{16}	$D_2(4)$; $S_4(4)$; $2C_{2h}(4)$; $C_{2v}(4)$; $3C_2(8)$; $C_s(8)$
$I4/mmm$	D_{4h}^{17}	$2D_{4h}(2)$; $D_{2h}(4)$; $D_{2d}(4)$; $C_{4v}(4)$; $C_{2h}(8)$; $4C_{2v}(8)$; $C_2(16)$; $3C_s(16)$
$I4/mcm$	D_{4h}^{18}	$D_4(4)$; $D_{2d}(4)$; $C_{4h}(4)$; $D_{2h}(4)$; $C_{2h}(8)$; $C_4(8)$; $2C_{2v}(8)$; $2C_2(16)$; $2C_s(16)$

APPENDIX 1 (Continued)

Space group		
Hermann–Mauguin symbols[1]	Schoenflies symbols[2]	Site symmetry[2]
$I4_1/amd$	D_{4h}^{19}	$2D_{2d}(4)$; $2C_{2h}(8)$; $C_{2v}(8)$; $2C_2(16)$; $C_s(16)$
$I4_1/acd$	D_{4h}^{20}	$D_2(8)$; $S_4(8)$; $C_i(16)$; $3C_2(16)$
$P23$	T^1	$2T$; $2D_2(3)$; $C_3(4)$; $4C_2(6)$
$F23$	T^2	$4T(4)$; $C_3(16)$; $2C_2(24)$
$I23$	T^3	$T(2)$; $D_2(6)$; $C_3(8)$; $2C_2(12)$
$P2_13$	T^4	$C_3(4)$
$I2_13$	T^5	$C_3(8)$; $C_2(12)$
$Pm3$	T_h^1	$2T_h$; $2D_{2h}(3)$; $4C_{2v}(6)$; $C_3(8)$; $2C_s(12)$
$Pn3$	T_h^2	$T(2)$; $2C_{3i}(4)$; $D_2(6)$; $C_3(8)$; $2C_2(12)$
$Fm3$	T_h^3	$2T_h(4)$; $T(8)$; $C_{2h}(24)$; $C_{2v}(24)$; $C_3(32)$; $C_2(48)$; $C_s(48)$
$Fd3$	T_h^4	$2T(8)$; $3C_{3i}(16)$; $C_3(32)$; $C_2(48)$
$Im3$	T_h^5	$T_h(2)$; $D_{2h}(6)$; $C_{3i}(8)$; $2C_{2v}(12)$; $C_3(16)$; $C_s(24)$
$Pa3$	T_h^6	$2C_{3i}(4)$; $C_3(8)$
$Ia3$	T_h^7	$2C_{3i}(8)$; $C_3(16)$; $C_2(24)$
$P\bar{4}3m$	T_d^1	$2T_d$; $2D_{2d}(3)$; $C_{3v}(4)$; $2C_{2v}(6)$; $C_2(12)$; $C_s(12)$
$F\bar{4}3m$	T_d^2	$4T_d(4)$; $C_{3v}(16)$; $2C_{2v}(24)$; $C_s(48)$
$I\bar{4}3m$	T_d^3	$T_d(2)$; $2D_{2d}(6)$; $C_{3v}(8)$; $S_4(12)$; $C_{2v}(12)$; $C_2(24)$; $C_s(24)$
$P\bar{4}3n$	T_d^4	$T(2)$; $D_2(6)$; $2S_4(6)$; $C_3(8)$; $3C_2(12)$
$F\bar{4}3c$	T_d^5	$2T(8)$; $2S_4(24)$; $C_3(32)$; $2C_2(48)$
$I\bar{4}3d$	T_d^6	$2S_4(12)$; $C_3(16)$; $C_2(24)$
$P432$	O^1	$2O$; $2D_4(3)$; $2C_4(6)$; $C_3(8)$; $3C_2(12)$
$P4_232$	O^2	$T(2)$; $2D_3(4)$; $3D_2(6)$; $C_3(8)$; $5C_2(12)$
$F432$	O^3	$2O(4)$; $T(8)$; $D_2(24)$; $C_4(24)$; $C_3(32)$; $3C_2(48)$
$F4_132$	O^4	$2T(8)$; $2D_3(16)$; $C_3(32)$; $2C_2(48)$
$I432$	O^5	$O(2)$; $D_4(6)$; $D_3(8)$; $D_2(12)$; $C_4(12)$; $C_3(16)$; $3C_2(24)$
$P4_332$	O^6	$2D_3(4)$; $C_3(8)$; $C_2(12)$
$P4_132$	O^7	$2D_3(4)$; $C_3(8)$; $C_2(12)$
$I4_132$	O^8	$2D_3(8)$; $2D_2(12)$; $C_3(16)$; $3C_2(24)$
$Pm3m$	O_h^1	$2O_h$; $2D_{4h}(3)$; $2C_{4v}(6)$; $C_{3v}(8)$; $3C_{2v}(12)$; $3C_s(24)$
$Pn3n$	O_h^2	$O(2)$; $D_4(6)$; $C_{3i}(8)$; $S_4(12)$; $C_4(12)$; $C_3(16)$; $2C_2(24)$
$Pm3n$	O_h^3	$T_h(2)$; $D_{2h}(6)$; $2D_{2d}(6)$; $D_3(8)$; $3C_{2v}(12)$; $C_3(16)$; $C_2(24)$; $C_s(24)$

APPENDIX 1 (Continued)

Space group		
Hermann–Mauguin symbols[1]	Schoenflies symbols[2]	Site symmetry[2]
$Pn3m$	O_h^4	$T_d(2)$; $2D_{3d}(4)$; $D_{2d}(6)$; $C_{3v}(8)$; $D_2(12)$; $C_{2v}(12)$; $3C_2(24)$; $C_s(24)$
$Fm3m$	O_h^5	$2O_h(4)$; $T_d(8)$; $D_{2h}(24)$; $C_{4v}(24)$; $C_{3v}(32)$; $3C_{2v}(48)$; $2C_s(96)$
$Fm3c$	O_h^6	$O(8)$; $T_h(8)$; $D_{2d}(24)$; $C_{4h}(24)$; $C_{2v}(48)$; $C_4(48)$; $C_3(64)$; $C_2(96)$; $C_s(96)$
$Fd3m$	O_h^7	$2T_d(8)$; $2D_{3d}(16)$; $C_{3v}(32)$; $C_{2v}(48)$; $C_s(96)$; $C_2(96)$
$Fd3c$	O_h^8	$T(16)$; $D_3(32)$; $C_{3i}(32)$; $S_4(48)$; $C_3(64)$; $2C_2(96)$
$Im3m$	O_h^9	$O_h(2)$; $D_{4h}(6)$; $D_{3d}(8)$; $D_{2d}(12)$; $C_{4v}(12)$; $C_{3v}(16)$; $2C_{2v}(24)$; $C_2(48)$; $2C_s(48)$
$Ia3d$	O_h^{10}	$C_{3i}(16)$; $D_3(16)$; $D_2(24)$; $S_4(24)$; $C_3(32)$; $2C_2(48)$
$P3$	C_3^1	$3C_3$
$R3$	C_3^4	C_3
$P\bar{3}$	C_{3i}^1	$2C_{3i}$; $2C_3(2)$; $2C_i(3)$
$R\bar{3}$	C_{3i}^2	$2C_{3i}$; $C_3(2)$; $2C_i(3)$
$P3m1$	C_{3v}^1	$3C_{3v}$; $C_s(3)$
$P31m$	C_{3v}^2	C_{3v}; $C_3(2)$; $C_s(3)$
$P3c1$	C_{3v}^3	$3C_3(2)$
$P31c$	C_{3v}^4	$2C_3(2)$
$R3m$	C_{3v}^5	C_{3v}; $C_s(3)$
$R3c$	C_{3v}^6	$C_3(2)$
$P312$	D_3^1	$6D_3$; $3C_3(2)$; $2C_2(3)$
$P321$	D_3^2	$2D_3$; $2C_3(2)$; $2C_2(3)$
$P3_112$	D_3^3	$2C_2(3)$
$P3_121$	D_3^4	$2C_2(3)$
$P3_212$	D_3^5	$2C_2(3)$
$P3_221$	D_3^6	$2C_2(3)$
$R32$	D_3^7	$2D_3$; $C_3(2)$; $2C_2(3)$
$P\bar{3}1m$	D_{3d}^1	$2D_{3d}$; $2D_3(2)$; $C_{3v}(2)$; $2C_{2h}(3)$; $C_3(4)$; $2C_2(6)$; $C_s(6)$
$P\bar{3}1c$	D_{3d}^2	$3D_3(2)$; $C_{3i}(2)$; $2C_3(4)$; $C_i(6)$; $C_2(6)$
$P\bar{3}m1$	D_{3d}^3	$2D_{3d}$; $2C_{3v}(2)$; $2C_{2h}(3)$; $2C_2(6)$; $C_i(6)$
$P\bar{3}c1$	D_{3d}^4	$D_3(2)$; $C_{3i}(2)$; $2C_3(4)$; $C_i(6)$; $C_2(6)$
$R\bar{3}m$	D_{3d}^5	$2D_{3d}$; $C_{3v}(2)$; $2C_{2h}(3)$; $2C_2(6)$; $C_s(6)$
$R\bar{3}c$	D_{3d}^6	$D_3(2)$; $C_{3i}(2)$; $C_3(4)$; $C_i(6)$; $C_2(6)$
$P\bar{6}$	C_{3h}^1	$6C_{3h}$; $3C_3(2)$; $2C_s(3)$
$P\bar{6}m2$	D_{3h}^1	$6D_{3h}$; $3C_{3v}(2)$; $2C_{2v}(3)$; $3C_3(6)$

APPENDIX 1 (Continued)

Space group: Hermann–Mauguin symbols[1]	Space group: Schoenflies symbols[2]	Site symmetry[2]
$P\bar{6}c2$	D_{3h}^2	$3D_3(2)$; $3C_{3h}(2)$; $3C_3(4)$; $C_2(6)$; $C_s(6)$
$P\bar{6}2m$	D_{3h}^3	$2D_{3h}$; $2C_{3h}(2)$; $C_{3v}(2)$; $2C_{2v}(3)$; $C_3(4)$; $3C_s(6)$
$P\bar{6}2c$	D_{3h}^4	$D_3(2)$; $3C_{3h}(2)$; $2C_3(4)$; $C_2(6)$; $C_s(6)$
$P6$	C_6^1	C_6; $C_3(2)$; $C_2(3)$
$P6_2$	C_6^4	$2C_2(3)$
$P6_4$	C_6^5	$2C_2(3)$
$P6_3$	C_6^6	$2C_3(2)$
$P6mm$	C_{6v}^1	C_{6v}; $C_{3v}(2)$; $C_{2v}(3)$; $2C_s(6)$
$P6cc$	C_{6v}^2	$C_6(2)$; $C_3(4)$; $C_2(6)$
$P6_3cm$	C_{6v}^3	$C_{3v}(2)$; $C_3(4)$; $C_s(6)$
$P6_3mc$	C_{6v}^4	$2C_{3v}(2)$; $C_s(6)$
$P6/m$	C_{6h}^1	$2C_{6h}$; $2C_{3h}(2)$; $C_6(2)$; $2C_{2h}(3)$; $C_3(4)$; $C_2(6)$; $2C_s(6)$
$P6_3/m$	C_{6h}^2	$3C_{3h}(2)$; $C_{3i}(2)$; $2C_3(4)$; $C_i(6)$; $C_s(6)$
$P622$	D_6^1	$2D_6$; $2D_3(2)$; $C_6(2)$; $2D_2(3)$; $C_3(4)$; $5C_2(6)$
$P6_122$	D_6^2	$2C_2(6)$
$P6_522$	D_6^3	$2C_2(6)$
$P6_222$	D_6^4	$4D_2(3)$; $6C_2(6)$
$P6_422$	D_6^5	$4D_2(3)$; $6C_2(6)$
$P6_322$	D_6^6	$4D_3(2)$; $2C_3(4)$; $2C_2(6)$
$P6/mmm$	D_{6h}^1	$2D_{6h}$; $2D_{3h}(2)$; $C_{6v}(2)$; $2D_{2h}(3)$; $C_{3v}(4)$; $5C_{2v}(6)$; $4C_s(12)$
$P6/mcc$	D_{6h}^2	$D_6(2)$; $C_{6h}(2)$; $D_3(4)$; $C_{3h}(4)$; $C_6(4)$; $D_2(6)$; $C_{2h}(6)$; $C_3(8)$; $3C_2(12)$; $C_s(12)$
$P6_3/mcm$	D_{6h}^3	$D_{3h}(2)$; $D_{3d}(2)$; $C_{3h}(4)$; $D_3(4)$; $C_{3v}(4)$; $C_{2h}(6)$; $C_{2v}(6)$; $C_3(8)$; $C_2(12)$; $2C_s(12)$
$P6_3/mmc$	D_{6h}^4	$D_{3d}(2)$; $3D_{3h}(2)$; $2C_{3v}(4)$; $C_{2h}(6)$; $C_{2v}(6)$; $C_2(12)$; $2C_s(12)$

Note: The site symmetry is described by several symbols. In the example C_{2h}^2: $4C_i(2)$; $C_s(2)$, the space group C_{2h}^2 contains four sets of sites having a point symmetry C_i with two equivalent sites per set and one set of sites having a point symmetry C_s with two equivalent sites per set.

The data above are presented for a crystallographic unit cell. The number of equivalent sites per set (the number in the parentheses) or the molecules per crystallographic unit cell (Z) must be converted to one in terms of a primitive Bravais cell. This may be done by dividing the number in parentheses by the cell multiplicity. For A-, B-, C- or I-type lattices, divide by two, and for F-type lattices divide by four. All other types of lattices may be considered to be primitive.

Correlation Tables for the Species of a Group and Its Subgroups[3]

C_4	C_2
A	A
B	A
E	$2B$

C_6	C_3	C_2
A	A	A
B	A	B
E_1	E	$2B$
E_2	E	$2A$

D_2	C_2	C_2	C_2
A	A	A	A
B_1	A	B	B
B_2	B	A	B
B_3	B	B	A

D_3	C_3	C_2
A_1	A	A
A_2	A	B
E	E	$A+B$

			C_2'	C_2''
D_4	C_4	C_2	C_2	C_2
A_1	A	A	A	A
A_2	A	A	B	B
B_1	B	A	A	B
B_2	B	A	B	A
E	E	$2B$	$A+B$	$A+B$

D_5	C_5	C_2
A_1	A	A
A_2	A	B
E_1	E_1	$A+B$
E_2	E_2	$A+B$

		C_2'	C_2''				C_2'	C_2''
D_6	C_6	D_3	D_3	D_2	C_3	C_2	C_2	C_2
A_1	A	A_1	A_1	A	A	A	A	A
A_2	A	A_2	A_2	B_1	A	A	B	B
B_1	B	A_1	A_2	B_2	A	B	A	B
B_2	B	A_2	A_1	B_3	A	B	B	A
E_1	E_1	E	E	B_2+B_3	E	$2B$	$A+B$	$A+B$
E_2	E_2	E	E	$A+B_1$	E	$2A$	$A+B$	$A+B$

		$\sigma(zx)$	$\sigma(yz)$
C_{2v}	C_2	C_s	C_s
A_1	A	A'	A'
A_2	A	A''	A''
B_1	B	A'	A''
B_2	B	A''	A'

C_{3v}	C_3	C_s
A_1	A	A'
A_2	A	A''
E	E	$A'+A''$

		σ_v	σ_d		σ_v	σ_d
C_{4v}	C_4	C_{2v}	C_{2v}	C_2	C_s	C_s
A_1	A	A_1	A_1	A	A'	A'
A_2	A	A_2	A_2	A	A''	A''
B_1	B	A_1	A_2	A	A'	A''
B_2	B	A_2	A_1	A	A''	A'
E	E	B_1+B_2	B_1+B_2	$2B$	$A'+A''$	$A'+A''$

C_{5v}	C_5	C_s
A_1	A	A'
A_2	A	A''
E_1	E_1	$A'+A''$
E_2	E_2	$A'+A''$

		σ_v	σ_d	$\sigma_v \rightarrow \sigma(zx)$			σ_v	σ_d
C_{6v}	C_6	C_{3v}	C_{3v}	C_{2v}	C_3	C_2	C_s	C_s
A_1	A	A_1	A_1	A_1	A	A	A'	A'
A_2	A	A_2	A_2	A_2	A	A	A''	A''
B_1	B	A_1	A_2	B_1	A	B	A'	A''
B_2	B	A_2	A_1	B_2	A	B	A''	A'
E_1	E_1	E	E	B_1+B_2	E	$2B$	$A'+A''$	$A'+A''$
E_2	E_2	E	E	A_1+A_2	E	$2A$	$A'+A''$	$A'+A''$

C_{2h}	C_2	C_s	C_i
A_g	A	A'	A_g
B_g	B	A''	A_g
A_u	A	A''	A_u
B_u	B	A'	A_u

C_{3h}	C_3	C_s
A'	A	A'
E'	E	$2A'$
A''	A	A''
E''	E	$2A''$

C_{4h}	C_4	S_4	C_{2h}	C_2	C_s	C_i
A_g	A	A	A_g	A	A'	A_g
B_g	B	B	A_g	A	A'	A_g
E_g	E	E	$2B_g$	$2B$	$2A''$	$2A_g$
A_u	A	B	A_u	A	A''	A_u
B_u	B	A	A_u	A	A''	A_u
E_u	E	E	$2B_u$	$2B$	$2A'$	$2A_u$

C_{5h}	C_5	C_s
A'	A	A'
E_1'	E_1	$2A'$
E_2'	E_2	$2A'$
A''	A	A''
E_1''	E_1	$2A''$
E_2''	E_2	$2A''$

C_{6h}	C_6	C_{3h}	S_6	C_{2h}	C_3	C_2	C_s	C_i
A_g	A	A'	A_g	A_g	A	A	A'	A_g
B_g	B	A''	A_g	B_g	A	B	A''	A_g
E_{1g}	E_1	E''	E_g	$2B_g$	E	$2B$	$2A''$	$2A_g$
E_{2g}	E_2	E'	E_g	$2A_g$	E	$2A$	$2A'$	$2A_g$
A_u	A	A''	A_u	A_u	A	A	A''	A_u
B_u	B	A'	A_u	B_u	A	B	A'	A_u
E_{1u}	E_1	E'	E_u	$2B_u$	E	$2B$	$2A'$	$2A_u$
E_{2u}	E_2	E''	E_u	$2A_u$	E	$2A$	$2A''$	$2A_u$

D_{2h}	D_2	$C_2(z)$ C_{2v}	$C_2(y)$ C_{2v}	$C_2(x)$ C_{2v}	$C_2(z)$ C_{2h}	$C_2(y)$ C_{2h}	$C_2(x)$ C_{2h}	$C_2(z)$ C_2	$C_2(y)$ C_2	$C_2(x)$ C_2	$\sigma(xy)$ C_s	$\sigma(zx)$ C_s	$\sigma(yz)$ C_s
A_g	A	A_1	A_1	A_1	A_g	A_g	A_g	A	A	A	A'	A'	A'
B_{1g}	B_1	A_2	B_2	B_1	A_g	B_g	B_g	A	B	B	A'	A''	A''
B_{2g}	B_2	B_1	A_2	B_2	B_g	A_g	B_g	B	A	B	A''	A'	A''
B_{3g}	B_3	B_2	B_1	A_2	B_g	B_g	A_g	B	B	A	A''	A''	A'
A_u	A	A_2	A_2	A_2	A_u	A_u	A_u	A	A	A	A''	A''	A''
B_{1u}	B_1	A_1	B_1	B_2	A_u	B_u	B_u	A	B	B	A''	A'	A'
B_{2u}	B_2	B_2	A_1	B_1	B_u	A_u	B_u	B	A	B	A'	A''	A'
B_{3u}	B_3	B_1	B_2	A_1	B_u	B_u	A_u	B	B	A	A'	A'	A''

D_{3h}	C_{3h}	D_3	C_{3v}	$\sigma_h \rightarrow \sigma_v(zy)$ C_{2v}	C_3	C_2	σ_h C_s	σ_v C_s
A_1'	A'	A_1	A_1	A_1	A	A	A'	A'
A_2'	A'	A_2	A_2	B_2	A	B	A'	A''
E'	E'	E	E	A_1+B_2	E	$A+B$	$2A'$	$A'+A''$
A_1''	A''	A_1	A_2	A_2	A	A	A''	A''
A_2''	A''	A_2	A_1	B_1	A	B	A''	A'
E''	E''	E	E	A_2+B_1	E	$A+B$	$2A''$	$A'+A''$

D_{4h}	D_4	$C_2' \rightarrow C_2''$ D_{2d}	$C_2'' \rightarrow C_2'$ D_{2d}	C_{4v}	C_{4h}	C_2' D_{2h}	C_2'' D_{2h}	C_4	S_4	C_2' D_2	C_2'' D_2
A_{1g}	A_1	A_1	A_1	A_1	A_g	A_g	A_g	A	A	A	A
A_{2g}	A_2	A_2	A_2	A_2	A_g	B_{1g}	B_{1g}	A	A	B_1	B_1
B_{1g}	B_1	B_1	B_2	B_1	B_g	A_g	B_{1g}	B	B	A	B_1
B_{2g}	B_2	B_2	B_1	B_2	B_g	B_{1g}	A_g	B	B	B_1	A
E_g	E	E	E	E	E_g	$B_{2g}+B_{3g}$	$B_{2g}+B_{3g}$	E	E	B_2+B_3	B_2+B_3
A_{1u}	A_1	B_1	B_1	A_2	A_u	A_u	A_u	A	B	A	A
A_{2u}	A_2	B_2	B_2	A_1	A_u	B_{1u}	B_{1u}	A	B	B_1	B_1
B_{1u}	B_1	A_1	A_2	B_2	B_u	A_u	B_{1u}	B	A	A	B_1
B_{2u}	B_2	A_2	A_1	B_1	B_u	B_{1u}	A_u	B	A	B_1	A
E_u	E	E	E	E	E_u	$B_{2u}+B_{3u}$	$B_{2u}+B_{3u}$	E	E	B_2+B_3	B_2+B_3

D_{4h} (cont.)	C_2, σ_v C_{2v}	C_2, σ_d C_{2v}	C_2' C_{2v}	C_2'' C_{2v}	C_2 C_{2h}	C_2' C_{2h}	C_2'' C_{2h}	C_2 C_2	C_2' C_2	C_2'' C_2
A_{1g}	A_1	A_1	A_1	A_1	A_g	A_g	A_g	A	A	A
A_{2g}	A_2	A_2	B_1	B_1	A_g	B_g	B_g	A	B	B
B_{1g}	A_1	A_2	A_1	B_1	A_g	A_g	B_g	A	A	B
B_{2g}	A_2	A_1	B_1	A_1	A_g	B_g	A_g	A	B	A
E_g	B_1+B_2	B_1+B_2	A_2+B_2	A_2+B_2	$2B_g$	A_g+B_g	A_g+B_g	$2B$	$A+B$	$A+B$
A_{1u}	A_2	A_2	A_2	A_2	A_u	A_u	A_u	A	A	A
A_{2u}	A_1	A_1	B_2	B_2	A_u	B_u	B_u	A	B	B
B_{1u}	A_2	A_1	A_2	B_2	A_u	A_u	B_u	A	A	B
B_{2u}	A_1	A_2	B_2	A_2	A_u	B_u	A_u	A	B	A
E_u	B_1+B_2	B_1+B_2	A_1+B_1	A_1+B_1	$2B_u$	A_u+B_u	A_u+B_u	$2B$	$A+B$	$A+B$

D_{4h}	σ_h	σ_v	σ_d	
(cont.)	C_s	C_s	C_s	C_i
A_{1g}	A'	A'	A'	A_g
A_{2g}	A'	A''	A''	A_g
B_{1g}	A'	A'	A''	A_g
B_{2g}	A'	A''	A'	A_g
E_g	$2A''$	$A'+A''$	$A'+A''$	$2A_g$
A_{1u}	A''	A''	A''	A_u
A_{2u}	A''	A'	A'	A_u
B_{1u}	A''	A''	A'	A_u
B_{2u}	A''	A'	A''	A_u
E_u	$2A'$	$A'+A''$	$A'+A''$	$2A_u$

					$\sigma_h \rightarrow \sigma(zx)$		σ_h	σ_v
D_{5h}	D_5	C_{5v}	C_{5h}	C_5	C_{2v}	C_2	C_s	C_s
A_1'	A_1	A_1	A'	A	A_1	A	A'	A'
A_2'	A_2	A_2	A'	A	B_1	B	A'	A''
E_1'	E_1	E_1	E_1'	E_1	A_1+B_1	$A+B$	$2A'$	$A'+A''$
E_2'	E_2	E_2	E_2'	E_2	A_1+B_1	$A+B$	$2A'$	$A'+A''$
A_1''	A_1	A_2	A''	A	A_2	A	A''	A''
A_2''	A_2	A_1	A''	A	B_2	B	A''	A'
E_1''	E_1	E_1	E_1''	E_1	A_2+B_2	$A+B$	$2A''$	$A'+A''$
E_2''	E_2	E_2	E_2''	E_2	A_2+B_2	$A+B$	$2A''$	$A'+A''$

								$\sigma_h \rightarrow \sigma(xy)$								
		C_2'	C_2''			C_2''	C_2'	$\sigma_v \rightarrow \sigma(yz)$			C_2'	C_2''	σ_v	σ_d		
D_{6h}	D_6	D_{3h}	D_{3h}	C_{6v}	C_{6h}	D_{3d}	D_{3d}	D_{2h}	C_6	C_{3h}	D_3	D_3	C_{3v}	C_{3v}	S_6	D_2
A_{1g}	A_1	A_1'	A_1'	A_1	A_g	A_{1g}	A_{1g}	A_g	A	A'	A_1	A_1	A_1	A_1	A_g	A
A_{2g}	A_2	A_2'	A_2'	A_2	A_g	A_{2g}	A_{2g}	B_{1g}	A	A'	A_2	A_2	A_2	A_2	A_g	B_1
B_{1g}	B_1	A_1''	A_2''	B_2	B_g	A_{2g}	A_{1g}	B_{2g}	B	A''	A_1	A_2	A_2	A_1	A_g	B_2
B_{2g}	B_2	A_2''	A_1''	B_1	B_g	A_{1g}	A_{2g}	B_{3g}	B	A''	A_2	A_1	A_1	A_2	A_g	B_3
E_{1g}	E_1	E''	E''	E_1	E_{1g}	E_g	E_g	$B_{2g}+B_{3g}$	E_1	E''	E	E	E	E	E_g	B_2+B_3
E_{2g}	E_2	E'	E'	E_2	E_{2g}	E_g	E_g	A_g+B_{1g}	E_2	E'	E	E	E	E	E_g	$A+B_1$
A_{1u}	A_1	A_1''	A_1''	A_2	A_u	A_{1u}	A_{1u}	A_u	A	A''	A_1	A_1	A_2	A_2	A_u	A
A_{2u}	A_2	A_2''	A_2''	A_1	A_u	A_{2u}	A_{2u}	B_{1u}	A	A''	A_2	A_2	A_1	A_1	A_u	B_1
B_{1u}	B_1	A_1'	A_2'	B_1	B_u	A_{2u}	A_{1u}	B_{2u}	B	A'	A_1	A_2	A_1	A_2	A_u	B_2
B_{2u}	B_2	A_2'	A_1'	B_2	B_u	A_{1u}	A_{2u}	B_{3u}	B	A'	A_2	A_1	A_2	A_1	A_u	B_3
E_{1u}	E_1	E'	E'	E_1	E_{1u}	E_u	E_u	$B_{2u}+B_{3u}$	E_1	E'	E	E	E	E	E_u	B_2+B_3
E_{2u}	E_2	E''	E''	E_2	E_{2u}	E_u	E_u	A_u+B_{1u}	E_2	E''	E	E	E	E	E_u	$A+B_1$

D_{6h} (cont.)	C_2' C_{2v}	C_2'' C_{2v}	C_2 C_{2h}	C_2' C_{2h}	C_2'' C_{2h}	C_3	C_2 C_2	C_2' C_2	C_2'' C_2	σ_h C_s	σ_d C_s
A_{1g}	A_1	A_1	A_g	A_g	A_g	A	A	A	A	A'	A'
A_{2g}	B_1	B_1	A_g	B_g	B_g	A	A	B	B	A'	A''
B_{1g}	A_2	B_2	B_g	A_g	B_g	A	B	A	B	A''	A'
B_{2g}	B_2	A_2	B_g	B_g	A_g	A	B	B	A	A''	A''
E_{1g}	A_2+B_2	A_2+B_2	$2B_g$	A_g+B_g	A_g+B_g	E	$2B$	$A+B$	$A+B$	$2A''$	$A'+A''$
E_{2g}	A_1+B_1	A_1+B_1	$2A_g$	A_g+B_g	A_g+B_g	E	$2A$	$A+B$	$A+B$	$2A'$	$A'+A''$
A_{1u}	A_2	A_2	A_u	A_u	A_u	A	A	A	A	A''	A''
A_{2u}	B_2	B_2	A_u	B_u	B_u	A	A	B	B	A''	A'
B_{1u}	A_1	B_1	B_u	A_u	B_u	A	B	A	B	A'	A''
B_{2u}	B_1	A_1	B_u	B_u	A_u	A	B	B	A	A'	A'
E_{1u}	A_1+B_1	A_1+B_1	$2B_u$	A_u+B_u	A_u+B_u	E	$2B$	$A+B$	$A+B$	$2A'$	$A'+A''$
E_{2u}	A_2+B_2	A_2+B_2	$2A_u$	A_u+B_u	A_u+B_u	E	$2A$	$A+B$	$A+B$	$2A''$	$A'+A''$

D_{6h} (cont.)	σ_v C_s	C_i
A_{1g}	A'	A_g
A_{2g}	A''	A_g
B_{1g}	A''	A_g
B_{2g}	A'	A_g
E_{1g}	$A'+A''$	$2A_g$
E_{2g}	$A'+A''$	$2A_g$
A_{1u}	A''	A_u
A_{2u}	A'	A_u
B_{1u}	A'	A_u
B_{2u}	A''	A_u
E_{1u}	$A'+A''$	$2A_u$
E_{2u}	$A'+A''$	$2A_u$

$D_{\infty h}$	A_1'	A_2'	E_1'	E_2'	E_3'	A_1''	A_2''	E_1''	E_2''	E_3''
$C_{\infty v}$	A_1	A_2	E_1	E_2	E_3	A_2	A_1	E_1	E_2	E_3

D_{2d}	S_4	$C_2 \rightarrow C_2(z)$ D_2	C_{2v}	C_2 C_2	C_2' C_2	C_s
A_1	A	A	A_1	A	A	A'
A_2	A	B_1	A_2	A	B	A''
B_1	B	A	A_2	A	A	A''
B_2	B	B_1	A_1	A	B	A'
E	E	B_2+B_3	B_1+B_2	$2B$	$A+B$	$A'+A''$

D_{3d}	D_3	C_{3v}	S_6	C_3	C_{2h}	C_2	C_s	C_i
A_{1g}	A_1	A_1	A_g	A	A_g	A	A'	A_g
A_{2g}	A_2	A_2	A_g	A	B_g	B	A''	A_g
E_g	E	E	E_g	E	A_g+B_g	$A+B$	$A'+A''$	$2A_g$
A_{1u}	A_1	A_2	A_u	A	A_u	A	A''	A_u
A_{2u}	A_2	A_1	A_u	A	B_u	B	A'	A_u
E_u	E	E	E_u	E	A_u+B_u	$A+B$	$A'+A''$	$2A_u$

D_{4d}	D_4	C_{4v}	S_8	C_4	C_{2v}	C_2 C_2	C_2' C_2	C_s
A_1	A_1	A_1	A	A	A_1	A	A	A'
A_2	A_2	A_2	A	A	A_2	A	B	A''
B_1	A_1	A_2	B	A	A_2	A	A	A''
B_2	A_2	A_1	B	A	A_1	A	B	A'
E_1	E	E	E_1	E	B_1+B_2	$2B$	$A+B$	$A'+A''$
E_2	B_1+B_2	B_1+B_2	E_2	$2B$	A_1+A_2	$2A$	$A+B$	$A'+A''$
E_3	E	E	E_3	E	B_1+B_2	$2B$	$A+B$	$A'+A''$

D_{5d}	D_5	C_{5v}	C_5	C_2	C_s	C_i
A_{1g}	A_1	A_1	A	A	A'	A_g
A_{2g}	A_2	A_2	A	B	A''	A_g
E_{1g}	E_1	E_1	E_1	$A+B$	$A'+A''$	$2A_g$
E_{2g}	E_2	E_2	E_2	$A+B$	$A'+A''$	$2A_g$
A_{1u}	A_1	A_2	A	A	A''	A_u
A_{2u}	A_2	A_1	A	B	A'	A_u
E_{1u}	E_1	E_1	E_1	$A+B$	$A'+A''$	$2A_u$
E_{2u}	E_2	E_2	E_2	$A+B$	$A'+A''$	$2A_u$

D_{6d}	D_6	C_{6v}	C_6	D_{2d}	D_3	C_{3v}	D_2	C_{2v}	S_4	C_3	C_2 C_2
A_1	A_1	A_1	A	A_1	A_1	A_1	A	A_1	A	A	A
A_2	A_2	A_2	A	A_2	A_2	A_2	B_1	A_2	A	A	A
B_1	A_1	A_2	A	B_1	A_1	A_2	A	A_2	B	A	A
B_2	A_2	A_1	A	B_2	A_2	A_1	B_1	A_1	B	A	A
E_1	E_1	E_1	E_1	E	E	E	B_2+B_3	B_1+B_2	E	E	$2B$
E_2	E_2	E_2	E_2	B_1+B_2	E	E	$A+B_1$	A_1+A_2	$2B$	E	$2A$
E_3	B_1+B_2	B_1+B_2	$2B$	E	A_1+A_2	A_1+A_2	B_2+B_3	B_1+B_2	E	$2A$	$2B$
E_4	E_2	E_2	E_2	A_1+A_2	E	E	$A+B_1$	A_1+A_2	$2A$	E	$2A$
E_5	E_1	E_1	E_1	E	E	E	B_2+B_3	B_1+B_2	E	E	$2B$

D_{6d} (cont.)	C_2' C_2	C_s
A_1	A	A'
A_2	B	A''
B_1	A	A''
B_2	B	A'
E_1	$A+B$	$A'+A''$
E_2	$A+B$	$A'+A''$
E_3	$A+B$	$A'+A''$
E_4	$A+B$	$A'+A''$
E_5	$A+B$	$A'+A''$

S_4	C_2
A	A
B	A
E	$2B$

S_6	C_3	C_i
A_g	A	A_g
E_g	E	$2A_g$
A_u	A	A_u
E_u	E	$2A_u$

S_8	C_4	C_2
A	A	A
B	A	A
E_1	E	$2B$
E_2	$2B$	$2A$
E_3	E	$2B$

T	D_2	C_3	C_2
A	A	A	A
E	$2A$	E	$2A$
F	$B_1+B_2+B_3$	$A+E$	$A+2B$

T_h	T	D_{2h}	S_6	D_2	C_{2v}	C_{2h}	C_3	C_2
A_g	A	A_g	A_g	A	A_1	A_g	A	A
E_g	E	$2A_g$	E_g	$2A$	$2A_1$	$2A_g$	E	$2A$
F_g	F	$B_{1g}+B_{2g}+B_{3g}$	A_g+E_g	$B_1+B_2+B_3$	$A_2+B_1+B_2$	A_g+2B_g	$A+E$	$A+2B$
A_u	A	A_u	A_u	A	A_2	A_u	A	A
E_u	E	$2A_u$	E_u	$2A$	$2A_2$	$2A_u$	E	$2A$
F_u	F	$B_{1u}+B_{2u}+B_{3u}$	A_u+E_u	$B_1+B_2+B_3$	$A_1+B_1+B_2$	A_u+2B_u	$A+E$	$A+2B$

T_h (cont.)	C_s	C_i
A_g	A'	A_g
E_g	$2A'$	$2A_g$
F_g	$A'+2A''$	$3A_g$
A_u	A''	A_u
E_u	$2A''$	$2A_u$
F_u	$2A'+A''$	$3A_u$

T_d	T	D_{2d}	C_{3v}	S_4	D_2	C_{2v}	C_3	C_2	C_s
A_1	A	A_1	A_1	A	A	A_1	A	A	A'
A_2	A	B_1	A_2	B	A	A_2	A	A	A''
E	E	A_1+B_1	E	$A+B$	$2A$	A_1+A_2	E	$2A$	$A'+A''$
F_1	F	A_2+E	A_2+E	$A+E$	$B_1+B_2+B_3$	$A_2+B_1+B_2$	$A+E$	$A+2B$	$A'+2A''$
F_2	F	B_2+E	A_1+E	$B+E$	$B_1+B_2+B_3$	$A_1+B_1+B_2$	$A+E$	$A+2B$	$2A'+A''$

O	T	D_4	D_3	C_4	$3C_2$ D_2	C_2, $2C_2'$ D_2	C_3	C_2	C_2
A_1	A	A_1	A_1	A	A	A	A	A	A
A_2	A	B_1	A_2	B	A	B_1	A	A	B
E	E	A_1+B_1	E	$A+B$	$2A$	$A+B_1$	E	$2A$	$A+B$
F_1	F	A_2+E	A_2+E	$A+E$	$B_1+B_2+B_3$	$B_1+B_2+B_3$	$A+E$	$A+2B$	$A+2B$
F_2	F	B_2+E	A_1+E	$B+E$	$B_1+B_2+B_3$	$A+B_2+B_3$	$A+E$	$A+2B$	$2A+B$

O_h*	O	T_d	T_h	D_{4h}	D_{3d}
A_{1g}	A_1	A_1	A_g	A_{1g}	A_{1g}
A_{2g}	A_2	A_2	A_g	B_{1g}	A_{2g}
E_g	E	E	E_g	$A_{1g}+B_{1g}$	E_g
F_{1g}	F_1	F_1	F_g	$A_{2g}+E_g$	$A_{2g}+E_g$
F_{2g}	F_2	F_2	F_g	$B_{2g}+E_g$	$A_{1g}+E_g$
A_{1u}	A_1	A_2	A_u	A_{1u}	A_{1u}
A_{2u}	A_2	A_1	A_u	B_{1u}	A_{2u}
E_u	E	E	E_u	$A_{1u}+B_{1u}$	E_u
F_{1u}	F_1	F_2	F_u	$A_{2u}+E_u$	$A_{2u}+E_u$
F_{2u}	F_2	F_1	F_u	$B_{2u}+E_u$	$A_{1u}+E_u$

* To find correlations with smaller subgroups, carry out the correlation in two steps; for example, if the correlation of O_h with C_{2v} is desired, use the above table to pass from O_h to T_d and then employ the table for T_d to go on to C_{2v}.

BIBLIOGRAPHY

1. International Tables for X-Ray Crystallography, Vol. 1 (N. F. M. Henry and K. Lonsdale, Eds.), Kynoch Press, Birmingham, England (1965).
2. R. S. Halford, *J. Chem. Phys.* **14,** 8 (1946).
3. D. M. Adams, *Metal–Ligand and Related Vibrations*, St. Martin's Press, New York (1968).

Appendix 2

PROCEDURE USED FOR FACTOR GROUP ANALYSIS OF SOLIDS

In making low-frequency assignments for solids, it becomes imperative that factor group analysis procedures be applied for the molecule whenever possible. Appendix 2 in conjunction with Appendix 1 will outline these procedures. Several examples will serve to illustrate the method.

Certain limitations in the method are apparent. It is necessary to know the space group of the solid, and for this x-ray data must be available. The arrangement of the atoms in the Bravais unit cell is helpful. If these data are unavailable, the analysis cannot be easily performed.

In order to determine the number of lattice modes that will be found, it is necessary to calculate the group characters for various representations. These are listed as follows and follow the previous nomenclature[1,2]:

(1) The unit cell modes (internal and lattice) based on $3n$ Cartesian coordinates are given by the group character

$$\chi_i(n_i) = \mu_R(\pm 1 + 2 \cos \varphi) \tag{A-2-1}$$

(2) The acoustic modes are given by

$$\chi_i(T) = \pm 1 + 2 \cos \varphi \tag{A-2-2}$$

(3) The translatory lattice modes are given by

$$\chi_i(T') = (\mu_R(S) - 1)(\pm 1 + 2 \cos \varphi) \tag{A-2-3}$$

(4) The rotatory lattice modes are given by

$$\chi_i(R') = [\mu_R(S - P)]\, \chi_i(P) \tag{A-2-4}$$

where μ_R is the number of atoms invariant under the symmetry operation R; $\mu_R(S)$ is the number of structural groups remaining invariant under symmetry operation R; $\mu_R(S-P)$ is the number of polyatomic groups remaining invariant under an operation R; P is the number of monatomic groups; and φ is the angle of rotation corresponding to the symmetry operation R. Plus and minus signs stand, respectively, for proper and improper rotations;

$\chi_i(P) = (1 \pm 2\cos\varphi)$ for nonlinear polyatomic groups, $(\pm 2\cos\varphi)$ for operations $C(\varphi)$ and $S(\varphi)$ in a linear polyatomic group, and 0 for operations $C_2(\varphi)$ and δ_v in a linear polyatomic group.

EXAMPLE 1—$NaNO_3$

Sodium nitrate[3] belongs to the D_{3d}^6 ($R\overline{3}c$) space group having a rhombohedral or pseudohexagonal crystal structure. There are two molecules per unit cell. Figure 1 illustrates the Bravais unit cell. Table I contains the character table for the D_{3d}^6 space group, and subsequently the table is developed to satisfy equations (A-2-1) to (A-2-4) for the different operations in the group. The operations for the nitrate group may be expressed as follows:

$2S_6$	(1) (2) (3,4) (5,9,7,8,6,10)
	(1) (2) (3,4) (5,10,6,8,7,9)
$2C_3$	(1) (2) (3) (4) (8,10,9) (5,7,6)
	(1) (2) (3) (4) (8,9,10) (5,6,7)

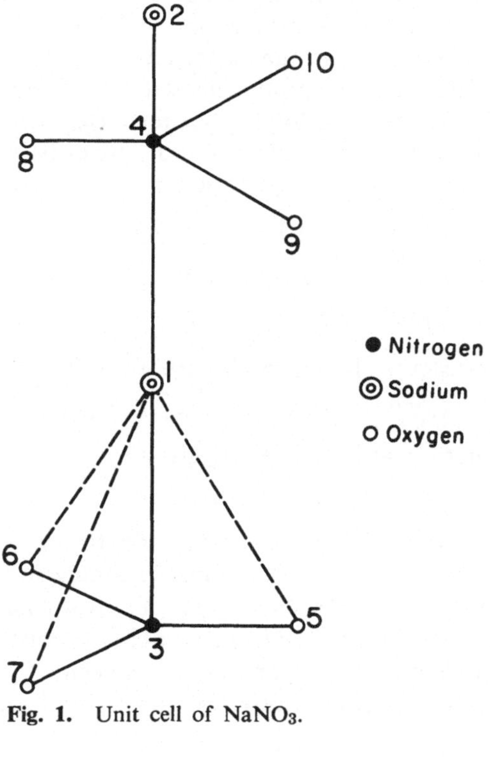

Fig. 1. Unit cell of $NaNO_3$.

Table I. Character Table and Distribution of Unit Cell Modes in $NaNO_3$

$D_{3d}^6(R\bar{3}c)$	E	$2C_3$	$3C_2$	i	$2S_6$	$3\sigma_d$	n_i	T	T'	R'	n_i'	Activity IR	Activity R
A_{1g}	1	1	1	1	1	1	1	0	0	0	1	ia	a
A_{1u}	1	1	1	−1	−1	−1	2	0	1	0	1	ia	ia
A_{2g}	1	1	−1	1	1	−1	3	0	1	1	1	ia	ia
A_{2u}	1	1	−1	−1	−1	1	4	1	1	1	1	a	ia
E_g	2	−1	0	2	−1	0	4	0	1	1	2	ia	a
E_u	2	−1	0	−2	1	0	6	1	2	1	2	a	ia
φ_R	0°	120°	180°	180°	60°	0°							
$\cos\varphi_R$	1	−0.5	−1	−1	0.5	1							
$\pm 1 + 2\cos\varphi_R$	3	0	−1	−3	0	1							
μ_R	10	4	4	2	2	0							
$\mu_R(S)$	4	2	2	2	2	0							
$\mu_R(S-P)$	2	0	2	0	0	0							
$\chi_i(n_i)$	30	0	−4	−6	0	0							
$\chi_i(T)$	3	0	−1	−3	0	1							
$\chi_i(T')$	9	0	−1	−3	0	−1							
$\chi_i(R')$	6	0	−2	0	0	0							

n_i = total unit cell modes (lattice + internal); T = acoustic modes; T' = translatory lattice modes; R' = rotatory lattice modes; n_i' = internal modes of polyatomic unit.

$i = S_2 = S_6^3$

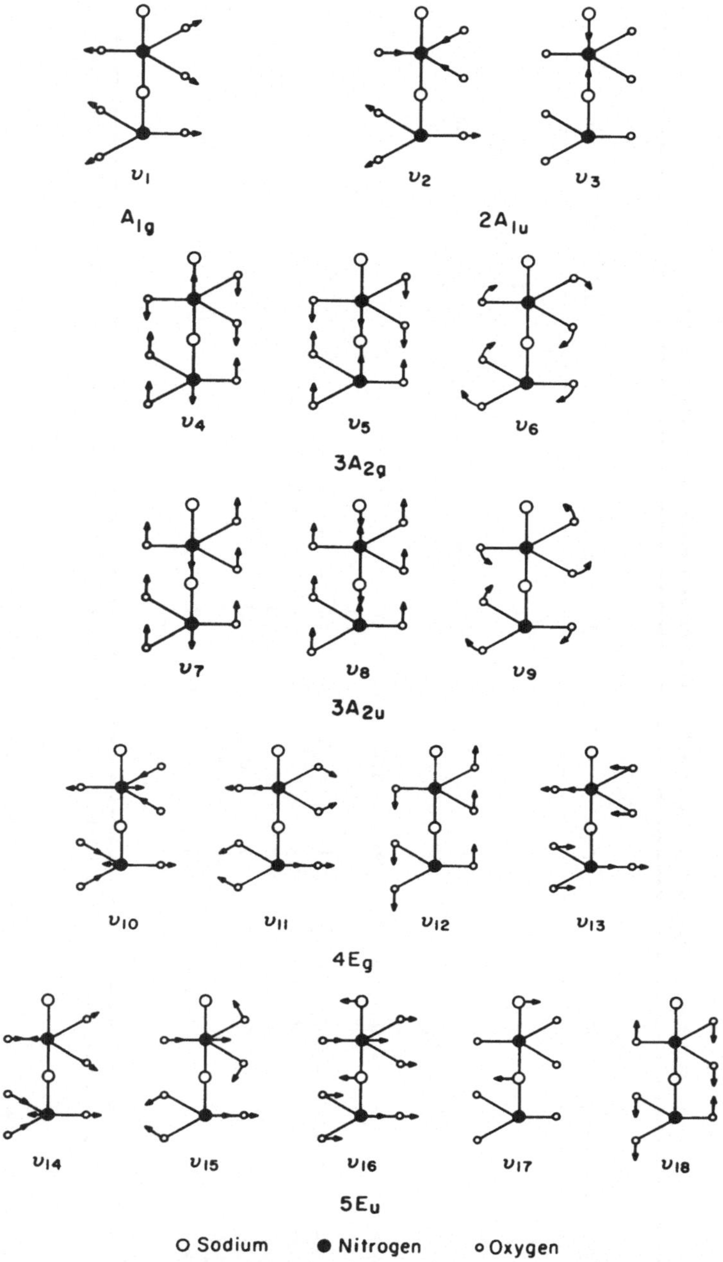

Fig. 2. Vibration modes of $NaNO_3$. Only one component of the degenerate pairs of E_g and E_u is shown. (Courtesy of American Institute of Physics, New York. From Nakagawa and Walter.[6])

i	(1) (2) (3,4) (5,8) (6,9) (7,10)
$3\sigma_v$	(1,2) (3,4) (5,8) (6,10) (7,9)
(glide)	(1,2) (3,4) (5,10) (6,9) (7,8)
	(1,2) (3,4) (5,9) (6,8) (7,10)
$3C_2$	(1,2) (3) (4) (5) (8) (6,7) (9,10)
	(1,2) (3) (4) (6) (9) (5,7) (8,10)
	(1,2) (3) (4) (7) (10) (5,6) (8,9)

Having obtained the characters of irreducible representations for $\chi_i(n_i)$, $\chi_i(T)$, $\chi_i(T')$, and $\chi_i(R')$, one is now ready to develop the right-hand part of the table and calculate n_i, T, T', R', and n for each vibration species. To calculate n_i, use is made of the reduction formula

$$N_i = \frac{1}{N_g} \Sigma n_e \chi_i(R) \chi_i(n_i) \qquad \text{(A-2-5)}$$

where N_i is the number of times an irreducible representation appears in a vibration species; N_g is the number of elements in the group; n_e is the number of elements in each operation class; $\chi_i(R)$ is the character of the vibration species for the space group involved; and $\chi_i(n_i)$ is the character for the total unit cell modes. For T one uses the same reduction formula but replaces $\chi_i(n_i)$ with $\chi_i(T)$, etc.

The development of Table I shows that from the expected 30 degrees of freedom, 12 internal frequencies are calculated for NO_3^- and 18 lattice vibrations for $NaNO_3$. The different modes which correspond to the irreducible representations of the D_{3d} space group are $1A_{1g} + 2A_{1u} + 3A_{2g} + 4A_{2u} + 4E_g + 6E_u$. All gerade ($g$) modes are Raman active, and all ungerade (u) are infrared active. The A_{2g} and A_{1u} modes are forbidden. Figure

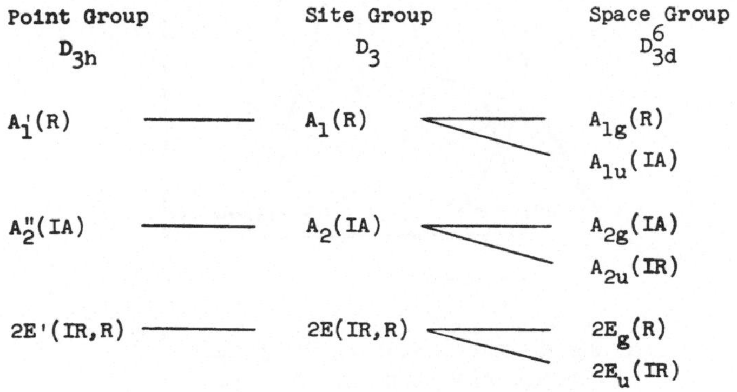

Fig. 3. Correlation chart for NO_3^- vibrations.

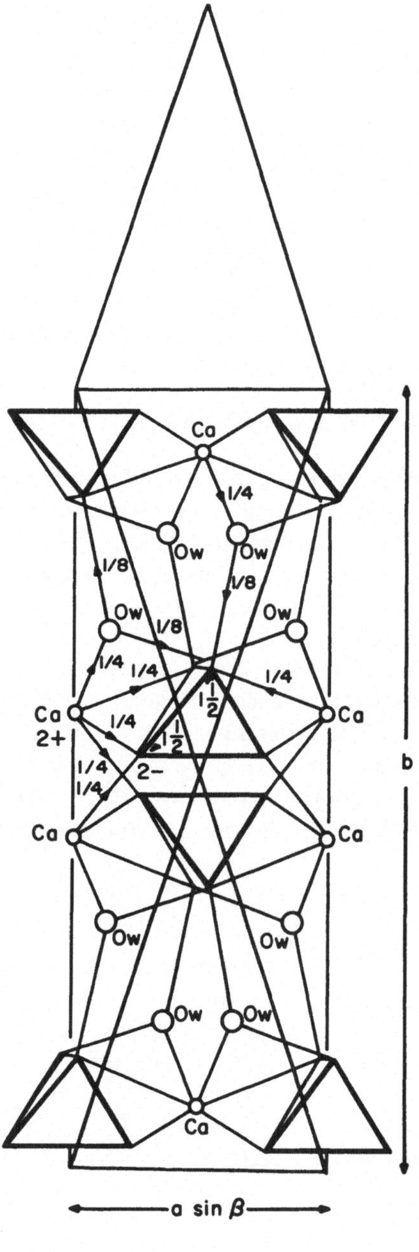

Fig. 4. Bravais unit cell for gypsum. (Courtesy of Plenum Press, New York. From Mitra and Gielesse.[4])

Table II. Character Table and Distribution of Unit Cell Modes for Gypsum

									n_i'		Activity	
C_{2h}^6 $(C2/c)$	E	C_2	i	σ_h	n_i	T	T'	R'	SO_4^{2-}	H_2O	IR	R
A_g	1	1	1	1	17	0	5	4	5	3	ia	a
A_u	1	1	−1	−1	17	1	4	4	5	3	a	ia
B_g	1	−1	1	−1	19	0	7	5	4	3	ia	a
B_u	1	−1	−1	1	19	2	5	5	4	3	a	ia
φ	0°	180°	180°	0°								
$\cos\varphi$	1	−1	−1	1								
$\pm 1 + 2\cos\varphi$	3	−1	−3	1								
μ_R	24	4	0	0								
$\mu_R(S)$	8	4	0	0								
$\mu_R(S-P)$	6	2	0	0								
$\chi_i(n_i)$	72	−4	0	0								
$\chi_i(T)$	3	−1	−3	1								
$\chi_i(T')$	21	−3	3	−1								
$\chi_i(R')$	18	−2	0	0								

2 shows all the modes for $NaNO_3$ corresponding to those calculated by space group analysis.

Figure 3 illustrates the correlation chart for $NaNO_3$ based on a site symmetry of D_3 and a point group of D_{3h} for NO_3. The selection of the proper site group is sometimes difficult. In using Halford's method,[5] an acceptable site group must be a subgroup of both the space group and the molecular group. In general, the site group will be of lower order than the molecular group. Further elimination can be made on the basis of the number of molecules contained in the unit cell (see Appendix 1—Site Symmetries). In some cases it is necessary to resort to a study of the map of the unit cell to determine the site group. Having selected a site symmetry and knowing the point group symmetry, a correlation chart is possible. Reference to Appendix 1—Correlation Tables allows one to select the proper species for the group and subgroup.

EXAMPLE 2—$CaSO_4 \cdot 2H_2O$

Gypsum,[4] $CaSO_4 \cdot 2H_2O$, belongs to the space group C_{2h}^6 ($C2/c$) with two molecules per Bravais unit cell. Figure 4 shows the Bravais unit cell

Table IIIA. Character Table and Distribution of Normal Modes for an XY_4 Molecule in a T_d Point Group

							Activity	
T_d	E	$8C_3$	$6\sigma_d$	$6S_4$	$3S_4^2 = 3C_2$	n	IR	R
A_1	1	1	1	1	1	1	ia	a
A_2	1	1	−1	−1	1	0	ia	ia
E	2	−1	0	0	2	1	ia	a
F_1	3	0	−1	1	−1	0	ia	ia
F_2	3	0	1	−1	−1	2	a	a

Table IIIB. Character Table and Distribution of Normal Modes for an XY_4 Molecule in a C_2 Point Group

				Activity	
C_2	E	$C_2(y)$	n	IR	R
A	1	1	5	a	a
B	1	−1	4	a	a
μ_R	5	2			
$\Xi(R)$	9	0			
φ	0°	180°			
$\cos\varphi$	1	−1			
$\pm 1 + 2\cos\varphi$	3	−1			

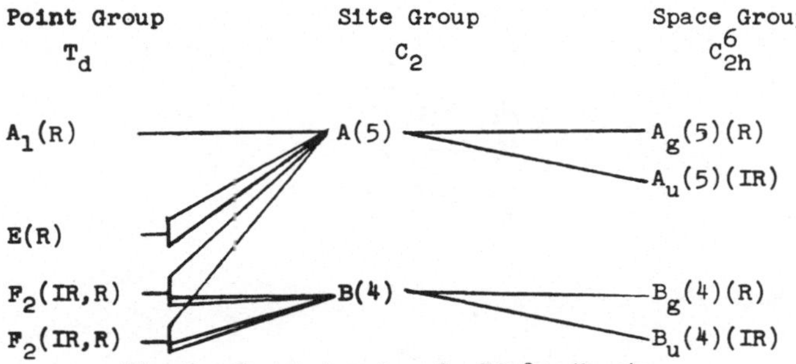

Fig. 5. Correlation chart for SO_4^{2-} vibrations.

for gypsum. Table II shows the character table and the distribution of the unit cell modes. In order to determine the number of internal modes belonging to SO_4^{2-} and H_2O, the lattice is first considered to be composed only of $CaSO_4$, and the water molecules are ignored. Next the lattice is regarded as consisting only of water, ignoring the $CaSO_4$.

Seventy-two degrees of freedom are expected for gypsum, 39 of which are optical lattice modes and 30 of which are internal modes associated with SO_4^{2-} and H_2O.

The internal modes of vibrations for the SO_4^{2-} in gypsum may be considered by the site group method of Halford.[5] Free sulfate ions have a tetrahedral symmetry. Table IIIA contains the character table for a T_d molecule. There are four modes for the free ion: A_1 and E modes are Raman active, and the two F_2 vibrations are active in both the infrared and Raman.

For the gypsum molecule the C_2 crystal axis coincides with a C_2 axis for each SO_4^{2-} group. Thus, the motions of the sulfate ion in crystal may be considered to occupy a C_2 site symmetry. The character table and distribution of normal modes of an XY_4 molecule in a C_2 site group are given in Table IIIB. Since there are two SO_4^{2-} per unit cell, each mode may split into two, depending on whether it is gerade (g) or ungerade (u) to the point of inversion of the unit cell. This is shown in the correlation chart in Fig. 5 for which the correlation tables in Appendix 1 may be used. These tables are helpful in determining which species of a subgroup are to be correlated with the species of the main group. For example, the A_1, E, and F_2 species of the T_d group are found to correlate with the species of the C_2 group as follows:

T_d		C_2
A_1	$\longrightarrow$	A
E	$\longrightarrow$	$2A$
F_2	$\longrightarrow$	$A+2B$

The species of the C_{2h} group correlate with the species of the C_2 group as follows:

C_{2h}		C_2
A_g	$\longrightarrow$	A
B_g	$\longrightarrow$	B
A_u	$\longrightarrow$	A
B_u	$\longrightarrow$	B

BIBLIOGRAPHY

1. S. Bhagavantam and T. Venkatarayudu, *Proc. Indian Acad. Sci.* **A9,** 224 (1939).
2. J. R. Ferraro and J. S. Ziomek, *Introductory Group Theory and Its Application to Molecular Structure*, Plenum Press, New York (1969).
3. D. W. James and W. H. Leong, *J. Chem. Phys.* **49,** 5089 (1968).
4. S. S. Mitra and P. J. Gielesse, *Prog. Infrared Spectry.* **2,** 47 (1964).
5. R. S. Halford, *J. Chem. Phys.* **14,** 8 (1946).
6. I. Nakagawa and J. L. Walter, *J. Chem. Phys.* **51,** 1389 (1969).

INDEX